Sustainable Development for the Americas

Resilience and Sustainability in Civil, Mechanical, Aerospace and Manufacturing Engineering Systems

Series Editor
Mohammad Noori
Cal Poly San Luis Obispo
Published Titles

Reliability and Safety of Cable-Supported Bridges
Naiwei Lu, Yang Liu, and Mohammad Noori

Reliability-Based Analysis and Design of Structures and Infrastructure
Ehsan Noroozinejad Farsangi, Mohammad Noori, Paolo Gardoni, Izuru Takewaki, Humberto Varum, Aleksandra Bogdanovic

Seismic Analysis and Design using the Endurance Time Method
Homayoon E. Estekanchi and Hassan A. Vafai

Thermal and Structural Electronic Packaging Analysis for Space and Extreme Environments
Juan Cepeda-Rizo, Jeremiah Gayle, and Joshua Ravich

Sustainable Development for the Americas
Science, Health, and Engineering Policy and Diplomacy
E. William Colglazier, Hassan A. Vafai, Kevin E. Lansey, with the administrative assistance of Molli D. Bryson

Series Editor Bio: Prof. Mohammad Noori is a professor of mechanical engineering at California Polytechnic State University, San Luis Obispo. He received his BS (1977), his MS (1980) and his PhD (1984) from the University of Illinois, Oklahoma State University and the University of Virginia respectively; all degrees in Civil Engineering with a focus on Applied Mechanics. His research interests are in stochastic mechanics, non-linear random vibrations, earthquake engineering and structural health monitoring, AI-based techniques for damage detection, stochastic mechanics, and seismic isolation. He serves as the executive editor, associate editor, the technical editor or a member of editorial boards of 8 international journals. He has published over 250 refereed papers, has been an invited guest editor of over 20 technical books, has authored/co-authored 6 books, and has presented over 100 keynote and invited presentations. He is a Fellow of ASME, and has received the Japan Society for Promotion of Science Fellowship.

For more information about this series, please visit: https://www.routledge.com/Resilience-and-Sustainability-in-Civil-Mechanical-Aerospace-and-Manufacturing/book-series/ENG

Sustainable Development for the Americas

Science, Health, and Engineering Policy and Diplomacy

Edited by
Honorary Editor: E. William Colglazier
Hassan A. Vafai
Kevin E. Lansey

With the administrative assistance of Molli D. Bryson

CRC Press
Taylor & Francis Group
Boca Raton London New York

CRC Press is an imprint of the
Taylor & Francis Group, an informa business

First edition published 2022
by CRC Press
6000 Broken Sound Parkway NW, Suite 300, Boca Raton, FL 33487-2742

and by CRC Press
2 Park Square, Milton Park, Abingdon, Oxon OX14 4RN

CRC Press is an imprint of Taylor & Francis Group, LLC

Library of Congress Cataloging-in-Publication Data
Names: Colglazier, E. William (Elmer William), editor. |
Vafai, Hassan A., editor. | Lansey, Kevin E., editor.
Title: Sustainable development for the Americas: science, health and engineering policy
and diplomacy / edited by: honoraray editor: E. William Colglazier, Hassan A. Vafai,
Kevin E. Lansey, with the administrative assistance of Molli D. Bryson
Description: First edition. | Boca Raton: CRC Press, 2022. | Includes index.
Identifiers: LCCN 2021020838 (print) | LCCN 2021020839 (ebook) |
ISBN 9781032115719 (hardback) | ISBN 9781032115726 (paperback) |
ISBN 9781003220503 (ebook)

ISBN: 978-1-032-11571-9 (hbk)
ISBN: 978-1-032-11572-6 (pbk)
ISBN: 978-1-003-22050-3 (ebk)

DOI: 10.1201/9781003220503

Typeset in Times
by Newgen Publishing UK

Contents

Foreword by the Senior Vice President for Academic Affairs and Provost at The University of Arizona

It has been almost one year since we hosted the 2019 *Science, Health and Engineering Policy and Diplomacy; Sustainable Development for the Americas* meeting here in Tucson, Arizona. I had the great pleasure of delivering the welcome address, during which I noted the importance and timeliness of science and policy discussions and resolutions. The global pandemic struck just a few months later, shaking nations and communities around the world. Now, in the midst of the COVID-19 pandemic, we must re-commit to gathering as global citizens to decide on science policies that support healthy communities and healthy environments around the world.

It is no coincidence that this meeting took place at The University of Arizona, a public, land-grant, research-intensive, and student-centered university located a mere 75 miles from Mexico's border. UArizona is designated as an Hispanic Serving Institution (HSI) and was recognized in 2019 as one of only nine universities to be recognized with the Seal of Excelencia from the national organization Excelencia in Education. We have long engaged internationally with Latin American universities and communities to advance understanding and policy on border issues such as water, immigration, trade, and related topics. In the science and engineering disciplines, we routinely collaborate broadly with the global community to advance knowledge and develop new technologies to support sustainable communities.

The need for international collaboration with science as the core framework for setting policies and solving problems is now more important than ever. Well-known examples of the use of this framework in policy settings include:

- The Montreal Protocol, which banned the use of ozone depleting contaminants in cars, refrigerators, and air conditioners;
- Paris Climate Agreement, a global agreement to combat dangerous temperature increases; and
- CERN, the United Organization for Nuclear Research, a research organization that collaboratively operates the world's largest particle physics lab.

The 2019 meeting was unprecedented in terms of the sheer numbers of experts on this topic, who came to UArizona's campus to present, educate, ideate, and connect. This foreword serves as a companion piece to a textbook in which multiple presenters from that meeting cover many of the conference topics.

Academic and policy discourse cannot ignore the recent trends toward diminished public trust in science, which is compounded by a growing distrust in media sources. Access to information and the ability to determine the quality of said information are an enormous challenge today, for even the most discerning eye. The lack of trust in science is compounded by a general misunderstanding of the scientific method as a process, rather than a set of settled theories or facts. Mistrust is perpetuated in online forums by real and perceived conflicts between cultures, religious beliefs, and within science communities, nationally and internationally. At the 2019 meeting, the focus may have been on partnership and diplomacy between the Americas, but our efforts must be global and inclusive to be impactful.

When scientists partner with economists, political scientists, sociologists, communications professionals, and humanists, we can develop policies which encompass cultural sensitivity, competency, and capital. Partnerships across the disciplines can deliver informed and strategically impactful solutions that reflect the cultural, sociological, socioeconomic, and political differences between our countries while capitalizing on the strengths of those differences.

I am grateful to the key contributors at the conference, who inspired this compilation of papers, and I thank them for their pioneering efforts in this important discipline and practice. Many thanks to William Colglazier, for serving as the honorary chair. Special thanks also are owed to Hassan Vafai, my engineering colleague at UArizona, and to our esteemed conference panelists, Cherry Murray, Ronit Prawer, and Vaughan Turekian. It is my great hope that the collaborations that were fostered and nurtured at the 2019 meeting will yield real and lasting positive outcomes for humanity and the environment in the years ahead.

Liesl Folks, *Ph.D., M.B.A.*

Preface

This book is a collection of selected papers and transcribed panel discussions presented at The University of Arizona's second Science, Health and Engineering Policy and Diplomacy Conference, held on October 23–25, 2019 in Tucson, Arizona. This conference was organized by the Science, Health and Engineering Policy and Diplomacy Initiative (SPDI) at The University of Arizona, and was sponsored by the Richard Lounsbery Foundation and the World Academy of Sciences, along with the following departments from The University of Arizona: the Office of the Provost, the Department of Civil and Architectural Engineering and Mechanics, the Udall Center for Studies in Public Policy, the College of Engineering, the Office of Global Initiatives, and the College of Science.

As a follow up to the 2017 conference, *Science Diplomacy with a Focus on the Americas*, this meeting was intended to convene a broad audience related to fundamental work and application of science policy and diplomacy. The conference placed an emphasis on sustainable development for the Americas, which is a particular strength of The University of Arizona as a certified Hispanic-Serving Institution (HSI). The proceedings for the 2017 conference were published by Momentum Press as a three-volume series, and this volume builds upon that foundation.

This book provides a spectrum of perspectives from some of the most renowned practitioners within the realm of science policy and diplomacy. Speakers range from Nobel laureates to esteemed scientists, scholars to early career scientists and graduate students, from around the world—including Africa, the Middle East, China, Europe, Latin and Central America, as well as the United States—and from various sectors including local and federal governments, non-profits, research and development firms, academia, and the National Academies. The diversity of experiences and variety of viewpoints offer valuable insights into a range of aspects of science and diplomacy.

The two-day conference convened over a dozen distinguished scholars to present keynote presentations and participate in moderated panel discussions. After a call for papers, over 150 papers were received, reviewed by the scientific committee, and some 60 papers accepted for presentation at the conference, in 15 break-out sessions. These sessions examined global sustainability challenges and the science diplomacy approaches that can deliver solutions for the 21st century.

On the first night of the conference, the honorary conference chairman, Dr. E. William Colglazier, moderated a panel discussion under the theme "Overcoming Social, Economic, and Environmental Challenges for the Americas and the World: How Can Science, Technology, and Innovation Help Find Solutions?" Drs. Cherry Murray, Ronit Prawer, and Vaughan Turekian shared their perspectives and experiences in science diplomacy and how it can advance diplomatic relations as well as scientific advancement. In the next session, held the following morning, Dr. Colglazier moderated a panel discussion titled "Sustaining International Scientific Cooperation Among Nations: Why It Matters." Drs. Peter Agre and Ronit Prawer gave specific examples of how science diplomacy can help to thaw icy relations with adversarial regions, while also discussing obstacles to conduct collaborative research, and how those might be minimized.

The following plenary sessions sampled the history of science diplomacy from Ivan the Terrible to the present day and explored the importance of sustained international scientific cooperation, building community resilience, and the role of political will in sustainability and diplomatic efforts. Presentations examined topics such as global change, arid environments, water security and governance, transboundary conflict and cooperation, urban and rural resilience, and public health. They scrutinized topics that directly impact decision-making—new technologies, big data, and changing international norms—and provided case studies from various geographic regions through the lens of science diplomacy, including the U.S.-Mexico border, the Gulf of California, South America, Europe, the Middle East, Central and South Asia, China, and others. Other sessions focused on the role of educational cooperation, developing networking capacity, and maximizing global impact for careers in science diplomacy.

Throughout the conference, there was an underlying theme of the importance of collaboration. Through the expertise and experiences shared, it was clear that building a network of people, organizations, and countries engaged in science diplomacy is crucial for mutual growth and to overcome conflicting political stances.

Some of the key recommendations of the conference included:

- In times of political and international uncertainty, international scientific collaboration can serve as an invaluable diplomatic tool.
- Even small science diplomatic efforts can influence myriad issues, from overfishing to human rights negotiations to global carbon emission reduction.
- Equally important to beneficial scientific impact is building the infrastructure and communication networks to share information and initiate partnerships.
- Short-term research collaborations can morph into lasting international partnerships, especially between national health agencies. These long-term relationships can survive through shifting political landscapes and should be prioritized as the ultimate goal.
- It is critical that we develop more resilient materials and sustainable practices across the gamut of industries and disciplines, as we face increasingly complex global challenges.
- Sustainability efforts require a great deal of engagement and political will, from local communities to state departments. Science diplomats—from experts and scientists to spokespersons and ambassadors—can help facilitate at all levels and bring in valued resources from technology sharing, capacity building, and knowledge exchanges.
- It is crucial that we encourage and lead younger generations into science diplomacy early in their careers via robust, interdisciplinary training programs and outreach education.

The sessions and subsequent question and answer periods were transcribed, edited, and appropriate references were added by volunteer rapporteurs. The papers were thoroughly reviewed by a scientific committee and approved by the speakers for inclusion in the publication. The result is a comprehensive view from those who have significantly impacted the field.

The publication of these proceedings is the main resolution from the conference, as well as the commitment to organize subsequent similar events to continue to build upon the collaborations and mechanisms in place to further expand the reach of science in diplomacy, science policy, and diplomacy for science.

Hassan A. Vafai, *Corresponding Editor*

Acknowledgments

We are grateful to the persons and institutions that have made the Sustainable Development for the Americas conference and the publication of these proceedings possible. Many University of Arizona (UA) entities on campus provided financial and in-kind support for the conference: the Office of the Provost, College of Engineering, Department of Civil & Architectural Engineering & Mechanics, College of Science, Office of Global initiatives, Udall Center for Studies in Public Policy, Institute of the Environment. We would also like to acknowledge the support we have received from the Richard Lounsbery Foundation and the World Academy of Sciences. In addition, we especially would like to express our thanks to Dr. Jeff Goldberg, Dean Emeritus of the College of Engineering, who recognizes the value of science and engineering diplomacy and played a pivotal financial and intellectual role in advancing the conference forward.

Further, we would like to convey our utmost sincerest thanks to Dr. Leisl Folks, Senior Vice President for Academic Affairs and Provost (UA), for her support of our endeavor and providing remarks at the conference's opening session. Also, unlimited appreciation goes to Dr. Elizabeth Cantwell, Senior Vice President for Research and Innovation (UA), for her support and remarks at the concluding session of the conference. Furthermore, we express our gratitude to Dr. David Hahn, Craig M. Berg Dean, College of Engineering (UA), for his support and remarks at the conference.

The Sustainable Development for the Americas conference, convened at UA, was a seminal event. However, it would not have been realized without the dedication and commitment of the Honorary Conference Chairman, Dr. E. William Colglazier, and we thank him for his support. As a recognized leader in the field with a comprehensive awareness of current activities, trends, and historical context, Dr. Colglazier was instrumental in establishing the program and assembling the eminent group of keynote speakers to achieve that goal. He spent many hours with us in conversations on conference content and structure as well as connecting with potential speakers. From the opening session and throughout the course of the conference, he provided a comprehensive perspective on the contributions, as well as value and future of the SPDI.

The second group that guided the intellectual direction of the conference was the members of the Scientific Committee of the Conference that consisted of UA faculty and administrators: James Buizer, Institute of the Environment; Kacey Ernst, Department of Epidemiology; Andrea Gerlak, School of Geography and Development; Jeffrey Goldberg, Dean Emeritus, College of Engineering; John Hildebrand, Department of Neuroscience; Ryan Jones, Latin America Programs and Grants; Monica Ramirez-Andreotta, Department of Environmental Science; Joaquin Ruiz, Dean, College of Science; Christopher Scott, Udall Center for Studies in Public Policy; Alex Thome, Center for Innovation in Brain Science; Robert Varady, Udall Center for Studies in Public Policy; and Jane Zavisca, Associate Dean, Research and Graduate Studies.

We also thank the distinguished keynote speakers and the scientists from different parts of the world who presented papers at different sessions for investing their time to join us at the UA to describe the results of experience and research in prepared talks and participate in the wide-ranging discussions that followed their presentations. Particularly, we appreciate their efforts and significant time spent to collaborate with the conference rapporteurs and our staff to convert their talks to this proceeding. The enthusiasm and professionalism of this group elevated the conference quality to an influential event.

Further, we acknowledge the work done by the team of rapporteurs, including supporting conference activities and transcribing, researching, and condensing the talks for clarity. This outstanding team, led by Molli Bryson, comprised a self-identified multidisciplinary group of graduate and undergraduate UA students interested in learning and, perhaps, developing a career in science diplomacy consisted of:

Rachel Abraham
Talia Anderson
Adriana Arcelay
Kinzie Bailey
Charles Devine
Mercury Fox
Stella Heflin
John Keck
Tanner Muscarella
Amber Petrie
Brandon Strange
Singleton Thibodeaux-Yost
Susan Washko
and Laura Werthmann.

Leading up to and following the conference, the UA hosted four seminars to highlight the SPDI. We thank each of them: Dr. John Boright, Executive Director of International Affairs of the U.S. National Academies of Sciences, Engineering and Medicine; Dr. Frances Colón, CEO, Jasperi Consulting, and Former Deputy Science and Technology Adviser to the Secretary of State; Dr. Kathleen Harder, Director, Center for Design in Health and Human Factors and Ergonomics Graduate Program Chair, University of Minnesota; Dr. Mahlet Mesfin, Former Director of AAAS Center for Science Diplomacy.

Additionally, we thank Professor Mohammad Noori for his valuable advice and recommendations in the fields of Sustainability and Resiliency for the planning of the conference. Also, we would like to express our gratitude to Dr. Josef Clements and Ms. Lisa Wilford from Taylor and Francis Group for their valuable advice and recommendations regarding the quality and content presentation of the book. Also, we would like to acknowledge Mr. Navin Prathath, manager at Newgen KnowledgeWorks Pvt Ltd, and his colleagues; Ms. Jeno Jay Kattakal; and Mr. Sedhu Madhavan for their accuracy and patience in preparing the manuscripts for publication.

The extraordinary efforts of Ms. Molli Bryson as the lead administrative conference organizer and her team are greatly appreciated. In this position, she assisted with planning, managing the logistics, and the administration of other conference events, along with the following individuals:

Kian Alavy, Global Initiatives, UA
Ally Beaulieu, Udall Center for Studies in Public Policy, UA
Neha Gupta, Hydrology and Atmospheric Sciences, UA
Liliana Salas, School of Information, UA
Sehrish Shabaz Cloudhary, School of Information, UA
and Diana Zamora-Reyes, Hydrology and Atmospheric Sciences, UA

Finally, our gratitude goes to the staff of the UA Department of Civil and Architectural Engineering and Mechanics for fitting this out-of-the-ordinary task into their already at-capacity workload. Of particular note, we thank Ms. Therese Lane, Senior Business Manager, for her efforts in financial oversight, and Ms. Jeanne Yordy, Associate Accountant, for her assistance throughout this long process. Also, we would like to express our gratitude to Dr. Shelley McGrath, Manager, Evaluation and Communications for the Office of the University Provost; Lewis Humphreys, Senior Software and IT Licensing Manager, Tech Launch Arizona; and Bob Sleeper, Senior Licensing Manager, Tech Launch Arizona, for their contributing efforts and persistent support.

Hassan A. Vafai, *Corresponding Editor*

About the Editors

HONORARY EDITOR

E. William Colglazier is Editor-in-Chief of *Science and Diplomacy* and Senior Scholar in the Center for Science Diplomacy at the American Association for the Advancement of Science (AAAS). He works there to advance knowledge and practice on science policy and science diplomacy and to support international collaboration in science and technology. From 2016 to 2018 he co-chaired the ten-member group appointed by the UN Secretary General to advise on science, technology, and innovation for achieving the 17 Sustainable Development Goals of the UN 2030 Agenda. He served as the Science and Technology Adviser to the Secretary of State from 2011 to 2014. As Executive Officer of the National Academy of Sciences and National Research Council for 17 years, he oversaw the studies that provide independent, objective scientific advice on public policy issues.

He received his Ph.D. in theoretical physics from the California Institute of Technology in 1971, and subsequently worked at the Stanford Linear Accelerator Center, the Institute for Advanced Study in Princeton, the Center for Science and International Affairs at Harvard's Kennedy School of Government, and the University of Tennessee. While at Harvard, he also served as Associate Director of the Program in Science, Technology, and Humanism of the Aspen Institute. As Professor of Physics in Tennessee, he directed research centers dealing with environmental, energy, and waste management issues. In 1976–1977, he was an AAAS Congressional Science Fellow.

In 2015, Colglazier received from the American Physical Society the Joseph A. Burton Forum Award for "outstanding contributions to the public understanding or resolution of issues involving the interface of physics and society" and from the Government of Japan the Order of the Rising Sun, Gold Rays with Neck Ribbon, for "contributing to science and technology exchange and mutual understanding between Japan and the United States."

CORRESPONDING EDITOR

Hassan A. Vafai has held position of professorship in engineering at different universities including Sharif University of Technology, Washington State University, and The University of Arizona. His research interests include the fields of structural mechanics, structural dynamics, offshore structures, and earthquake engineering. He was founder and editor-in-chief of *Scientia*, a peer-reviewed international journal of science and technology. Throughout his careers, he has received numerous awards for his contribution to the promotion of science and technology worldwide including being elected "Eternal Spirit in Science," the highest honor by the National Academy of Sciences of Iran conferred on outstanding scientists; an honorary doctorate by the Senatus Academicus of Moscow Region State Institution of Higher Education; and the "Order of Palm Academicques" awarded by the Ministry of Education, Research and Technology of France.

CO-EDITOR

Kevin E. Lansey is Professor of the Department of Civil Engineering and Engineering Mechanics. His interests are related to water supply planning, water–energy links, and water distribution systems design, operation, and monitoring. He has published over 100 journal articles, ten book chapters, and two textbooks in those areas. For his contributions, he was awarded the ASCE Huber Civil Engineering Research Prize for Young Researchers in 2002.

WITH THE ASSISTANCE OF

Molli D. Bryson currently serves as the Environmental Program Coordinator for the Udall Center for Studies in Public Policy at The University of Arizona. Since earning her bachelor's degree in English, her professional interest has centered around science communications, especially in the realm of environmental diplomacy. She coordinates communications for the department, including website management, marketing, and cataloguing publications, as well as assisting in coordinating an array of events ranging from internal lecture series to large, international conferences. Her experience working with prestigious individuals from around the world has guided her career toward the subject of science editing and researching how best to communicate science to the public.

List of Contributors

SPEAKER BIOGRAPHIES

PETER AGRE

Before serving as 2009–2010 president of the American Association for the Advancement of Science, Agre was professor of medicine and biological chemistry at the Johns Hopkins School of Medicine and vice chancellor for science and technology at Duke Medical Center. He directs the Johns Hopkins Malaria Research Institute and is program director of the Southern Africa International Center of Excellence for Malaria Research in Zimbabwe and Zambia. He chaired the Committee on Human Rights of the National Academy of Sciences and in 2003 was awarded the Nobel Prize in chemistry.

LAUREN ALEXANDER AUGUSTINE

Lauren Alexander Augustine is the executive director of the Gulf Research Program of the National Academies of Sciences, Engineering, and Medicine (NASEM). Prior to accepting her position at NASEM, Augustine served as director of the National Academies' Resilient America Program, which supports communities' efforts to build resilience to extreme events using science and diverse stakeholder engagement. Developing disaster and community resilience is an important component of the Gulf Research Program's mission, and Augustine's knowledge and experience in these areas will be valuable in her new role. She earned her B.S. in applied mathematics and systems engineering and her M.S. in environmental planning and policy from the University of Virginia, and her Ph.D. from Harvard University in an interdisciplinary program that combined hydrology, geomorphology, and ecology.

LIESEL FOLKS

Liesl Folks is the Senior Vice President and Provost at The University of Arizona. She is the chief academic officer of the university and oversees all academic programs and units exclusive of the health sciences. She is responsible for all affairs related to the academic mission of the University including the faculty, academic programs, and related budgeting. As a member of the President's senior leadership team, her role includes the implementation of the strategic plan for the institution, and ensuring that the University demonstrates excellence as it pursues its missions of delivering an outstanding education that prepares students for the 4th Industrial Revolution, conducting research and discovery that addresses society's grand challenges, defining what it means to be a global university, living and breathing the University's core values to enable institutional excellence, as well as advancing the land grant mission in the State of Arizona. As the chief academic officer, Dr. Folks leads 16 colleges and 24 schools as well as academic affairs, faculty affairs, libraries, University Information Technology Services, UA Global, and student services. She also holds the rank of professor in the Department of Electrical and Computer Engineering.

Prior to serving in this role, Dr. Folks was the dean of the School of Engineering and Applied Sciences at the University of Buffalo where she built a reputation as a champion for women and underrepresented minorities in science and engineering degree programs. As an internationally recognized expert in the fields of magnetic materials and devices, nanoscale metrology, and spin-electronic devices, Dr. Folks holds 14 U.S. patents and has authored more than 60 peer-reviewed papers. She also serves on numerous boards and committees and recently became chair of the triennial congressionally mandated National Academies Review of the National Nanotechnology Initiative. Prior to working in higher education, she spent 16 years in research and development at IBM Almaden Research Center, Hitachi Global Storage Technologies, and Western Digital. Dr. Folks earned a master's degree in business administration from Cornell University after completing both a doctorate and Bachelor of Science degrees in physics from the University of Western Australia.

Cherry A. Murray

Cherry Murray grew up traveling abroad as her father was a diplomat. She obtained both a B.S. and a Ph.D. in physics from the Massachusetts Institute of Technology, after which she spent 27 years at Bell Labs in New Jersey, then became deputy director for science and technology at Lawrence Livermore National Lab for five years, then was dean of engineering and applied sciences at Harvard for six years, took two years of government service leave as a presidential appointee as Director of the U.S. Department of Energy Office of Science. She is now Professor of Physics and Deputy Director of Research of the Biosphere 2 Institute at The University of Arizona. Her research interests have varied from experimental condensed matter and surface physics to nanotechnology, innovation, research and development of telecommunications networks, national security and science and technology policy. Her current interests include international science policy, research, development, and innovation to sustain human civilization on future Earth.

Ronit Prawer

Prawer is the Director of the U.K. Government's Science and Innovation Network (SIN) for the Eastern USA, where she works to grow and strengthen the bilateral relationship between Great Britain and the US on research and innovation. Before arriving in the US, Prawer served for three years as the Science Attaché to the British Ambassador in Tel Aviv, where she established the UK's first Science Diplomacy projects in the region: STREAM, a multi-lateral water research program and platform for scientists from across the Middle East to work together with British scientists to tackle critical water issues; and GROWTH, a unique scholarship scheme bringing together young Palestinian and Israeli scientists. She hails from Melbourne, Australia, where she trained in genetics and worked to identify the genes underlying male infertility. In addition to her Genetics qualification, she also holds a degree in English Literature and French from the University of Melbourne.

Léna Salamé

Léna is a lawyer specialized in conflict management. She served in the UN system for 17 years as the strategic and operational coordinator of its program on water cooperation (PCCP). She is now leading the development of the Universities' Partnership for Water Cooperation and Diplomacy (UPWCD). Because of her professional achievements she is credited for having played a central role in the development and promotion of the current trend of hydro-diplomacy.

Glenn Schweitzer

Schweitzer has promoted international scientific cooperation and science diplomacy for over 50 years, working with scientists in countries such as Russia and Iran. He has been director of the Office for Central Europe and Eurasia of the National Academies since 1985, except for 1992–1994, when he went to Moscow at the request of the State Department to establish and lead the International Science and Technology Center. Earlier in his career he was first science officer at the U.S. Embassy in Moscow, and was instrumental in establishing the Office of the Science and Technology Advisor to the U.S. Secretary of State.

Vaughan Turekian

Since his 2015 appointment as science and technology adviser to the U.S. Secretary of State, Turekian has advised on international environment, science, technology, and health matters affecting U.S. foreign policy. Before that he was chief international officer at the American Association for the Advancement of Science and director of the association's Center for Science Diplomacy. From 2002 to 2006, he was special assistant and adviser to the State Department undersecretary for global affairs, consulting on sustainable development, climate change, environment, energy, science, technology, and health. Earlier in his career he was program director for the committee on global change research at the National Academy of Sciences.

Zhishen Wu

Zhishen Wu is a professor at Southeast University, China and Ibaraki University, Japan. He received his B.S. and M.S. from Southeast University, China and his Ph.D. from Nagoya University, Japan. His research expertise includes structural, concrete, and maintenance engineering and advanced composite materials. He is the author or co-author of over 600 papers in refereed journals and international conference proceedings including over 50 keynote or invited papers. He also holds 50 authorized patents of inventions. He is chairman of the China Chemical Fibers Association Committee on Basalt Fibers, and is the Vice President of the International Society for Structural Health Monitoring of Intelligent Infrastructure (ISHMII). He is also an elected follow of ISHMII, ASCE, Japan Society of Civil Engineering (JSCE), and International Institute of FRP in Construction (IIFC). Wu serves as an editor for the *International Journal of Sustainable Materials and Structural Systems* and has led several important research projects in both Japan and China.

Overview

This book titled *Sustainable Development for the Americas: Science, Health and Engineering Policy and Diplomacy* is a collection of selected papers and transcribed panel discussions presented at The University of Arizona's second Science, Health and Engineering Policy and Diplomacy Conference, held on October 23–25, 2019 in Tucson, Arizona. This successful conference was organized by the Science, Health and Engineering Policy and Diplomacy Initiative (SPDI) at The University of Arizona, and was sponsored by the Richard Lounsbery Foundation and the World Academy of Sciences, along with the following departments from The University of Arizona: the Office of the Provost, the Department of Civil and Architectural Engineering and Mechanics, the Udall Center for Studies in Public Policy, the College of Engineering, the Office of Global Initiatives, and the College of Science.

Following the debut conference in 2017, *Science Diplomacy with a Focus on the Americas*, this meeting intended to convene a broad audience related to fundamental work and application of science policy and diplomacy, with an emphasis on sustainable development for the Americas, which is a particular strength of The University of Arizona as a certified Hispanic-Serving Institution (HSI). The proceedings for the 2017 conference were published by Momentum Press as a three-volume series, and this book builds upon that foundation.

This book delivers a spectrum of perspectives from some of the most renowned practitioners within the realm of science policy and diplomacy. Speakers range from Nobel laureates to esteemed journal editors to early career scientists and graduate students, as well as from various sectors including local and federal governments, non-profits, research and development firms, academia, the National Academies, and more. The diversity of experiences and variety of perspectives offer valuable insights in every aspect of science and diplomacy.

The two-day conference convened over a dozen distinguished scholars to present keynote presentations and participate in moderated panel discussions. After a call for proposals, over 60 speakers were invited to present their papers in one of 15 breakout sessions. These sessions examined the global sustainability challenges and the science diplomacy approaches that can deliver solutions for the 21st century.

The plenary sessions sampled the history of science diplomacy from Ivan the Terrible to the present day and explored the importance of sustained international scientific cooperation, building community resilience, and the role of political will in sustainability and diplomatic efforts. Presentations examined topics such as global change, arid environments, water security and governance, transboundary conflict and cooperation, urban and rural resilience, and public health. They scrutinized topics that directly impact decision-making—new technologies, big data, and changing international norms—and provided case studies from various geographic regions through the lens of science diplomacy, including the U.S.-Mexico border, the Gulf of California, South America, Europe, the Middle East, Central and South Asia, China, and others. Other sessions focused on the role of educational cooperation, developing your networking capacity, and maximizing global impact for careers in science diplomacy.

Throughout the conference, there was an underlying theme of the importance of collaboration. Through the expertise and experiences shared at the conference, it was clear that building a network of people, organizations, and countries engaged in science diplomacy is crucial for mutual growth and to help thaw icy political relations.

Some of the key recommendations of the conference included:

- In times of political and international uncertainty, international scientific collaboration can serve as a strong diplomatic tool.
- Even small science diplomatic efforts can influence myriad issues, from overfishing to human rights negotiations to global carbon emission reduction.
- Just as important is building the infrastructure and communication networks to share information and start partnerships.
- Short-term research collaborations can morph into longtime international partnerships, especially between national health agencies. These long-term relationships can survive through shifting political landscapes and should be prioritized as the ultimate goal.
- It is critical that we develop more resilient materials and sustainable practices across the gamut of industries and disciplines, as we face increasingly complex global challenges.
- Sustainability efforts require a great deal of engagement and political will, from local communities to state departments. Science diplomats—from experts and scientists to spokespersons and ambassadors—can help facilitate at all levels and bring in valued resources from technology sharing, capacity building, and knowledge exchanges.
- It is crucial that we encourage and lead younger generations into science diplomacy early on via robust, interdisciplinary training programs and outreach education.

The publication of these proceedings is the main resolution from the conference, as well as the commitment to organize subsequent similar events to continue to build upon the collaborations and mechanisms in place to further expand the reach of science in diplomacy, science policy, and diplomacy for science.

CONFERENCE CO-CHAIRS AND THE ORGANIZING COMMITTEE:

Honorary Conference Chair

E. William Colglazier, former Science and Technology Adviser to the U.S. Secretary of State and Editor-in-Chief of the American Association for the Advancement of Science journal *Science & Diplomacy*

CONFERENCE CO-CHAIRS

Kevin Lansey, Department of Civil and Architectural Engineering and Mechanics, The University of Arizona
Hassan A. Vafai, Department of Civil and Architectural Engineering and Mechanics, The University of Arizona

COMMITTEE MEMBERS

James Buizer, Institute of the Environment, The University of Arizona
Kacey Ernst, Department of Epidemiology, The University of Arizona
Andrea Gerlak, School of Geography and Development, The University of Arizona
Jeffrey Goldberg, Dean Emeritus, College of Engineering, The University of Arizona
John Hildebrand, Department of Neuroscience, The University of Arizona
Ryan Jones, Latin America Programs and Grants, The University of Arizona
Monica Ramirez-Andreotta, Department of Environmental Science, The University of Arizona
Joaquin Ruiz, Dean, College of Science, The University of Arizona
Christopher Scott, Udall Center for Studies in Public Policy, The University of Arizona
Alex Thome, Center for Innovation in Brain Science, The University of Arizona
Robert Varady, Udall Center for Studies in Public Policy, The University of Arizona
Jane Zavisca, Associate Dean, Research and Graduate Studies, The University of Arizona

CONFERENCE ORGANIZERS

Molli Bryson, Udall Center for Studies in Public Policy, The University of Arizona
Kian Alavy, Global Initiatives, The University of Arizona
Ally Beaulieu, Udall Center for Studies in Public Policy, The University of Arizona
Neha Gupta, Hydrology and Atmospheric Sciences, The University of Arizona
Liliana Salas Nino, School of Information, The University of Arizona
Sehrish Shabaz Cloudhary, School of Information, The University of Arizona
Diana Zamora-Reyes, Hydrology and Atmospheric Sciences, The University of Arizona

STUDENT RAPPORTEURS

Rachel Abraham, Talia Anderson, Adriana Arcelay, Kinzie Bailey, Charles Devine, Mercury Fox, Stella Heflin, John Keck, Tanner Muscarella, Amber Petrie, Brandon Strange, Singleton Thibodeaux-Yost, Susan Washko, and Laura Werthmann.

Conference Program

Wednesday, October 23, 2019

6:30–6:45pm	Welcome Address: **Liesl Folks**, Provost, University of Arizona
6:45–8:10pm	Introductory Remarks and Panel Discussion: *Overcoming Social, Economic, and Environmental Challenges for the Americas and the World: How Can Science, Technology, and Innovation Help Find Solutions* Honorary Conference Chair: **E. William Colglazier** Panelists: **Cherry A. Murray**, **Ronit Prawer**, and **Vaughan C. Turekian**
8:10–8:25pm	Public Question and Answer Period

Thursday, October 24, 2019

8:00–8:10am	Conference Overview: **Kevin Lansey**
8:10–8:20am	Welcome Address: **Jeffrey Goldberg**
8:20–9:15am	Panel Discussion: *Sustaining International Scientific Cooperation Among Nations: Why It Matters* Chair: **E. William Colglazier** Panelists: **Peter Agre** and **Ronit Prawer**
9:15–10:00am	Keynote address: *Legacy from Ivan the Terrible to the Nobel Laureates* by **Glenn Schweitzer**, introduced by **Joaquin Ruiz**
PARALLEL SESSION 1 10:30–12:00pm	**Hydrodiplomacy in the U.S.-Mexico Border Region** Session Chair: **Margaret Wilder** *The 1944 Water Treaty and the Future of U.S.-Mexico Environmental Governance* by **Stephen P. Mumme** *Evolving Together: Recent Trends in US-Mexico Water Diplomacy and Collaboration in the Colorado River Basin* by **Mariana Rivera-Torres** *Minute 319 and Minute 323 of the US-Mexico Water Treaty: When Science Matters and When it Doesn't* by **Karl W. Flessa** *The U.S.-Mexico Transboundary Aquifer Assessment Program (TAAP): Implementation, Evolution, and Lessons Learned* by **Elia M. Tapia** and **Sharon B. Megdal**
	Science Diplomacy in Times of Uncertainty Session Chair: **James Buizer** *The Value of Science in Multilateralism* by **Randy Burd** *The Shifting Political Landscape: Science as the Bellwether* by **Michelle Wyman**
	Global Impact through STEM and Health Session Chair: **Emily Halvorson-Otts** *Life on Edge: The Interface between the US and Latin America* by **Dolores Duran-Cerda**, **Ricardo Castro-Salazar**, **Hassan Vafai**, **Kevin Lansey**, **Ross Zimmerman**, and **Emily Halvorson-Otts** *Antibacterial Activity of Traditional Native American Medicines Against Streptococcus Pyogenes* by **Saffeyya-Grace Ibrahim** and **Cori Leonetti** Panel Discussion: *The Pimeria Alta Alliance: Opportunities and Challenges* Panelists: **Dolores Duran-Cerda** and **Steve Farley**

PARALLEL SESSION 2
1:15–2:45pm

Science, Technology and Health
Session Chair: **Kacey Ernst**
Updating the Nuremberg Code: The Need for International Healthcare Collaborations in Artificial Intelligence and Robotics Ethics by **Stephanie Zawada** and **Jerzy Rozenblit**
Region-Based T1-Weighted MRI Morphometric Changes in High School Collision Sport Athletes by **Pratik Kashyap**
Building Bridges: Practical Health-Science Diplomacy Tools Community Tools to Link Public Groups and City Decision-makers y **Ruth Larbey**
Antibacterial and Antioxidant Xanthones and Benzophenones from Garcinia Smeathmannii by **Hugues Fouotsa et al.**

Science for Science Policy
Session Chair: **Jane Zavisca**
Science in Legislation—Applying Scientific Peer Review and Rigor to Public Policymaking by **CJ Pommier**, **Josh Hoskinson**, **Ansel Andrews**, and **Henry Werchan**
Recent institutional transformations on the technological governance in Latin America by **Martin Peréz Comisso**
From Citizen Science to Policy by **Samragyee Gautam**
Forging Sister Wetlands in El Paso, Texas and Ciudad Juárez, Mexico: Urban Transboundary Conservation and Restoration on the U.S. by **Ana Córdova**, **Adriana Zuñiga-Terán**, and **John Sproul**

Role of Educational Cooperation
Session Chair: **Monica Ramirez-Andreotta**
Environmental Justice Promotion through Experiential Outdoor Education on the River – Part 2 – HEAL by **Dakota Goodman**
Silos and Spiders: What Measles Tells Us about Policy and Education by **Deb Hughes Hallett**
Autonomous Rural Development and Well-Being in the Upper Gulf of California Communities by **Osmar Villalobos**
A Model for Water Diplomacy by **Katherine Himes**

2:45–3:30pm

Keynote address: *Sustainability Enhancement of Infrastructures with Smart and Resilient Materials* by **Zhishen Wu**, introduced by **David W. Hahn**

PARALLEL SESSION 3
4:00–5:30pm

Science Policy Issues
Session Chairs: **Ryan Jones** and **Alex Thome**
Barriers to Entry (And Re-entry): A Comparative Analysis of Launching Practices Between the US, Canada, and Mexico by **Matthew Contursi** and **Timiebi Aganaba**
US Public Preferences in Geoengineering Research Funding and Governance by **Leah Kaplan** and **John P. Nelson**
Air Forces Southern Partnerships with South America by **Zoja Bazarnic**, **Lt. Col. Eric Alonsobernal**, and **Tyrone Barbery**
Influence of Solar Magnetic Activities on Rainfall and Surface Air Temperature in West African Sub-region by **Esther Hanson**, **Francisca N. Okeke**, and **Dean Pesnell**

40 years of Diplomacy: Science, Education, Fisheries, and Community Development: Intercultural Center for the Study of Desert and Oceans (CEDO Intercultural)
Session Chairs: **Nélida Barajas Acosta**, **Christine Flannagan**, and **Robert G. Varady**
Citizen Science to Governance: Building a Sustainable Future for Northern Gulf of California Communities by **Peggy J. Turk-Boyer**
Saving Vaquita: CEDO Science, Sustainable Fisheries and Smart Seafood Choices by **Sarah Mesnick**
Marine Biology in the Desert by **Katrina Mangin**

Science Diplomacy Education
Session Chair: **Andrea K. Gerlak**
Building Science Diplomacy at the University of Arizona by **Andrea K. Gerlak**
Teaching Science Diplomacy by **Hassan Vafai**
The Transdisciplinary Environmental Science for Society (TESS) Program by **Daniel Ferguson**
A Role for the Universities' Partnership for Water Cooperation and Diplomacy by **Léna Salamé**

Friday, October 25, 2019

8:00–8:15am Welcome Address: **John Hildebrand**

8:15–9:00am Keynote: *Using Science to Build Community Resilience* by **Lauren Alexander Augustine**

PARALLEL SESSION 4
9:15–10:45am

Science and Diplomacy in the Middle East
Session Chair: **Julie Ellison-Speight**
Academia's Role in Science Diplomacy and Connecting Regions by **Sharon B. Megdal**
Arava Institute for Environmental Studies (Middle Eastern NGOs in Science Diplomacy) by **David Lehrer**
Science Diplomacy in Immigration Policy by **Megan Carney**
The Private Sector in Middle Eastern Scientific Diplomacy by **Mohamad Amin Saad**
HIVE: Harnessing Innovation through Virtual Exchange Program by **Julie Ellison-Speight**

Building Regional Science & Technology (S&T) Innovation Ecosystems to Deliver on the SDG's in the Americas
Session Chair: **Joel Cuello**
Tech Parks as Regional Innovation Catalyst through Public-Private-University Partnerships by **Carol Stewart**
Galvanizing Regional Innovation through University Technology Transfer by **Douglas Hockstad**
Building S&T Innovation Ecosystems in Mexico by **Justin Dutram**
Laws of Building S&T Innovation Ecosystems in Emerging Economies by **Joel Cuello**

	Bringing New Perspectives and Knowledge into Education Session Chair: **Dominic Boccelli** *Marginalized Knowledges and Education* by **Alberto Arenas** *Expanding Environmental Science Literacy to Include Cultural and Community Knowledge* by **Kristin Gunckel** *Speaking Back to 'Sustainable Development': Land Education and Indigenous Perspectives* by **Austin Cruz** *Data Sharing and Engineering Faculty: An Analysis of Selected Publications* by **Chris A. Wiley**
PARALLEL SESSION 5 11:00–12:30pm	**Sustainable Urban Development** Session Chair: **David Quanrud** *Framework of Urban Sustainability Indicators in Drylands* by **David Quanrud**, **Parastoo Parivar**, and **Hassan Vafai** *Sustainable Water Resources: Urban Scapes and Restoring the Commons* by **Itzchak Kornfeld** *Hydro-energy Cooperation in the Bangladesh-Bhutan-India-Nepal Region: Prospects for Transboundary Energy and Water Security in South Asia* by **Christopher Scott**, **Aditi Mukherji**, **Udisha Saklani**, and **Padmendra Shrestha** *The Central Role of Energy in the Urban Transition: Global Challenges for Sustainability* by **Joseph R. Burger**, **James H. Brown**, **John W. Day Jr.**, **Tatiana P. Flanagan**, and **Eric D. Roy**
	Panel Discussion: Connecting Your Science and Engineering Training with Careers in Policy and Diplomacy Session Chair: **Katherine Himes** Panelists: **Alex Thome** and **Ryan Jones**
	Hydrodiplomacy in Ambos Nogales: Lessons learned from binational water management Session Chair: **Adriana Zuñiga-Terán** *Governance, Cooperation Requirements, and Challenges for a Binational Solution to Watershed Management* by **Francisco Lara-Valencia** *Watershed Modeling in Ambos Nogales* by **Laura Norman** *The Role of Green Infrastructure in Watershed Management* by **Joaquin Murrieta** *Potential Sustainable Solutions to Water Resources Management Challenges* by **Stephanie Buechler**
12:45–1:30pm	Panel Discussion: *Networking for Science Diplomacy* Session Chair: **Christopher Scott** Panelists: **Nélida Acosta Barajas**, **Cherry A. Murray**, and **Robert G. Varady**
2:00–2:30pm	Concluding Talk: *Determine « X » in the Following Formula: (Trust + X^2) × Perseverance = Political Will* by **Léna Salamé**, introduced by **Elizabeth Cantwell**
2:30–3:00pm	Concluding Remarks, Best Paper Awards, and Participation Certificates: **E. William Colglazier**

1 Overcoming Social, Economic, and Environmental Challenges for the Americas and the World

How Can Science, Technology, and Innovation Help Find Solutions

DOI: 10.1201/9781003220503-1

SUMMARY

This chapter is a transcription of the opening plenary session for the conference. Honorary Chairman of the conference, Dr. E. William Colglazier, moderates a panel that includes Cherry A. Murray, Deputy Director of Research of the Biosphere 2 Institute; Ronit Prawer, Eastern U.S. Region Director for the U.K. Science and Innovation Network; and Vaughan C. Turekian, Executive Director of Policy and Global Affairs at the National Academies of Sciences, Engineering and Medicine. They provide their perspectives and experience in science diplomacy how it can advance diplomatic relations as well as scientific advancement. This is followed by a Question and Answer period with the audience.

SCIENCE DIPLOMACY AT THE UNIVERSITY OF ARIZONA

Liesl Folks
Senior Vice President and Provost, The University of Arizona

Welcome everybody! It's a great delight to see so many faces here for this important event. I would like to thank Kevin Lansey for inviting me to participate. I am very new to campus, still finding my way around, and still really discovering all the amazing things that go on around this wonderful research-intensive university.

It is a great pleasure to be here, particularly because if there is one thing of high importance to us right now, it is the role of policy in society. I want to particularly say a special thanks to the following people: Bill Colglazier, who is the honorary chair and organizing this evening's panel session, the keynote speakers and plenary session panelists, all the sponsors, and members of the organizing committee. It is a heavy burden organizing any kind of conference, so thank you for bringing this marvelous group together.

Given the theme of this session, which is *Overcoming Social, Economic, and Environmental Challenges for the Americas and the World: How Can Science, Technology, and Innovation Help Find Solutions,* I would like to take just a moment to highlight the important role that The University of Arizona has played in collaborations across the state of Arizona, of course, but also across the nation and the world. In fact, there is nothing we do anymore in science that doesn't ask of us to collaborate broadly. We have worked with European countries on particle physics to attempt to describe the origins of the universe. Speaking of the universe, we have partnered with corporations, universities, and NASA on the OSIRIS-REX mission to Mars, with the goal of helping us to better understand our origins and our destiny. We have engaged internationally with Latin American universities to research border issues such as water, immigration, and related topics.

The University of Arizona has demonstrated a record of expertise and an outstanding reputation in science, health, and engineering policy and diplomacy. We connect scientists with policy folks in areas such as climate science and climate change through the Udall Center, and more broadly through the Institute for Energy Solutions. The University's environment is conducive to these interdisciplinary collaborations. We work, for example, with the Navajo people in addressing food, energy, and water insecurity with faculty from different departments coming together to accomplish research goals and to extend that research into practice in our communities regionally and nationally. Importantly, we find that our work with our Native peoples here has relevance to work with Native peoples around the world. There are partnerships building even since I've arrived with people in my home country in Australia, which is an amazing connection when you think about it.

Introduction to Conference Chair

To discuss how we can expand the impact of science further, I would like to introduce the honorary conference chair, E. William Colglazier who will moderate the panel discussion this evening.

In 2016 William Colglazier was appointed to the group selected to promote the role of science, technology, and innovation in achieving the United Nations' "2030 Agenda for Sustainable Development" (United Nations General Assembly 2015). He was Science and Technology Advisor to the U.S. Secretary of State from 2011 to 2014, an executive officer of the National Academy of Sciences and the National Research Council. He is editor-in-chief of *Science and Diplomacy* and a senior scholar in the Center for Science Diplomacy at the American Association for the Advancement of Science. While at Harvard's Center for Science and International Affairs, he was the associate director of the Aspen Institute Program in Science, Technology, and Humanism. With that, I would like to ask Bill to come up to the stage and introduce his panelists so that we can kick off the discussion.

Thank you very much for being here, everybody.

FIGURES 1.1 Opening session of the SPD conference, The University of Arizona, October 23, 2019.

PANEL DISCUSSION

Honorary Chair

E. William Colglazier
Senior Scholar and Editor-in-Chief of Science & Diplomacy, AAAS Center for Science Diplomacy

Panelists

Cherry A. Murray
Deputy Director of Research, Biosphere 2 Institute; Professor of Physics, The University of Arizona

Ronit Prawer
Eastern U.S. Region Director, U.K. Science and Innovation Network

Vaughan C. Turekian
Executive Director of Policy and Global Affairs, National Academies of Sciences, Engineering and Medicine

SCIENCE, ENGINEERING, AND HEALTH DIPLOMACY: WHY IT MATTERS FOR AMERICA AND THE WORLD

E. William Colglazier
Honorary Chairman of the Conference

It is a great pleasure to be back at The University of Arizona. The conference that was held two years ago, *Science Diplomacy for the Americas*, was great. I am very glad the university is continuing with this series, and hope similar conferences continue to be held at least every two years. The title this year, which was the logo on the nice button in our bags, is *Sustainable Development for the Americas: Science, Health and Engineering Policy and Diplomacy*. Arizona is obviously a unique place being right next to the border, interacting not only with Mexico and Central America, but with all of Latin America as well as Canada.

To "kick off" the conference, we have three distinguished practitioners of science diplomacy, for our session this evening, titled *Science, Engineering, and Health Diplomacy: Why It Matters for America and the World*. I am asking each of the speakers to tell us about what they think about science diplomacy as a very broad concept and how people interpret it in different ways, and also why it matters to them and why they think it is important. I would like for them to speak about their experiences using scientific knowledge, expertise, and international collaboration to advance diplomacy among nations, and also to tell us about some of the things that they believe science diplomacy has accomplished in their own careers. What they have been involved in, what some of the challenges have been, what some of the obstacles are, and what lessons they have learned that we can use as we think about how science and technology, and really all academic fields and all areas of human knowledge, can be useful, not only for diplomacy but also for helping countries to accomplish their broad goals. After they make their remarks, I will come back with two questions, but will leave plenty of time to open the conversation up to all of you in the audience.

FIGURE 1.2 Honorary Conference Chair, E. William Colglazier, at the opening session of the SPD conference, The University of Arizona, October 23, 2019.

When I think about science diplomacy, the part that fascinates and interests me the most is when science and technology can actually lead the way in terms of reaching diplomatic objectives. Sometimes people in the military talk about the "tip of the spear" being the leading edge, the people that are out in front. There are some cases where I think engaging in science and technology internationally has done that and has had benefits for diplomacy and for international relations. I think that has occurred in several areas that I have observed during my own career, certainly during the height of the Cold War in the 1980s, the fact that American scientists and Soviet scientists conducted what were called Track II discussions, not formal government discussions, about arms control and nuclear weapons. Those discussions actually had a tremendous influence when there was a window of opportunity in terms of international relations between the U.S. and Soviet Union, and contributed to many of the arms control agreements in the 1980s and 1990s (Bureau of Arms Control 1987).

Currently the area where we all are concerned is the issue of climate change. One of the important partnerships where science diplomacy also was the leading edge was the creation of the Intergovernmental Panel on Climate Change in 1988, which brought together the world scientists that are focused on these problems, but also engaged with the policymakers and the decision-makers at the global level (Intergovernmental Panel on Climate Change n.d.). Their work and their partnership with the decision-makers helped lead the way to raising this issue up to the point where it actually could lead to progress on the diplomatic front at the global level.

In the case of dealing with different nations, one of the upcoming speakers at this conference is Dr. Glenn Schweitzer, a friend and colleague who has been heavily influential in the scientific engagement that was encouraged by both the U.S. Government and the Iranian Government, which has gone on for more than 15 years

(Schweitzer and Neureiter 2008). It was also very useful for diplomacy. Although this scientific dialogue never dealt with the hard, nuclear issues, it dealt with all other areas of science with collaborations, meetings, and workshops between the scientists in the two countries. That helped create the goodwill that led to some of the diplomatic achievements.

But I think as all of us know, politics is a more powerful force than science. And of course, in some of these areas I mentioned, there have been setbacks, and certainly in arms control agreements with the U.S. and Russia backing away; in environmental agreements with the U.S. withdrawing from the Paris Climate Accord (Hersher 2019); in bilateral agreements with the problems right now after the U.S. withdrew from the 2015 Iranian nuclear agreement, the Joint Comprehensive Plan of Action or JCPOA (BBC Press 2018). So it shows you that none of these diplomatic challenges are easy; sometimes you have to take the long view. I still remain an optimist that science and technology can make a difference, and we have to keep using it to be the tip of the spear for trying to advance diplomacy, conversation, and dialogue between nations to tackle some of the really tough issues that our societies deal with.

Introduction to Panel Members

So I am going to turn it over to my three colleagues. Again, they will talk about their own view of what science diplomacy is, why it matters to them, to international engagement, and what are some of the lessons that they have learned in their careers putting science diplomacy into practice.

Introduction to Cherry Murray

Our first speaker is Professor Cherry Murray. I have known Cherry for a long time. She is truly a distinguished scientist, member of the National Academy of Sciences. Recently, she was the Dean of Applied Science at Harvard and is now professor here at The University of Arizona. She is also Deputy Director of Research at Biosphere 2, which I had the distinct pleasure of touring this afternoon. It is truly a unique facility, doing really very interesting and cutting-edge science in specialized areas. She is also the co-chair of the InterAcademy Partnership (IAP) of the world's science academies, and so she may tell you a little bit about that. She also had experience at the highest levels of U.S. national labs, including Bell Labs. So, she has had a very long and distinguished career. I think The University of Arizona is lucky to be able to attract her back here.

Introduction to Ronit Prawer

Our second speaker is Dr. Ronit Prawer. She is the Eastern U.S. Region director of a terrific U.K. program that deals with both science and innovation internationally with offices and interactions in countries around the globe. Her office covers the eastern region of the U.S. and is located in Boston. The U.K. has been very effective at following events in other countries dealing with science, technology, innovation, and has capitalized very much on having these "eyes and ears" on the ground and in engaging with these ecosystems of innovation in various places of the world. She was

FIGURE 1.3 Panelists at the opening session of the SPD conference, The University of Arizona, October 23, 2019.

formerly a Science Counselor of the U.K. and Israel, so she has broad international experience.

Introduction to Vaughan Turekian

And our last speaker, Dr. Vaughan Turekian, is Executive Director of the Policy and Global Affairs Division of the U.S. National Academies of Science, Engineering and Medicine. He was also a former Science and Technology Adviser to the Secretary of State. He co-chairs the advisory committee at the United Nations that provides advice on the role of science, technology, and innovation (STI) for achieving the 17 Sustainable Development Goals, and helps to organize the annual Multi-stakeholder STI Forum at the United Nations. All three speakers have done truly impressive and amazing things in their careers. They have a unique and wise perspective on how science and technology can be a positive force in influencing human relations and relationships between countries.

WHAT IS SCIENCE DIPLOMACY?

Cherry Murray
Deputy Director for Research, Biosphere 2 Institute, The University of Arizona

I'll start out with, first off, my father was a diplomat, so I got into diplomacy from birth basically. I lived maybe two years in the U.S. Otherwise, I lived internationally until I went to college. I got very, very interested in physics, so after I got my Ph.D. at MIT, I went to Bell Labs. I was at Bell Labs for 27 years. Then I went to

Lawrence Livermore National Lab as deputy director. I was Senior Vice President at Bell Labs and Deputy Director for Science and Technology at Lawrence Livermore. Then I went to Harvard to become Dean. Then I went to the Department of Energy (DOE) to become the director of the Office of Science. For two years, I was a federal employee. I have now been at The University of Arizona since July.

So what is science diplomacy? I think Dr. Colglazier identified one part of what I think science diplomacy is, which is using scientists as the tip of the diplomatic sword. But I also think a lot of science policy is relevant to science diplomacy. So in trying to figure out what my charge was tonight, I decided I would just share a few things that I have done in science policy and policy for science.

Science Policy

Of over 25 National Academy studies, probably the most impactful one that I have been on was called Rising Above the Gathering Storm (National Academy 2007). This study was a slightly different study from the usual consensus study because it was hundreds of different people and workshops, but gathering together the rationale for why the U.S. (and other countries, because other countries have used this study) should follow its recommendations for science, general education, science education, support of young people—and young assistant professors in particular—in science, research and development portfolio, particularly fundamental science.

I also served on the President's Oil Spill Commission after the Deepwater Horizon accident, which had some ramifications, since it was in the Gulf of Mexico, with the country of Mexico (Obama 2010). It's now ten years since that incident. On the 14th of November, we're going to have a tenth anniversary meeting of the Commission. The Commission had a huge effect on the Bureau of Ocean Management. They completely reorganized how they were managing things. It turned out that before our report, none of the companies had any oil spill equipment that could easily be borrowed in an emergency (U.S. Congress House Committee on Natural Resources 2011). So a kind of emergency supply of equipment from different companies was made. The companies, or many of them, spent a very large amount of money trying to clean up the shore. Unfortunately, what Dr. Colglazier said is very true in this case, and much of this has been undone in the last ten years (Fears 2019).

Science for Policy

When I was at the DOE, I served on the National Science and Technology Council (NSTC)—which is, since we don't have a science ministry, we have a very large number of different agencies that fund science, as well as the National Science Foundation—getting them all together in a room to talk about various issues such as new initiatives. For example when I was at the DOE, we got exascale computing launched, funded semiconductor research, things like freedom of federal scientists, international relationships, and that sort of thing. I am on advisory committees for Germany, Singapore, and Japan advising either their energy research and development, or their national labs, or their science portfolio. All very, very interesting, I would say. I just got elected co-chair of the IAP. There are six co-chairs, so I'm one of six co-chairs of the IAP, which is a partnership of over 140 different national academies of science, engineering, and medicine. I am learning right now the power and impact of getting a network of academies working together.

Diplomacy for Science

At the time I was at the DOE, we began the diplomatic process of starting a new \$3–\$3.5 billion dollar class international science experiment in the U.S. The U.S. participates in international science abroad. We don't have science facilities in the U.S. that are this large that are international, very much like CERN, and that's what this is. And for that reason, we needed to negotiate a U.S.-U.K. science agreement because, believe it or not, we didn't have one. So we worked through it; Vaughan was at the State Department at the time, and working through the Office of Science and Technology Policy, we actually got an agreement. One story I have is that we really wanted to make sure that a certain amount of funding got put into the British government budget at a certain time. It turned out that Ernie Moniz, the U.S. Secretary of Energy, knew British Foreign Secretary Philip Hammond really well. So Ernie wrote him a little note saying, "you know, this is a fantastic experiment. It looks at the flavors of neutrinos, and we would really like to have that in the budget." And it got in the U.K. budget, and CERN provided a very large amount of in-kind in their five-year budget (U.S. Department of Energy 2015). We did, I think, 20 different international agreements and MOUs, including with Brazil, Chile, and Canada. The facility is now under construction and working well, very much modeled on the CERN model of having spokespeople for each part of the experiment that are international. And this is the first U.S.-based truly international facility.

The other thing I did was along with the United Nations Conference of Parties, otherwise known as COP24. There was an Energy Ministerial and my job was to stimulate investment and research in energy R&D, in particular, sustainable energy R&D. So we began what was called Mission Innovation, a partnership of 26 different nations around the world who committed to increasing their funding by a factor of two on renewable energy research and development.

Science for Diplomacy

When I was at Lawrence Livermore National Lab, we had the very last of the lab-to-lab Track II collaborations, where the weapons labs of the former Soviet Union and of the U.S. had scientist-to-scientist meetings, which were extremely helpful for things like nuclear forensics of material that could possibly be stolen and is found someplace. At the time, the Nunn-Lugar Cooperative Threat Agreement was coming to an end. That was the agreement that basically helped the Russian scientists locate all of their special nuclear materials, which you would not want to get into the hands of terrorists when the Soviet Union fell apart (Kuiper, n.d.). The U.S. Congress actually provided a considerable amount of money for the U.S. national labs to help them find and locate the materials and sequester them. I also was on several Russian Academy of Sciences-National Academy of Sciences arms control meetings at the International Atomic Energy Agency (IAEA). However, and unfortunately, these conversations have not been continued in the last few years.

So very briefly, some key insights and lessons learned throughout my experience:

Communication is always important. First of all, who is your audience? So whom do you want to encourage to do something? If you know your audience, then you need to find the right spokesperson. It may not be you. It may not actually be those elite scientists. For the more populist-leaning people, the spokesperson needs to be someone whom they actually trust.

Active listening! Figure out what the audience wants. What is your target? Who are you trying to help? What actually are their problems? You need to listen to them.

The same message needs to be given over and over. So, for example, just because you have testified to some Congressional Committee on a certain subject it doesn't mean that your information is in the record and you don't have to say anything again. No, staffers turnover, Congresspeople turnover. You need to say it constantly. You need to be in touch with staffers and Congresspeople or whoever else you want to reach.

Use systems thinking. Societal problems, such as how to deal with managing our Earth, are extremely complex. In fact, they are complex systems, so you need to use systems thinking. That means there *will* be unintended consequences of whatever you do. So try to figure out what they are and monitor what's happening. *Good is always better than perfect*, because you're not going to get perfect.

Have patience. Circumstances change. So when something can't be solved or politically is not on the table, it can be in, say, five years or ten years. As I said, politicians and officials change and compromise is often a solution, and that's what politicians do all the time. Scientists don't like to compromise. And there're some things you shouldn't compromise on, such as ethics, but compromise where maybe you're not getting everything that you think is going to be important for whatever. *Doing something is better than doing nothing*. And with that, I will step down.

HOW DOES SCIENCE DIPLOMACY ADVANCE DIPLOMACY AMONG NATIONS?

Ronit Prawer
Eastern U.S. Region Director, U.K. Science and Innovation Network

First and foremost, a huge thank you to Bill Colglazier and to Hassan Vafai for inviting me to what I'm sure are going to be two days of very rigorous discussions, which I'm looking forward to immensely. I have to say that when I was looking at the program, I felt like a kid in a candy store! I want to be in every single one of the parallel sessions, so if anyone is the technological entrepreneur who has invented a Hermione Granger-style time-turner, I'm in the market for one.

A couple of years ago, Neil deGrasse Tyson very hopefully tweeted "The good thing about science is that it's true whether or not you believe it" (@neiltyson, June 14, 2013). If you'll permit me another favorite aphorism, the satirical journalist Sir

David Frost is frequently quoted as having pronounced that "*diplomacy* is the art of letting someone else have your way." I often tell my team that if you combine those two helpful adages you might say that *Science Diplomacy* is the art of subjecting governments to objective truths and then helping them navigate the implications.

As Bill mentioned, I am a "practicing" science diplomat, rather than a student of the discipline per se. In my current role as Regional Director for the British Foreign Office's Science and Innovation Network, I represent the U.K. government here in the U.S., overseeing our bilateral scientific relationship, and my team works to nurture and promote that relationship. Unlike my co-panelists and mentors here on the stage, I have not yet escaped the civil service, though one day I hope I'll join them and find out what it is like on the other side.

On that note, it's important that I give you a brief disclaimer before we dive in. Today I'll speak today of personal experiences as a science diplomat, and how it has felt for me as an individual to be working at the intersection of science, government, and society, a nexus which you will continue to examine across this conference. My comments therefore will reflect my own personal opinions and experiences, and not those of the British government.

What I thought I would do today is tell you just one story. It is the story of a seminal moment in my career which illustrated to me the power and potential of science diplomacy to change the course of international relations. Dr. Colglazier's "homework" question to us panelists today was to ask: "How does science diplomacy advance diplomacy among nations?" I'd like to answer that question for you through the lens of this particular experience.

But before I do that, I realize that you don't have to be an astute linguist to have noticed that I am in fact Australian, rather than British. I won't go into the long and winding road of *how* I got here, but it might be worth telling you a little bit about *why* an irreverent Antipodean would want to serve Queen and Country in the British Civil Service, particularly seeing as Bill alluded to this in his introduction. Let me spend just a quick moment on *British* science diplomacy and give you a bit more detail on the network that Bill mentioned.

The answer as to why an Aussie would be working for the Brits is of course, in part, the world-class U.K. science that I have the honor of representing. But let's go beyond the scientific community, and more specifically to science diplomacy. The U.K. has an inspirational and extensive history of science diplomacy and has been, I would say, a global pioneer in both its practice and its scholarship.

British Science Diplomacy

In the 1700s, the Royal Society appointed its first Foreign Secretary, Philip Zollman, and his role was actually to maintain relationships and share expertise with scientists abroad (Massarella 1992). Now, just for context, it took another 60 years before the British government appointed its first Secretary of State for Foreign Affairs. So in effect, you might say that the U.K. government was doing science diplomacy before it was doing any other kind of diplomacy (Poliakoff 2015). Fast-forward 150 years and the U.K. sent its first accredited government scientific representative abroad, to the U.S. in fact, which was the beginning of this special relationship in science. That gentleman's role was the precursor of mine, and he was also—fun fact—the grandson of Charles Darwin. So you could say I'm basically related.

The U.K.'s commitment to science diplomacy is not just Downton Abbey-style historical; it's actually extremely current. Fast forward to 2010. That same Royal Society authored what is probably the seminal piece of work on science diplomacy of the decade, together with AAAS and some of my colleagues on this panel (The Royal Society 2010).

I want to emphasize that the U.K.'s attitude towards science diplomacy isn't just like a landmark report here or there or a particular action or activity. It is actually a shift in *modus operandi*. The U.K. government was early to appreciate the changing nature of diplomacy and to adapt its foreign service accordingly.

Here is a case in point: when countries send Ambassadors abroad, they send them with key staff members "attached" to them (hence the French term *attachés*). These are the technical experts assigned to the staff of a diplomatic mission. They might include a political attaché, an economic attaché, a cultural attaché, or military attachés if there is a bilateral military relationship. Rather than having science as a subset of other portfolios as many other countries do, the U.K. recognized really early that it would be good to invest in specialized science attachés. If you think about it, this makes complete sense and is in fact critical level of specialization, because we know not everyone speaks "science." An economic attaché might have the language perhaps to talk about IP [Intellectual Property], but not PIs. The military attaché might be able to intelligently discuss security, but it would take a different skill set to discuss food security. The sort of person you wanted in the negotiation room on nuclear waste, or microplastics, or quantum cryptography is someone with a scientific background.

So, in the late 1990s, the U.K. started to include a science attaché in this critical team dedicated to the bilateral scientific relationship. These science attachés make up a network that I manage in the U.S.—the U.K. Science and Innovation Network, or SIN for short (and yes, that is definitely the best acronym in government—we are the "SINners"). It is the first of its kind globally, and now numbers 100 offices in 50 countries all over the world (UK Science & Innovation Network 2015).

In 2013, I was approached by the British Ambassador to Israel, who said to me, "I want science to be the foundation of what the wider bilateral relationship can be," and I hope you can now see why I very quickly said, "I'm in!" And I have been working with the British Civil Service ever since. That actually brings me to the story that I want to share with you today.

Science Diplomacy in the Middle East

When I began my role as science attaché Israel, I was introduced to a brilliant Palestinian scientist through a friend of my family, Hassan. Hassan was talking to me about his students, and at some point during our conversation, it dawned on me that his students were *all* abroad, in labs all around the world. Being globally connected of course is not uncommon for a scientist, but this was pretty extreme. So I asked him about it, and Hassan explained to me that if you graduated with an M.A. in a STEM discipline in the OPTs, you had three choices. Either you leave the university with your M.A. and go into industry or science-adjacent industry; or you stay on at the University to teach, or you head off to do a Ph.D. abroad, mainly in the U.S. and Europe. This, he explained to me, was because Palestinian universities had limited capacity to award doctoral degrees in the exact sciences. Alison Abbott wrote about this issue in *Nature* last year in a column on capacity building in Palestine (Abbott 2018). As many of you here are aware, a Ph.D. is a half-decade proposition. So a lot of promising young Palestinian scientists left academia, rather than leaving their homes and their communities in order to undertake this long process.

Barely ten kilometers away in Israel's world-class laboratories, I was having a very different conversation with Israeli PIs. The booming high-tech industry was growing, and for many of these companies, an M.A. graduate was the perfect employee, and the expertise of Ph.D. was often better achieved while on the job. The high-tech industry was devouring M.A. students. It turned out that Israeli labs and academics were actually desperate for high-quality M.A. graduates to come and do Ph.Ds. with them.

So, call it optimism, call it naiveté, stubbornness, or call it science diplomacy, but I knew what I wanted to do: I was going to bring Palestinian students to Israeli labs. A simple question of supply and demand, right?

Over the next year, I faced challenges ranging from funding to border crossings to academic boycott. An exhaustive list would be, well, exhausting and impossible. At one point, lab members in Tel Aviv couldn't even continue a Skype meeting with me, because they were in shelters under fire from militants next door to their partner labs in Gaza.

But I was determined. (And young and stubborn.) I put a lot of professional and personal capital on the line. I cajoled, I convinced. You can see my gray hairs, which I believe are from those initial years. But what I did absolutely pales in comparison to the bravery of the young Palestinian and Israeli scientists who risked everything from political backlash to promotion prospects, to even personal safety in order to be part of the program.

With their help, we established the U.K.'s first science diplomacy program between Israel and Palestine, and under that program, the U.K. funds five exceptional M.A. graduates from Palestinian universities and they complete their full Ph.Ds. in Israeli labs. We named that program GROWTH. It stands for Graduate Research Opportunities in Water Technology and Health because those are the topics that the students address (British Council, n.d.).

Let me give you an aside, just to give you a flavor of the intensity of establishing the scheme. At one point early on I said to my team, "guys, I've been calling this program the Israel-Palestine Ph.D. Program for Graduates (or something very long-winded). We're having a meeting with some funders and I want a better name, so let's get some ideas on the board!" And they said "Ron, we've got it. We'll call it Research in Israel and Palestine—R.I.P. for short!" They were of course being tongue in cheek, but there was a lot of uncertainty and a lot of skepticism that they were referencing with their humor. Knowing that earlier draft I'm sure you'll appreciate how proud I am of the name GROWTH—we got there eventually.

So, because I did not have enough gray hairs at that point, we also undertook to establish a second program in the region, and this is one that people have alluded to already, which is basically on transboundary water resources.

As many of you know, the Middle East is one of the most water-poor regions of the world. (I'm looking forward to attending some talks on this later in the program). Of course, we all know that water does not care about our politics or arbitrary political borders. It doesn't say, "Okay, everybody, this appears to be the 1948 green line, I guess we will stop here and turn around." So we established a program called STREAM: Science, Technology and Research Exchange Across MENA (UK Science & Innovation Network 2018). It is a funding instrument whereby one U.K. partner, an Israeli partner, and one partner from the wider MENA region can jointly apply for funding to work together on solving some of the most critical water issues of the region.

To tell you all the truth, I was a little worried we would have a lukewarm response to the initial Call for Proposals – after all, working across borders is extremely difficult, and the Call for Proposals explicitly demanded genuine collaboration. But the response was overwhelming and we received a huge number of excellent submissions. Last I checked, the program currently involves Jordan, Egypt, Morocco, the West Bank and Gaza, Lebanon, and more. It's a wonderful reminder that curious and courageous minds will always find a way to work together. Science diplomacy means doing all we can to support them.

Let me close with this: people often ask me if I have a favorite souvenir from my time in the Middle East, and indeed I do. It's a WhatsApp group. It's called "Oxford Friends," and is named that for this reason. While we were establishing the steering committees to help us administer the trilateral water program, we convened a set of discussions in Oxford so that participants could have some conversations in a quiet place away from their hometowns, which, shall we say, are not quiet.

In this WhatsApp group, people from all over the Middle East exchange data, they exchange blessings for festivals, and they check in on each other when the going gets rough in the area. I understand some of what is written in this WhatsApp group because it's in English, and in Hebrew, and in Arabic. I left Israel three years ago, but I still see my phone light up with blessings and brotherhood in that group, so that is unquestionably my favorite souvenir.

I bring this story to you as one of the examples of a project of science diplomacy really moving the needle in the region, and hopefully provides a satisfactory answer to Bill's question. Quietly, away from the headlines, young scientists are working hard in a spirit of collaboration and optimism. It might not happen tomorrow, but will certainly change the wider picture in the region. So on that note of optimism, perhaps, I'll turn over the time to Dr. Turekian.

HOW CAN SCIENCE DIPLOMACY AFFECT CHANGE?

Vaughan Turekian
Executive Director of Policy and Global Affairs, National Academies of Sciences, Engineering and Medicine

It is a real great pleasure to be back here at The University of Arizona. I had the wonderful opportunity to visit, once when I was younger, and then again two years ago I had an opportunity to come when Professor Vafai and Dr. Colglazier invited me to the Science and Diplomacy Conference in 2017. Back then, I was actually on the way of leaving my position at the State Department, during a time of transition, and it was an interesting conversation that took place there. What I *really* appreciated was the fact that you held this conference in February, and there is no greater draw to come to Arizona than to come in February. So I think we should all work on that as a formal output from science and diplomacy here; that we will all come to Arizona in February.

And Dr. Murray, it was so interesting hearing that your father was a diplomat and how that impacted you. My father was a scientist, and it actually also impacted my science diplomacy. I remember well, actually, as a very young boy—I was about five years old—and my father was going to Caltech to visit a very famous fellow geochemist. His name was Pat Patterson, Claire (Pat) Patterson. Pat was known as the person that had used lead isotopes to get a more exact date for the age of the Earth (Smith 2015). And there I was, sort of a young, very bored child sitting in this room while my father and Pat were sitting there, and they were screaming at each other! These are two titans of geochemistry just yelling at each other. Oh, my goodness. It was just… reverberating. I rushed out, as you would if your parent is yelling at somebody. I come out, with a little tear in my eye and said, "Dad, why? Why do you visit this man in California when you hate each other?" And he turned without a moment of hesitation and said, "Vaughan, Pat is one of my dearest friends. We were arguing about whether lead isotopes could be used to date some structure in the West Coast."

Thinking back to that moment, it dawned on me that one of the great things about science is that it allows us to disagree with each other around substance. And when we think about science diplomacy, what we are doing oftentimes, particularly when we think about high diplomacy, all those issues of great sort of national and international intrigue and bringing science into it, is saying that there may be times when we might not agree with the country, but we're going to find the things around science that we can work on together to build some sort of collaboration built around the science, the "truth" of science, whether we like it or not.

The United States and Cuba

I have been so fortunate through my own career to have been exposed to some of the great exploits of science diplomacy. We will hear about some of them this week, including from Dr. Peter Agre. If you've not heard Peter recount his stories of science diplomacy, they are amazing. One that strikes me, and I don't mean to "kill Hamlet in the first act" here and jump on his story, but it was one of my favorite moments of science diplomacy and is actually one of my proudest moments: the U.S.–Cuba relationship. Now, it's obviously changed in the last little bit. It's a relationship that changes every 63 years or every three. But what was interesting about that relationship was, very early on I had been involved with efforts to try to find ways to bring U.S. and Cuban scientists together. There was not much going on in the sort of the early 2000s around U.S.-Cuba relationships, but there was some advantage to having some scientific conversations.

At the time, I was involved in the Center for Science Diplomacy at AAAS. So you may have noticed, but Bill Colglazier and I basically work the same jobs at different times. So before Bill was at the Center for Science Diplomacy, I was at the Center for Science Diplomacy. He was actually the Science Adviser to the Secretary of State, and then I was, so it's always sort of a "who's up first" type of thing.

So in 2009, we went down to Cuba and Peter Agre was with us, which was a big deal; we brought a Nobel Prize winner with us. We stayed at the Hotel Nacional, which is an iconic hotel. It was just a remarkable place. It was this iconic, beautiful, gorgeous structure of this sort of old wood and was amazing, though it's pretty dilapidated now. So when we went, there were no Americans allowed, really. In fact, you had to get your own currency, sort of an American version of the Cuban currency called a CUC (or a Cuban convertible peso). We were really sort of one of the few Americans that are government, but we're *not* government because government officials aren't allowed down in Cuba at this point. So we're checking out of the hotel and we see a bunch of Americans, so we turn to them and say, "Who are you?" They say, "We're with the FAA [Federal Aviation Administration]. We're the air traffic team from Miami." This is 2009, right, and we say, "Well, what are you doing here?" They said, "Well, you know, we're not really that important," because you couldn't be up to a certain level in politics *and* go to Cuba, "we're the weenies; we're the technical people." So we said, "we're the weenies, too! We're the scientists." So we actually had a very good conversation. We're having this sort of nerdy conversation around things like operations and weather management, all these different issues. Sure enough, we leave and we're on the same charter jet. So when you would go to Cuba at that time, you took a chartered plane. There's no commercial aircraft. You go and the plane goes up and it goes right back down, 90 miles to Florida. We get on the charter flight. We see the air traffic controller group from Miami get on and they walk to the back of the plane. Then as we're taking off and leaving Cuban airspace, the charter pilot comes on the overhead speaker and he says "we're now leaving Cuban airspace, so to the Miami air traffic controllers in the back, your colleagues at Havana Air Traffic Control want me to tell you that they loved your visit. They learned a lot of technical things and they can't wait to meet you in Miami and teach you how to salsa."

Well, this bizarre world is happening. Now I'm interested, right? So we get back and I end up doing a few more things with Cuba, with a number of others, and we're looking for solid things to work on, science things to work out, not garbage. Real good science. What can we work on in health and a good health care system? What can we do on environment? There's a lot of shared stuff between the two countries. What can we work on in understanding atmospheric transport? And so I end up going into government, right around the same time that President Obama is looking to change the relationship with Cuba and looking for all sorts of things to do. So actually, that relationship came in environment, in health, and a number of other things, but there was a building of science into that. But that's not the story that really grabbed me.

The Science and Technology Adviser Office at the State Department sits right next to the Office of Cuban Affairs. This went from being the sleepiest office in the State Department to now this hub of activity. The President is now changing the relationship with Cuba. The President announces he's going to go down to Cuba and we get invited next door to watch the President's visit. We had done some of the negotiations around the U.S.–Cuba relationship, in no small part, actually, because Fidel Castro's son, Fidelito Castro, was the science advisor to the government of Cuba.

So there was a lot of effort to try to bring more science into the official relationship (Stone 2018).

So we're sitting in the Office of Cuban Affairs, sitting and watching. And suddenly you see this iconic picture of Air Force One, this massive airplane, coming in and flying over the old city of Havana. And I turned to my deputy and I said, "How much do you want to bet those air traffic controllers in Havana who are guiding in Air Force One, are the very same air traffic controllers who were learning from their U.S. counterparts about what to do?" You think about this: in that moment of time, we were seeing the continuity, the continuity of culture, the continuity of history, and the continuity of people trying to find ways to work together beyond the divisions that separate them. So we were seeing in that moment what happens when you actually bring people together, and technical people together, to try together to solve a problem of common interest. Whether it is trying to determine whether or not you can date some geological structure in California or land the most sophisticated airplane in the world, ultimately, science diplomacy is about bringing people together to solve a common problem.

QUESTIONS AND ANSWERS

William Colglazier

Now I am going to ask our panelists a question. Looking more towards the future, I was going to mention that with all the challenges that are occurring in the world right now, both on the technical side and on the policy side, I was going to ask whether the panelists were optimists or pessimists, but I think as you heard from them, all three are optimists. One good thing about people that are involved in science and technology, even when they're involved with diplomacy and in politics, they really are optimistic. They really do believe that science can make a difference for positive, not only for our country but also for all countries.

But anyone looking at the dystopian side, we see tremendous challenges facing our country. The political divisions do the same thing: increasing divisions between countries. I've mentioned that some of the successes for me in science diplomacy have been rolled back in recent times. We have a lot of challenges at the global level, at the national level, politically. And then you look at science and technology. It's developing incredibly fast. I mean, it has created tremendous opportunities, tremendous new possibilities for people. The digital revolution, robotics, what's occurring in the life sciences, you name it. Amazing things have occurred both in companies, some of which have emerged from almost nothing and in just ten years' time became some of the largest companies in the world. But at the same time, we've recognized also with all these new disruptive technologies that they can also bring along a number of challenges that we have to deal with. One of the things that all countries have to face is how do we maximize the upside? How do we deal with and negate the downside? I mean, just for example, look at social media. Now we have the era of fake news and using these new technologies for doing things that actually, in some cases, have brought about great divisions in society.

So my questions, which I'll just sort of just group together:

Question #1

What do the panelists recommend to young people, young scientists, engineers, and health professionals who want to make a difference working on these issues, applying what they've learned and using their expertise to actually improve the situation in the world, their country, and around the world?

Question #2

Considering all these things that are on the negative side, in terms of politics inside countries or between countries, with some of these enormous challenges coming about from these advancing technologies, which also have a disruptive side. What exactly is their solution for how science and technology can make the biggest difference to help us chart a positive path through both the political challenges and the technological challenges?

Each panelist will give their thoughts to the matter; then we'll open it up to all of you in the audience.

Cherry Murray: I am actually very optimistic that scientists can deal with some of the unintended consequences of new technologies, because I see them rallying together in workshops, British Royal Academy, U.S. National Academy of Sciences, and the Chinese Academy, for example, chartered an international workshop to talk about gene therapy in humans. It was at that workshop in Hong Kong when the Chinese scientist revealed that he had genetically modified two human babies. And there was an absolute uproar at that conference in the scientists, a completely international group of scientists, and they were quite clear that there are some ethical guidelines that need to be made internationally. (Stein 2018)

Very similar things are happening with A.I. I was at the Science and Technology Society Forum in Kyoto, and there was a session on cybersecurity. I was talking to the chief technology officer of Symantec cybersecurity firm, and he was talking about maybe we should even have a conference and workshop to think about what's ethical in publishing research on cybersecurity, because he has deep fakes, video tapes that people have openly published, the software and the ability to use eight minutes of video taken off the Web from YouTube, to create an entire speech that looks extremely convincing. Is that a good thing to publish? It's like, is it a good thing to publish the militarization of anthrax? I don't think so, but at least scientists can get in the room and talk about it, along with policymakers and ethicists, as well. I think that is more and more important, because science is going, and technology leaps because of science, it's going much faster than governments can keep up. There is no way that regulations are going to keep up with this. The scientist community has to regulate itself on the ethical principles.

Vaughn Turekian: I think those are great examples. I'm going to shift this a little bit because I think one of the things that I know I think about a lot right now is, and it has implications for the way I think The University of Arizona is going to address science diplomacy and why it is so important that you are actually thinking about this and moving in this direction of developing your programs. One of the big challenges is that we are living in a time where, and it's not just in the U.S. or in the U.K., it's truly around the world that we're seeing a real rise in populism and nationalism. That is a political reality. The reasons are very complicated. But that is, in fact, something that we've seen around the world, and in Latin America we are seeing it in similar fashion. You can go from Brazil to Mexico and points in between.

The reason why I think this is really critical and why it's so important for those that are sort of early in the thinking in this space is that even as the political system itself wants to move and migrate towards nationalism and populism, the very issues, like those that Dr. Murray mentioned, that are actually going to drive the system have a global imprint on them. What does that mean? Climate change, sustainability, emerging technologies. They are global by definition, almost global-scale issues, and yet they're coming into a political spectrum which is focusing much more nationally and in a populist way. Not to mention the diversifying of funding. The system itself has changed. In 1945, Vannevar Bush produces Science the Endless Frontier (Bush 1945). It defines the basic underpinning of the U.S. and in many ways the global science enterprise as one that is driven by basic research funded by governments. That assumption is not there anymore.

What it takes now is a whole new generation of thinking of what are the right structures, what does the right architecture for global science look like? Where does science diplomacy merge with international science cooperation? They're different, but they're not so different right now because they actually are about informing and driving science policy at a global scale.

We've got to look at this from all the different sort of pieces of it, deconstruct what was sort of assumed, reconstruct a new system and do it pretty quickly. There are a number of us who've been around this space can think about some of it, but it really requires new thinking because the very basic technologies, the very basic forms of communication and interaction, that are going to be driving are going to be the ones that are actually much more in the domain of the next generation.

Ronit Prawer: I was going to say something totally different, but Vaughan, as usual, has made me think. One of the key elements I think we need to consider when we talk about what science diplomacy looks like going forward *sounds* incredibly boring. But I want to share it with you anyway, because it's something I've been thinking about. The issue I'm talking about is monitoring and evaluation (or 'M&E') for science diplomacy. Don't all fall asleep quite yet!

Let me put it this way: as you heard today, each of the three of us has different stories of the way science diplomacy worked to change something, move the needle, change a metric. We've talked about how it helped reduce toxic waste, or helped engender U.S.-Cuba rapprochement. What we are hearing is the way in which we communicate the impact of science diplomacy to all of our stakeholders, be they governments, funders, philanthropists, big technology companies, or other partners. And, as you heard, the way we do that is deeply qualitative.

So, when I want to tell the government that science diplomacy is doing a good job, that it's important to continue to invest in it; or when I want to tell the British taxpayer that having science attachés in cities and states all across the world is good for the U.K. and good for science and the world, how do I communicate that to them? Is it a pounds and pence type proposition? Do I say, well, as a result of having this person here, this company opened up in the U.K. and that equals five more high-skilled jobs which is worth more to the U.K. economy than what you pay me, so it's worth it? Should I be quantifying it in that way? And what is the difference between the qualitative and quantitative metrics?

Now, let me tell you a small story. Every government likes to think that it operates like an actuary. For example, if you are thinking about whether or not you as a government should fund the next cohort of GROWTH scholars, do you think to yourself, "if this person from Israel is working with this person for Palestine, what is the percentage decrease in likelihood that they will have animosity between them and what is the percentage decrease in likelihood of violence? And how much does violence cost my economy, and is it therefore with the investment?" At the end of the day, do you just plug in the numbers and make an actuarial decision?

Here is my theory: deep down, few people are motivated by numbers, even though governments like to think they are. What governments are motivated by, and I can tell you this firsthand, is when they come into a lab and they go up to meet one of the Palestinian scientists and that person is holding a baby. And they say, "Oh did you just have a baby?" And the researcher replies, "no, this is my friend's baby, but it's Jewish prayer time, so she's gone off to do her prayers, and I'm just holding her kid while she does that, and then she looks after my kid on the days that I need to bring him to the lab." And in that moment of humanity and connection, in seeing compassion for others happen firsthand through science, that minister or that funder is convinced, and the numbers are less important.

What I'm trying to say is there is so much about qualitative impact that we need to figure out how to communicate. That isn't to say that quantitative metrics aren't also important, but we have to be careful about too much reliance on them. Metrics motivates behavior. A great example is a few years back when people started looking at the number of women on corporate boards, and organizations scrambled. It was all about "quick! Find a woman and put her on the board," rather than listening deeply to why that culture of excluding females wasn't working.

When you look at how metrics motivate behavior, so how are we—governments and companies and corporations and philanthropists—measuring the impact of science diplomacy? And is there some way to kind of standardize that and how to think about that more deeply? I am writing a policy paper about this for the U.K. government, so if anyone has ideas, I'd love to chat more!

Question

Bill Craig: My name is Bill Craig, I'm a local here in Tucson. I was a teacher and now I'm just interested in all of the things you have to do, because they really do matter. My question has to do with unexpected consequences in science. We deal with a lot of the aftermath because people put things out there and then those unexpected consequences cost all of us. And the cost of reduction of that is brutal. Wouldn't it be better off if we would do a more thorough job of testing in the beginning, and then not have to deal with quite so many of these unexpected consequences? Even on a small scale?

Answer

Cherry Murray: Yes. Some of them, though, you really cannot predict. But this is why monitoring what you're doing is very important. And yeah, you can do it at a small scale. Let me think of an example. The unintended consequences of Facebook are people are split into their own group and don't talk to anyone who disagrees with them.

Craig: I'm thinking more chemical engineering.

Murray: Well, chemical engineering is slightly different, since you don't have to deal as much with the social aspect, but the premise is similar. I was just going to say, I don't think anyone had any inkling, even with a small-scale Facebook trial, that this would happen. You just have to monitor it over time. For chemical engineering, it may be a little bit easier because it's chemistry and not a social science. But agreed. This is why we have clinical trials of medicines. The real unintended consequences of technology are when society starts using them at large and they become global scale. I mean, there *will* be unintended consequences of, for example, automatic cars. It will happen even though they're trying very hard to make sure that there will be fewer accidents in the cars and stuff or whatever. There will be some unintended consequences. There is an unintended consequence of having more efficient engines in your cars. People drive more. There are unintended consequences of plastics.

Question

Singleton Thibodeux: One of the questions I've had as an early career scientist, is becoming more involved in science and policy and communicating that and how you communicate that, which is difficult because as scientists, we're not always trained as communicators, but something that constantly comes up in how we communicate its ethics, and what a gray area that is in reducing something that is a very complex topic in discussion, into that is digestible in a small bite of information, but yet doesn't necessarily represent the true picture. I was hoping that you could speak a little bit more to how to define ethics in the science community, which currently is not something that is very well defined.

Answer

Murray: That is a really tough question, and I think it needs to be defined by the science community. Early on with DNA sequencing, there was called an Asilomar conference, which was a retreat of the people who were doing sequencing of DNA to start thinking about what is ethics in this case (Berg 2008). Similarly, what is ethics in the use of artificial intelligence and machine learning? There are some very unethical things that can be done and we would like to stop them or at least be aware of them. I think the scientists and engineers and medical doctors who are dealing with these technologies need to be thinking about them as they are utilizing them or developing them. This is why when I was dean of engineering, we had ethics in every classroom, engineers in particular. People's lives depend on things that engineers do, and therefore whatever they are doing, they have to keep that in mind. And that is also true of scientists developing new technologies.

Turekian: That is a really good, very tough question. One thing in thinking about this issue is the role of science in terms of our own perspectives as science advisors, when we served in that role. What you begin to realize whenever you're in that position of having somebody senior that you provide advice to, especially in a democracy, is they are hearing your science advice here [in their peripheral vision]. This is the part that says, "watch out, this could be this or no, this isn't a problem." But in a democracy, they are *hearing* this from us, but they are *listening* to a broader set of public.

And so it becomes incumbent on not only universities, but in every way of engagement to recognize that a technically literate society is actually helpful to this conversation because the scientific community might think it's setting ethics; the reality is they're not. They're setting it up to a point, but only up until a point of public outcry. So there is a flip. The science community could say "we think this is fine," but the public could step in and say, "no it's not, stop it." And that's the flip. The physics community saw that, I think in the 1970s, in many ways, and it showed that interaction is important. And it happens at all levels.

You also have to recognize then that we assume there are people that are interested in science and not interested in science. This simply isn't true; it's a false dichotomy. Everybody, by definition, is interested in some version of science, because if you're a farmer, if you're worrying about seasonal changes in your rainfall, you know that field better than anybody else. And so it becomes a challenge of how we think about understanding. What's the demand and then what's the supply of information and knowledge. Everybody is not thinking about everything all the time.

Prawer: I think that the only thing I would add to what my colleagues have said is that ethics are subjective. My perception of what is morally correct might be different to the person sitting next to me; let alone the conceptions of what is within ethical norms in other science cultures. I recently convened an 'A.I. and Ethics' meeting together with The Royal Society and the American Academy of Arts and Sciences, and the meeting looked at the Future of Work as it intersects with A.I. And even with two learned societies whose fellowship is genuinely interwoven, there are *still* subtle cultural differences in the way we perceive progress, and the extent to which we let market forces—or those forces driving the marketplace of ideas rather—how freely do we let those forces progress in the enterprise of trying to find new knowledge.

In order to have a productive meeting on "A.I. and Ethics," what we really needed to do was to talk to the researchers who are at the very forefront of A.I. and Machine Learning sciences and say, "I know you're the expert, but this person is a social scientist, this one is a philosopher, and this one is a historian, and the arts matter in this conversation, and you might be an A.I. expert, but you are not an A.I. *and* ethics expert."

We can have incredibly productive conversations when we as scientists really listen to people who aren't wired the same way as we are and when we interface with disciplines which function differently. For example, it was amazing watching A.I. researchers at this meeting talking to historians who are looking *backwards*, not looking *forwards* like we are trained to do, and hearing the conversation. "What do we learn from history?" they asked. "What do we learn from the third industrial revolution that we can apply to the fourth?" "How can we approach this socio-technological upheaval in a way that is productive?"

Finally, I think clarity is really important. That goes for people (we need to know what your ethics are, and you have to be very vocal about them) but it also applies to societies at large. To this end, have a look at the U.K.'s A.I. and Ethics center; it's actually called the Center for Data Ethics and Innovation. The Center is really taking a very profound look at this, with the aim of having a set of stated principles about our ethics around this topic, statements that we can be clear about and that we can be loud about. I think when you are clear and loud about your principles, you are more likely to be able to uphold them.

Colglazier: I think one very positive note here is that many areas of science are now thinking about this. I have to organize a session for the American Physical Society, and one of the sessions they want is on science and ethics, on science and integrity, which means that, of course, there are values and ethics in doing science. I get the impression they are much more concerned about the interaction with society in terms of ethics, particularly because in the case of A.I., in the case of gene editing, these are uncharted waters. But Vaughan's ultimate point that society is going to make a decision in the end, it is the scientists' responsibility to tell them about the implications that are occurring with these new technologies.

Glenn Schweitzer: I'm going to comment on your observation about the importance of evaluating the impact of science diplomacy efforts. For the last 20 years, I've been managing the largest U.S.-Iran science engagement program that exists (Schweitzer 2008). I'd like to give a loud shout out to The University of Arizona for setting the pace on how you document meetings of scientists. If you look at their publications, they are, I would say, by a factor of 10, better than anyone else's effort I've ever seen to document short-term engagement in a long-term fashion. If you want to know something about evaluating science diplomacy, turn to The University of Arizona.

Question

Unknown: In the news recently, the Swedish teenager who came over from Sweden on a solar powered sailboat to give her message to the United Nations that we should stop the burning of fossil fuel. Her slogan was "United behind the science" (Ortiz 2019). What would you tell Greta is a united science and what is the diplomacy needed to make that real?

Answer

Prawer: Greta is the climate prophetess. I think that's such a timely question, and speaks to something that Vaughan brought up earlier which is the erosion of the relationship between science and society. Many are concerned that public trust in science has decreased. We actually also convened an interesting meeting on the Truth, Trust and Expertise agenda, a project run by the European Academies of Science and which I encourage people to check out.

(ALLEA, n.d.)

I think one of the big problems inherent in getting people to "unite behind science" is around science editing. Not science *journalism*, but science *editing*. Let's say you were the Science Editor of *The New York Times*. How do you decide what stories to commission, stories which will interest the public? And at the same time, how do you compete with the pseudoscience clickbait that is everywhere? You know, we've all seen those headlines like "These top 10 foods will give you cancer," and the like.

The public is constantly fed science that is not what we, those of us in this room, would consider "good science," and that substandard information is widening the rift between science and society. It is harder for society to unite behind this science, and that contributes to the erosion of trust. I don't envy science editors, but I think this is an area that we really need to support. If you are a science editor, not only do you have to commission stories that are good science, not only do you have to make them dramatic enough to compete with clickbait, but you also need to fact-check them very carefully, because as you know, when someone retracts a story, that can be front page news, even when the story itself wasn't.

So when I think about Greta Thunberg's message, "uniting behind science," I was thinking about ways in which I can concretely help people to do that, and my answer is, it is something around supporting the enterprise of science editing. I don't quite know yet what that looks like, but I think, especially when you're a science journalist or if you're in science communication, you're going to get snapped up by a big company that will pay you millions of dollars to do their science communication or science PR. So I think we need to look at how do we keep good people in science journalism in the public arena, and how do we use that to allow the public to unite behind science, and then help them listen to Greta's message.

Question

Unknown: Speaking back to Ronit's story of the Israeli-Palestinian scientists, this made me remember something that I thought about years ago when I ran a conference in Greece and had a paper submitted by Greeks and Turks both from Cyprus. The woman was Greek. She had lived in Northern Cyprus before the invasion. She had worked with somebody from the Turkish side who had basically lived in her town to recreate the curriculum. This was before the U.N. didn't unite the two halves, when they thought it was going to unite, and they wanted a united curriculum between the Greeks and the Turks. They couldn't meet in either of their countries. They had to meet in foreign airports. They did a whole study whereupon the Turk wasn't allowed to the conference because he said that his address was in Northern Cyprus. So she gave the talk in halting English because she had hoped he would give the talk, but it was a triumph. These things are seared into our brains. I mean, yours is certainly seared into yours. Mine seared into mine.

My question is, are those [smaller interactions] better, or is it big programs with bigger and maybe easier things that we should be doing? There are tiny things like this which have an enormous impact on us and the people directly involved. Do they change the world? Or is it better to have big programs between big countries where people can go more easily?

Answer

Prawer: My answer would be that you have to start small. When there is such animosity, the only way to do this kind of work is out of the limelight inherent in large, showy initiatives, at least for a pilot phase. A lot of the students in our program never had their names published; never had a photo taken, and I think this is very familiar to a lot of people. I'm going to refer you to Vaughan, because I think the Cuba example is a really good example of when you start with something that doesn't feel huge and then all of a sudden, it's on a national platform.

Turekian: There are a lot of examples where there's been the use of something small to begin and then to keep something happening. I think that the trick to all of these issues is to identify when you can take something and scale it quickly, and how do you do that? And that's where you need to actually have.

Glenn uses the example of the work that he and Professor Vafai have done to keep just *anything* going with Iran. I mean, think about how up until 40 years ago, some of the largest numbers of students that were coming to the U.S. were coming from Iran. And that has eroded over time. We need to keep *something* going together, because at the end of the day, just doing what you can do allows you to keep just that spark going.

I think your question there was a slightly different one from the small to big one. That is, if you were going to invest a marginal dollar, where do you invest the marginal dollar? Do you invest in one of these sort of smaller activities, these person-to-person activities, or do you invest it in some massive facility, like CERN, something that is going to be the attractor and that could be the place that begins to bring people together?

Actually, and this may sound like a bit of a copout, but you actually need both, because the underlying conditions that allow for that big thing to actually be successful are that you have a scientist from one place looking at a scientist from another saying, "hey, do you want to cooperate?" I see this you know, and this is an oversimplification, but I see this with my own children. My own children will decide on the playground that they're going to find somebody else to say, "hey, do you want to be my friend?" Then, if enough of them are standing there, then they start saying, "well, you know, we're interested in playing baseball, and so are you." They've got a baseball team going now, but it actually requires somebody to say, "Hey, you're interested in something I'm interested in, so can we figure out a way to make that work?"

At a diplomacy level you actually need some of that. It can't be the only outcome, by the way. Think of geomorphology. You have the mass wasting events and the small events, and the big debate is, "what has a bigger impact on the landscape, the big stuff or the little stuff?" Have they figured that one out yet?

But the answer is, actually, you need both. And in fact, more and more of the theories actually say that you need the small stuff to lay the groundwork, so that when the big stuff comes in, you actually have something to actually push forward on.

Question

Alan Berman: My name is Alan Berman and I work locally for a non-profit called the Intercultural Center for Desert and Ocean Studies. My question was, are there two types of science diplomats, ones that focus on advocating for applying the results of science and others that focus on advocating for the scientific process? What made me think of that is Neil deGrasse Tyson's quote, "Science is true." That doesn't make sense to me because I see science as a process. It's a language. Many of the examples you gave are between people collaborating to speak the same language. But a language can't be true. However, for example, a lot of these results of climate science we could say are true enough to be applied. I'm wondering how you see yourselves as the people who are trying to actually convince people that it's worth applying these results of science, or just advocating for doing more science in general?

Answer

Prawer: I think you're right. We do both. We need to do both. The Royal Society and the AAAS 2010 paper on science diplomacy has three different categories of science diplomacy and it sounds like we need to add some more.

The U.K. government is fundamentally interested in evidence-based policy, so for me, one of my primary directives is to do a bit of language translation, to borrow your metaphor. I help government assimilate good science into its policymaking. Exactly as you suggest, I try to provide and translate good science—wherever it comes from around the world—to the U.K. government so that it can better do its job.

In terms of the "science in government," part of your question, our job as science diplomats is also to advocate for the scientific process. This comes back to the earlier question we discussed about interfaces, which is an interesting language question in itself. How, for example, do you talk to government about investing in quantum science? Before you get anywhere, you need to explain what "quantum" is to a politician whose training couldn't be further away. Fortunately, in the U.K., I don't have to advocate for doing more science. The U.K. government has consistently invested in science and fundamentally believes that it underlies prosperity, so that is a lot less of what my team and I do. But what we certainly spend time doing is the former of your two paradigms, and that is to advocate for the use of science for good governance, which is not always easy, but very worthwhile.

Murray: As you point out, science is a process and it is a cultural phenomenon among the scientists. Back in the days of Newton, we were not thinking of Einstein's gravitation, which was a totally new theory. The effects were a perturbation on what you would see in masses moving around except in the heavens, perhaps. But because Einstein added some to the force and mass equations of science, science changed. It's not saying Newton was wrong. He was, in fact, what would you call it, a first theory, zeroth order theory. Then you add some more perturbations to it, such as quantum, to make it more and more exact. So science is changing all the time, and what we think of as impossible to do in science is changing dramatically. It is a process and we must consider it that.

Turekian: I'll add a little bit of a different lens to this, which is the allusion to get the AAAS or Royal Society. If you really come down to some elements of this, when you think about science diplomacy, it really very much depends on who's doing it and how they're going to look at it. It will be a very different view from a foreign ministry then it will be from an university. It allows for that understanding that there's a breadth of the view of it. That said, it's all based upon the methodologies. Everyone has bought into the scientific method. We don't know where it's going to end up but it's going to be a method piece.

The other part is that we also need to recognize that when we look at science diplomacy, we have to look at the two different communities that are involved in it. Both sides need to see that there is a value to add, that it's in their interest to be engaged with the other person.

Science diplomacy starts by saying it's in the interest of Country A to work with Country B and vice versa. It is not just a one-way street, but actually is sort of the method. It is a method of interaction, of partnership, a "what's the incentive" structure. It is critically true when you actually talk about the two different communities that are leading it: the scientific community and the diplomatic community. If you look to a first approximation, scientists want access. If you're a scientist, you want access to other peoples' brains, you want access to other peoples' money, you want access to other peoples' samples, you want access to other peoples' data. So all this cooperation is driven because of that. If you think about the way that diplomats will look at things, first approximation, it's about influence. I want to influence the way that that country views my country. I want to influence the way that country makes some of its decisions that are beneficial to my country.

When we think about the merging of science diplomacy, the tension is from two very different cultures coming together. You have to find the place where the desire for the scientists to have access and the desire for the diplomats to have influence actually converge in a way that gets them to the third dimension of this chess game, that the other country sees it in their interest too, to have those two communities engage. If it's just me saying I want to influence the outcome and my scientists want your samples, science diplomacy is going nowhere. But if "okay, we're in this, we're dancing, we're seeing where this is going," *now* you've got action, and that's the way in which it's going to happen.

You can look at the examples of what Dr. Schweitzer and Dr. Vafai worked on in Iran, of what had happened in Cuba, a recent example with the U.S. in Syria, which actually fell apart in this case, and learn from these examples of why the sides need to actually come to some sort of equilibrium.

Question

Thibodeaux: I came here to University of Arizona to work on spacecraft operations for Mars, which is a dream job, but I am also heavily invested in a future and a career in science communication and diplomacy. However, I (and a lot of my fellow students and early career scientists) don't know how to do that. There's not a direct or clear path for us to engage. So maybe just for everybody at any level: What is your recommendation? How do we start to engage even more in becoming science diplomats in our local communities and beyond?

Answer

Prawer: My career path in science diplomacy was somewhat unusual, and I realize it doesn't happen that way for everyone. I think a lot of what we hear from young people who are interested in a career in science diplomacy is, "everyone who I talk to wants me to have science policy experience or science diplomacy experience, so how do I get a foot in the door?"

We designed a program, basically internships in our science offices. It is an internship program or I like to call it a "SINternship" program. It's tailored very much to the people that we work with. We have Ph.D. students join us, and we try to design projects which are flexible to work around lab time. We also try to make projects self-contained, so that students can have their name on a final product. We run this program specifically because as I mentioned earlier, the U.K. really cares about evidence-based policymaking. The more excellent scientists are involved in the process of science diplomacy and science communication, the more we will have that trust in the U.S. and the U.K.

Also, while we are here, I think that the words "Ph.D. student" are an oxymoron. You are a Ph.D. professional, so you are the actual expert in what you do when you are doing a Ph.D. program. I would say, if this is what you want to do, start now. There are educational organizations, volunteer organizations, that want people like you on their boards of advisors. They want you to come and give talks in their classrooms. They want you to come and explain your science to local government. Really put yourself out there. The first time you do it will be horrible. The second time you do it will probably also be horrible, but the only way to learn whether this career path is for you is to give it a shot. I think Ph.D. students often don't give themselves that platform, but remember that you are literally the expert on your Ph.D. subject, and you are entitled to market yourself that way. Go out and start!

Murray: The AAAS has internships and fellowships that are outstanding. At the Department of Energy, when I was director of the Office of Science, I had two

AAAS fellows in my office who are current fellows, and I had eight of them in the headquarters. There were a thousand federal employees in the Department of Energy, a seven-billion-dollar organization, but there were in the top, say hundred people in the front office, there were twelve former AAAS fellows. They really use that to get in to government. California and New York have similar programs and I'm told that so does Arizona for the Statehouse in Arizona (Arizona State Legislature 2020). Starting local I think would be an excellent idea. Talk about what you do in your research because you're the expert.

Turekian: I got started through AAAS fellowship with the State Department. In 2016, there were 300 Ph.D. scientists and engineers in the State Department, including 120 former AAAS fellows in the State Department. It's not a technical agency like DOE but it has actually a lot of technical talent. Now, the best lesson to be learned from that is actually a lesson you can take today: You don't build firehouses when there's a fire. You don't start science diplomacy, or scientists don't start communicating, when there's a problem. You actually have to be engaged in that conversation very early on, whatever institution you're in.

The power of the AAAS fellowship in a policy shop like the State Department is not that the scientist came sweeping in and said they could solve problems, it's that every day a scientist was sitting down the hall from a diplomat. For the first many years, the State Department rejected science like it was an organ being transplanted until they figured out, they're not leaving so we may still try to figure out how to use them, and they did over time. The science advisor's office up until recently was probably one of the great places of engaging into the department and realizing that you weren't just talking to a bunch of scientists; you were having access to a whole network of people, and you could be part of that network.

But then when you're here [at The University of Arizona], you can wander down the hall or across this beautiful campus and go find out what's going on in the international relations department. See what they're working on and say, "You know, you're working on this. I'm a real expert in water. You're interested in the Israeli-Palestinian issues. Maybe we should write a paper together."

Also, get a mentor! And people that are of a certain age, start mentoring! It's critical. One of my most important mentors said to me when I started the Science Adviser position, you have to occupy your space. And that's true in every part of life. If you want to do something, occupy your space, take a chance, do something different, go outside your comfort zone, and begin to engage differently. No one knows the path for this. We're all making it up. We have some things that didn't work, but we're all making this up. There are great opportunities to just say, "I'm going to go do this now and see what happens." You can do it when you're young, when you're old, you have kids, a mortgage and everything, then you start playing it as safe as you can, right down the middle.

Colglazier: Thank you all. One of the themes of these last suggestions was a great question about the small versus large, the small things actually provide a lot of the inspiration. You can start and do those small things, and sometimes they turn out to have a big impact when you scale with them.

REFERENCES

Abbott, A. 2018. "In the Palestinian territories, science struggles against all odds." *Nature* 563, 308–311. November 14, 2018. https://doi.org/10.1038/d41586-018-07350-9.

ALLEA, All European Academies. n.d. "Truth, Trust and Expertise." ALLEA website. Retrieved from https://allea.org/truth-trust-and-expertise/.

Arizona State Legislature. 2020. "Arizona Legislative Internships." Arizona State Legislature website. www.azleg.gov/azleginternships/.

BBC Press. 2018. "US Unleashes Sanctions on Iran, Hitting Oil, Banking and Shipping." *BBC News*, November 5, 2018. www.bbc.com/news/business-46092435.

Berg, P. 2008. "Asilomar 1975: DNA modification secured." *Nature* 455, 290–291. September 17, 2008. https://doi.org/10.1038/455290a.

British Council. n.d. "Israel-UK-Palestine GROWTH PhD Fellowship." Retrieved from www.britishcouncil.org.il/en/israel-uk-palestine-growth-phd-fellowship.

Bureau of Arms Control, Verification, and Compliance. 1987. "Treaty between the United States of America and the Union of Soviet Socialist Republics on the Elimination of their Intermediate-Range and Shorter-Range Missiles (INF Treaty)." U.S. Department of State Archives. Retrieved from https://2009-2017.state.gov/t/avc/trty/102360.htm#diplomaticnotes.

Bush, V. 1945. *Science the Endless Frontier*. A Report to the President by Vannevar Bush, Director of the Office of Scientific Research and Development, July 1945. www.nsf.gov/od/lpa/nsf50/vbush1945.htm.

Fears, D. 2019. "Trump Rolls Back Safety Rules Meant to Prevent Another Deepwater Horizon Spill." *Washington Post*, March 2, 2019. www.washingtonpost.com/climate-environment/2019/05/02/trump-rolls-back-safety-rules-meant-prevent-another-deepwater-horizon-spill/.

Hersher, R. 2019. "U.S. Formally Begins to Leave the Paris Climate Agreement." *NPR*, November 4, 2019. www.npr.org/2019/11/04/773474657/u-s-formally-begins-to-leave-the-paris-climate-agreement.

Intergovernmental Panel on Climate Change. n.d. "About the IPCC." www.ipcc.ch/about/.

Kuiper, K., ed. n.d. "Cooperative Threat Reduction." Retrieved from www.britannica.com/topic/Cooperative-Threat-Reduction.

Massarella, D. 1992. "Philip Henry Zollman, the Royal Society's First Assistant Secretary for Foreign Correspondence." *The Royal Society*, July 1, 1992. https://doi.org/10.1098/rsnr.1992.0023.

National Academy of Sciences, National Academy of Engineering, and Institute of Medicine. 2007. *Rising Above the Gathering Storm: Energizing and Employing America for a Brighter Economic Future*. Washington, DC: The National Academies Press. https://doi.org/10.17226/11463.

Obama, B. 2010. "National Commission on the BP Deepwater Horizon Oil Spill and Offshore Drilling." Executive Order 13543. Retrieved from https://obamawhitehouse.archives.gov/the-press-office/executive-order-national-commission-bp-deepwater-horizon-oil-spill-and-offshore-dri.

Ortiz, E. 2019. "Teen Climate Activist Greta Thunberg Tells Congress: 'Unite Behind the Science.' " *NBC News*, September 18, 2019. Retrieved from www.nbcnews.com/science/environment/climate-activist-greta-thunberg-tells-congress-unite-behind-science-n1055851.

Poliakoff, M. 2015. "The Royal Society, the Foreign Secretary, and International Relations." *Science and Diplomacy*. www.sciencediplomacy.org/letter-field/2015/royal-society-foreign-secretary-and-international-relations.

Schweitzer, G. and N. Neureiter. 2008. "Engaging Iran." *Science* 319: 5861, pp. 258. January 18, 2008. https://doi.org/10.1126/science.1154422.

Smith, D. 2015. "Getting the Lead Out." *Caltech News*, September 21, 2015. www.caltech.edu/about/news/getting-lead-out-47935.

Stein, R. 2018. "Chinese Scientist Says He's First to Create Genetically Modified Babies Using CRISPR." *NPR*, November 26, 2018. www.npr.org/sections/health-shots/2018/11/26/670752865/chinese-scientist-says-hes-first-to-genetically-edit-babies.

Stone, R. 2018. "Fidel Castro's Eldest Son, a Physicist, is Victim of Apparent Suicide." *Science*. February 4, 2018. https://doi.org/10.1126/science.aat2018.

The Royal Society. 2010. "New Frontiers in Science Diplomacy." January 2010. https://royalsociety.org/~/media/Royal_Society_Content/policy/publications/2010/4294969468.pdf

Tyson, Neil deGrasse (@neiltyson). 2013. "The Good Thing about Science." Twitter, June 14, 2013. https://twitter.com/neiltyson/status/345551599382446081?lang=en

UK Science & Innovation Network. 2015. U.K. Science and Innovation Network Report. https://assets.publishing.service.gov.uk/government/uploads/system/uploads/attachment_data/file/417600/bis-15-210-science-Innovation-network-report.pdf

UK Science & Innovation Network. 2018. "UK Science & Innovation Network Country Snapshot: Israel." Retrieved from https://assets.publishing.service.gov.uk/government/uploads/system/uploads/attachment_data/file/757883/SIN_Snapshot_-_Israel_2018.pdf.

United Nations General Assembly. 2015. "Transforming Our World: the 2030 Agenda for Sustainable Development." Resolution Adopted by the General Assembly on 25 September 2015. Session 70, Agenda Items 15 and 116. Retrieved from www.un.org/ga/search/view_doc.asp?symbol=A/RES/70/1&Lang=E

U.S. Congress House Committee on Natural Resources. 2011. "Final Report of the President's National Commission on the BP Deepwater Horizon Oil Spill and Offshore Drilling." Oversight Hearing before the Committee on Natural Resources, U.S. House of Representatives, One Hundred Twelfth Congress, First Session, Wednesday, January 26, 2011. Retrieved from www.govinfo.gov/content/pkg/CHRG-112hhrg63876/pdf/CHRG-112hhrg63876.pdf.

U.S. Department of Energy. 2015. "U.S.-CERN Agreement Paves Way for New Era of Scientific Discovery." *Energy News*, May 7, 2015. Retrieved from www.energy.gov/articles/us-cern-agreement-paves-way-new-era-scientific-discovery.

2 Sustaining International Scientific Cooperation among Nations

Why It Matters

DOI: 10.1201/9781003220503-2

SUMMARY

This chapter begins with a foreword by Jeffery Goldberg, Dean Emeritus of the College of Engineering at The University of Arizona, reflecting on the importance of international collaboration. Next is the transcription of the second plenary session for the conference, in which Honorary Chairman of the conference, Dr. E. William Colglazier, moderates a panel that includes Peter Agre, Nobel Laureate and Director of the Johns Hopkins Malaria Research Institute, and Ronit Prawer, Eastern U.S. Region Director for the U.K. Science and Innovation Network. Each shares their experiences about how science diplomacy can help thaw icy political relations as well as the obstacles of navigating science diplomacy in society. This is followed by a Question-and-Answer period with the audience.

Following this are two articles that were presented at the conference and subsequently reviewed by a scientific committee to be included in the proceedings: "Technological governance in Latin America: Institutional challenges to navigate emerging technologies (2010–2019)" by Martín Pérez Comisso, and "Science in Legislation: Applying Scientific Peer Review and Rigor to Public Policymaking" by CJ Pommier, Ansel Andrews, Josh Hoskinson, and Henry Werchan.

THE IMPORTANCE OF INTERNATIONAL COLLABORATION

Jeffrey Goldberg
Emeritus Dean, College of Engineering, The University of Arizona

From Wikipedia (corroborated by other sources), "Science diplomacy is the use of scientific collaborations **among nations** to **address common problems** and to **build constructive international partnerships**. Many experts and groups use a variety of definitions for science diplomacy. However, science diplomacy has become an umbrella term to describe a number of **formal or informal technical, research-based, academic or engineering exchanges**." In this note, I will argue why this idea is especially attractive to me, The University of Arizona (UArizona), and the U.S. as a whole.

I want to thank Drs. Kevin Lansey and Hassan Vafai of the UArizona for asking me to write this introductory note, for building the UArizona team, and for organizing the Fall 2019 conferences. We would not be moving these ideas forward without their leadership and hard work. I also want to thank Drs. E. William Colglazier, Peter Agre, and Ronit Prawer for speaking in the opening conference session, "Sustaining International Scientific Cooperation Among Nations: Why It Matters." They have been working with us since 2017 and their guidance and enthusiasm have been invaluable in helping the UArizona move forward.

The comments below are entirely my own and not those of UArizona. I am borrowing somewhat from my introductory remarks at the start of the conference. Also, I am writing this note in late June 2020 and the world has changed drastically since October 2019. We are now in the middle of a global COVID-19 pandemic with approximately 10 million people infected (tested positive) and approximately 500,000 deceased worldwide. It is clear that the pandemic has impacted everyday life globally in a way that has not been seen since the 1918 influenza outbreak. In the U.S., despite our significant wealth, we are approaching 2.5 million infected and 125,000 dead. The U.S. strategy of delaying any efforts for two months (January–February), closing the economy and attempting to ramp up testing/tracing/treatment/health-infrastructure for approximately two months (March–April), and re-opening the economy while trying to contain infections (May–June) has resulted in at best piecemeal success and significant failure. In late June, our infections are growing faster than they were in March (even when accounting for increased testing), and without treatment or vaccine, we seem to be in a steady state of over 40,000 new cases per day nationally. One can see clearly that our disjoint and inconsistent federal policy (state responsibility for a large majority of testing/tracing/treatment/supply chain and a different set of rules/policies for each state) and inconsistent messaging have led to large and dynamic differences in successes across the country. At this point, we cannot even convince 40% of the population on the scientific evidence for using face masks to reduce infection rate. We need to be able to do science diplomacy, not between nations, but between political factions!

In 2016, I was the Dean of Engineering at UArizona and Drs. Lansey and Vafai came to me with a different idea—running a conference and building university-wide expertise in science diplomacy. I did not know much about science (and engineering!) diplomacy at the time, but as I listened to their description, I was fascinated at the potential opportunities for students, faculty, and partners, as well as the overall benefit for the university and the nation. I knew of similar international program ideas (Fulbright Program, USAID project programs, offset projects for international defense contracts, and NSF international partnership programs such as PIRE and host of specific topic binational programs), but I was able to easily see the importance of such effort. My ideas centered around UArizona Engineering stakeholders: science diplomacy helps by providing the following:

- An opportunity for students to solve real problems and experience different cultures—something I never had or even realized existed when I was an undergraduate. All engineering colleges across the country are emphasizing international experiences and this is a key feature in ABET accreditation for colleges. The students often learn that technical problems are often only a piece of the overall problem. Students see that it is hard to understand the need/ problem if you do not understand the social environment, and it is difficult to implement solutions that are not culturally appropriate. I immediately think of the late Paul Pollock's book, Out of Poverty: What Works When Traditional Approaches Fail, when discussing cultural and technical appropriateness (Polak 2009). Having a college-wide program in science diplomacy can be an effective student-recruiting strategy.

- An opportunity for faculty to work on important theoretical and applied problems with international colleagues. I don't know a single engineering faculty member that does not want their work to have an impact. In these science diplomacy projects; one can quickly see impacts as even theoretical work is generally directed at a specific need. Just like engineering students, engineering faculty are interested in helping people live better lives. Finally, the funding that generally comes along with such projects can be used to support graduate students, summer salary, travel, and a host of other desirable items for faculty.
- An opportunity for the engineering college to build reputation internationally. I was a member of the American Society of Engineering Education's Deans Council as well as the International Federation of Engineering Education Societies Deans Council, and I thought that this was a willing and interested group for partnerships.
- An opportunity for all to work on an interdisciplinary team of engineers, scientists, medical professionals, social scientists, and others. As mentioned above, difficult problems often require solutions that engage more than technical expertise. I recall a water project our Engineers Without Borders students solved for a set of villages in West Africa. The solution was durable, reliable, low-cost, and satisfied the water transport needs. The solution also drastically changed the daily lives of the women and children of the villages—the impacts of which were never considered. Had we engaged a social scientist on the team, we almost surely would have had a better overall solution.
- An opportunity for communities that have problems to get help from experts to find appropriate solutions. The U.S. has expertise that can be shared, and this work can improve our reputation internationally. As a nation, we have not been perfect in such adventures and I am not denying that past imperialism or colonialism occurred when working with international partners; however, I believe that there can be true altruism to help those in need. In a recent Washington Post opinion article, E. J. Dionne states it far better than I can: "...Genuine American patriotism should never depend on false assertions of our perfection or a denial of our own sins. On the contrary, our desire to be what the prophet Isaiah called a 'light unto the nations' was always best invoked not as a boast but as a call to conscience" (Dionne 2020). Science diplomacy enables that call to conscience for individuals and organizations.

- An opportunity for the nation to build sustaining partnerships with trusted partners to cooperate on key global issues that cannot be solved by individual countries. It is clear that COVID-19 presents a perfect example of why such partnerships are needed. The need to share data honestly, work to design solutions, share products and materials fairly, and execute solution paths cannot be done without prior experience that generates trust and practice in working together. Science diplomacy partnerships can be a precursor to other partnerships that involve national economic growth, national security, and long-term trust. **I strongly believe that the long term for global success is not hundreds of individual countries all maximizing for their own benefit. It is not relationships that are only transactional with few, if any, long-term trusted partners.** If game theory has taught us anything, it is that gains from being in a stable coalition far exceed what any country can do individually. National success will be driven by the ability to work with long-term partners and not one-off transactional deals.

I want to close with comments on the current U.S. tactics to reduce H1B and J1 immigration, and reduce international graduate students studying in the U.S. All these ideas are weak economically and do not reduce unemployment even in the short term. There is no supply of unemployed citizens that can step into high-tech STEM positions that engage many of the 460,000 H1B and 216,000 J1 visa holders. Each year, 65,000 H1B visas are granted along with 20,000 additional for masters and Ph.D. graduates of U.S. institutions of higher learning (American Immigration Council 2020). This is a drop in the bucket when compared to the 40,000,000 now unemployed.

American Enterprise Institute economist James Pethokoukis wrote in his June 23 opinion article in *The Week*, "Foreign-born workers account for a fifth of STEM workers with a bachelor's degree and more than half of those with a Ph.D." (Pethokoukis 2020). In an April 2018 NBER Working Paper, Sari Pekkala Kerr and William Kerr found that immigrants account for about a quarter of U.S. entrepreneurship and innovation (Kerr and Kerr 2018). Immigration from Asia is a key driver, with Chinese and Indian ethnic inventors accounting for 22% of U.S. patents in 2018 versus fewer than 3% in 1975. It is impossible to have an economically vibrant America without global talent coming here and the predominant path of those individuals is through the U.S. research universities and the H1B and J1 visa programs. Shutting off these pipelines reduces our ability to do cutting-edge research, drive economic well-being, and help our international friends. Some of these students and companies eventually get back to home nations! I do not think that it is a difficult argument to show that the number one science diplomacy project worldwide is the U.S. efforts to educate the STEM graduate students of the world. The direct impact on millions of individuals is inestimable and that says nothing about the impact of those students on the economies and lifestyles of countries worldwide.

PANEL DISCUSSION

Honorary Chair:

E. William Colglazier
Senior Scholar and Editor-in-Chief of Science & Diplomacy, AAAS Center for Science Diplomacy

Panelists:

Peter Agre
Nobel Laureate; Director, Johns Hopkins Malaria Research Institute; Bloomberg Distinguished Professor, Johns Hopkins Bloomberg School of Public Health and Johns Hopkins School of Medicine

Ronit Prawer
Eastern U.S. Region Director, U.K. Science and Innovation Network

Preserving International Scientific Cooperation

E. William Colglazier
Honorary Chairman of the Conference

The subject of this session is *Sustaining International Scientific Cooperation among Nations: Why it Matters*. In the first session of this conference, we talked about science, engineering, and health diplomacy, about how the scientific knowledge and scientific values can help influence diplomacy, can help advance diplomacy, but also about how diplomacy and public policy can advance the scientific enterprise both at the national level and at the global level.

At this interface of science and policy, scientists have many roles. We heard from Dr. Cherry Murray last night what she has done in her remarkable career—serving inside government, serving outside government, at the National Academies, with universities, with national labs, with industry, serving at the highest levels of U.S. agencies. She has shown there are many ways that you can contribute. That is probably a lesson for all of us, especially for young people. In fact, some of the greatest science diplomats that I have known in my career were world-class scientists. They never served in the government, but devoted a lot of their time and efforts to using their scientific expertise to strengthen relations between countries especially in difficult times. Dr. Murray's career is a testament to how scientists can contribute at so many different levels, whether they are inside the government or outside the government. I might also mention, in terms of science diplomacy, there is science advancing diplomacy and diplomacy advancing science.

One of the remarkable things about science is that knowledge advances. There are a number of examples where the diplomats and the politicians and the policymakers were at an impasse and could not make progress. However, science was able to essentially "leap-frog" over the diplomatic impasses by creating new pathways, new options, that were not known in advance but once discovered enabled progress to be made.

Today, we are going to talk about international scientific cooperation. International scientific engagement, collaboration, and cooperation can be used to advance diplomacy, but also diplomacy can be used for advancing international scientific cooperation. That cooperation carried out with scientists in other countries is certainly essential to advancing the scientific enterprise, both nationally and globally.

American scientists have been used to being at the front of the pack for so many years. There is recognition now in almost all subfields of science that the U.S. has to be even more aggressive about international scientific cooperation. Very smart people are found in every other country, and many countries are making significant investments in science and building scientific capabilities. If a country wants to remain among the best in science, it has to engage with the best people, no matter where they are in the world.

There are unique research environments around the world. If a scientist wants to collect data and carry out experiments in locations that are not located domestically, that scientist will typically have to go internationally into those environments. I went on a tour of a remarkable facility, Biosphere 2, that is located here at The University of Arizona. It is a unique research environment, and is now being used by scientists coming from Germany, from Europe, and from Latin America. They are working there on scientific projects and doing things that they would not be able to do anywhere else in the world. You are able to see the rainforest environment outfitted with scientific equipment, the creation of an ocean with coral and fish, the landscape evolution project underway. Biosphere 2 represents a very unique resource, and one aspect of why it so unique is that it can be used by the scientific enterprise worldwide to advance our understanding of these ecosystems. International scientists are very interested in working there.

There are environments that American scientists do not have access to at home. What the diplomats can do is to remove some of the hurdles that come in the way of scientific engagement. Whether it is reducing problems with visas or sanctions or other types of regulations, there is much more that the diplomats can do to help the scientists.

Dr. Murray and I are on a committee of American Academy of Arts and Sciences that will be issuing three reports in 2020 and 2021. The reports are looking at international scientific cooperation. The first report talks about the imperative of international scientific cooperation for advancing the scientific enterprise, for our country and globally for all countries, and what are key issues that have to be focused upon. One of the other reports focuses on large-scale international projects, that is very large facilities that are too expensive for one country to carry out on its own. Although that issue certainly has been true in physics, but it is becoming more important in almost every field of science now and essential for making progress. So in some sense, the international scientific community needs to think carefully about how to set its priorities and how to develop these priority projects where one country might be the host but scientists from other countries are able to participate. That report is looking at the practical challenges, not only in setting priorities, but the management and governance, the financing, and all the issues associated with carrying out a multi-year large project. There are now examples of large scale scientific projects that are now becoming globalized with partner facilities located at sites in several countries.

The other working group is focused on the issues of collaboration between the advanced scientific countries and what you might call under-resourced or developing countries that are trying to increase their capabilities. One thing my colleague, Glenn Schweitzer, has always emphasized is that in international scientific cooperation, you have to emphasize what is the best science. Excellent science always has to be the hallmark. A lot that can be done with advanced countries working, collaborating, and helping to build the scientific capacity in under-resourced countries through international scientific cooperation, This other working group is very much focused on that important issue.

My hope is that, now with a year before the next U.S. administration, some of these reports on international scientific cooperation can be influential for stimulating new initiatives.

Introduction to Panel Members

Let me turn to our three experts today and the perspective that they are going to bring. Our original panelist, our Chinese colleague, Professor Zhishen Wu, was unable to travel to make it here today. He is able to deliver his talk via videoconferencing later this afternoon, so we are indeed grateful that he is able to do that. His absence does highlight one very important issue, and that is cooperation between U.S. scientists and Chinese scientists. My personal view is that for science and for our own country, we have to preserve cooperation and fundamental science research collaboration between American and Chinese scientists. There are now more roadblocks making some cases more difficult.

To me, one of the greatest diplomatic accomplishments of the U.S. has been training many excellent, smart people from other countries. Whether they stay here in the U.S. or they go home, it has benefitted our country and it has benefitted their home country. Right now, the latest data regarding the number of Chinese students that come to be educated at universities, roughly 80% stay in the U.S. and 20% go home.

The pressure coming from the U.S. government, from the FBI and others, has made international scientific cooperation face a more challenging environment. The balance between security and international scientific engagement has to be in the right place.

Introduction to Peter Agre

Dr. Peter Agre, a very distinguished Nobel Laureate. He is also one of the people I consider to be a true science diplomacy champion. He has been involved with Cuba, with Iran, with North Korea. He has just come back from Turkey. For those of you who read the news, you know there are challenging issues between the governments of the U.S. and Turkey. Peter's willingness to go to Turkey was in part to demonstrate at the leading technological university in Turkey, the "MIT" of Turkey, that American scientists are very much still interested in collaborating with scientists in Turkey, and also with engaging young and up-and-coming scientists in that country. We will hear his remarks about the importance of international scientific engagement and how it benefits both American scientists and scientists around the world, as well as being an advantage to diplomacy.

Introduction to Ronit Prawer

Our second speaker, who we also heard from in our earlier panel, Ronit Prawer. She has this incredible experience both in the Middle East applying science to foreign policy issues, and now with the U.K. government covering the Eastern U.S., from Boston on down. Her job entails learning from what happens in the U.S., and at the same time telling us about all the things that are going on in the U.K. regarding scientific innovation and how we can better connect the scientists and engineers and health professionals in both countries.

Let me ask first Peter to come up and speak for a few minutes, and then Ronit, and then we will convene as a panel to hear your input and advice, as well as answer questions as we are able.

Science Diplomacy through Medicine

Peter Agre
Nobel Laureate, Bloomberg Distinguished Professor and Director of the Johns Hopkins Malaria Research Institute

Good morning. It's a delight to be here. Tucson is so beautiful in the morning, in the autumn, with the sun coming up over the desert over the beautiful campus of The University of Arizona. I enjoyed having breakfast at a table with young scientists from this campus and from the University of Idaho. I think that, while there are clearly clouds on our international horizon, I think there is reason to be very optimistic.

I am sort of a late starter in the science diplomacy field. I was a medical doctor and medical researcher, always with an interest in international events and collaborations. By good fortune, I interned for the president of the American Association for the Advancement of Science, or AAAS. A new program for science diplomacy developed about 12 years ago. I was president ten years ago. The Center for Science Diplomacy, organized by Vaughan Turekian, who was here last night, Norman Neureiter, and led to a series of visits to countries where formal relations with the U.S. were minimal and certainly not friendly, but were opportunities for scientists to reach out and meet scientists from these countries. As Bill mentioned, we went to some pretty interesting places: repeated visits to North Korea, visits to Iran, multiple visits to Cuba, to Myanmar, Burma, and most recently to Turkey. The challenges are formidable, but I think the solution is really science and the scientists.

In a poll undertaken by the Zogby [International] firm in 2003, citizens of four Arab countries were asked about their impression of the U.S. in general. I think these were Morocco, Saudi Arabia, Lebanon, and the United Arab Emirates (Zogby 2002). The response to the question of their feelings toward the U.S. in general was very negative. 80% disapproved of the U.S. But when the same individuals were polled about their views of science and technology—and this is not just medical science, but all science, including engineering, technology—the response was completely different. 80% had a very *positive* view of the U.S. So this is clearly a door opener, and I think this is a door opener that our country really needs to take very seriously. It's an opportunity for scientists to meet scientists as individuals. I don't think science diplomacy is really a matter of our committee speaking to their committee. It's really about us as individuals.

For my day job, I direct the Malaria Research Institute at Johns Hopkins. In addition to a wonderful group of lab scientists in Baltimore, we have field operations in southern Africa, in Zambia, Zimbabwe, Congo, and now in Mozambique. So part of my time is spent in Africa. The challenges there are pretty significant, as you can imagine.

Challenges in Africa

Zimbabwe, a country which was once the richest in southern Africa, now is destitute because of an economic collapse, and this was brought about by political dealings with the Mugabe regime. Robert Mugabe led the liberation of Zimbabwe but stayed on for 37 years. I can't imagine any country having the same leadership for 37 years operating efficiently. How does one undertake clinical and epidemiological research in a place like Zimbabwe which is essentially totalitarian? I think the answer lies in the teachings of Niccolò Machiavelli: sometimes, the end does justify the means (Machiavelli 1981). By cooperating with them in a non-governmental organization in Harare, a biomedical research training center led by Shungu Munyati, we've been able to operate efficiently in Zimbabwe during a lot of considerable political instability. The Machiavellian angle is that Shungu Munyati was Robert Mugabe's niece. It was a connection that made things possible, and Shungu was a very decent scientist. It was a true scientific collaboration made possible.

Elsewhere in sub-Saharan Africa, there are other problems and governments are often unstable. But look at us in Washington, DC—I don't think we have an example which is clearly the answer for the world. But I think as scientists, we go forward, and some very good things are happening because of the collaborations. Malaria is still a major killer of small children in sub-Saharan Africa, but the workings of groups, like our group and many others, have led to progress, and the burden of Malaria has reduced, with the number of deaths reduced by about half. But that means we still have a lot of work to do. And this, of course, in almost every case is made possible by the individuals, many of them being Americans who work for years and decades in these locations.

Just briefly, I'd like to mention three of the wonderful trips that I participated in. The first two were with AAAS, with Vaughan Turekian and Norman Neureiter.

Visiting North Korea

The first was to North Korea. Just by show of hands, how many people here have been to North Korea? One. That's unusual; usually in audiences no hands go up. This was probably the last place an American would plan to visit, and it's made difficult and of course, presently it's forbidden by the U.S. State Department. They'll confiscate your passport if you go to North Korea. But it hasn't been such a rosy place for quite a long time. But, in 2009 with the inauguration of Barack Obama, there seemed to be a thaw. It was the workings of Vaughan Turekian and his group at AAAS that allowed a group of us—sponsored by the Lounsbery Foundation, by Maximillian Angerholzer—to visit Pyongyang and 15 different scientific institutes, research universities, and such, and it led to friendships which continue. I don't mean to imply that any great advances in the relations between our two countries have occurred, but there is a university outside of Pyongyang, known as the Pyongyang University of Science and Technology, which has a faculty of volunteer scientists, the language of instruction is English, and the enterprise is ongoing. And I think once the visa issues are relaxed, and I'm sure there will be a relaxation eventually, it will be an opportunity.

Visiting Iran

Another visit which I think of so fondly, and for which your colleague Hassan Vafai was instrumental, was a visit that Norman Neureiter, my wife, and I made to Iran six years ago. For a generation, the relations between Iran and the U.S. have been very frosty. Basically, a lot of angry rhetoric. But beneath the surface, there is some wisdom, and behind the scenes, some good will. The foreign minister of Iran at that time was a gentleman named Ali Akbar Salehi. His name has appeared in the news. He doesn't get a lot of visibility, and he doesn't seek visibility. But it was Professor Vafai who, with the invitation of Ali Akbar Salehi, organized a visit which I think was extremely positive.

Ali Akbar Salehi was actually not Iranian, but he was born in Iraq. He was educated at the American University of Beirut. He completed his Ph.D. at MIT. He has strong roots in the U.S. He has a brother who is a neurosurgeon in Chicago, he has another brother who is an engineer in Washington, DC, and his niece is an eye doctor at Johns Hopkins; she's my eye doctor, in fact. These are wonderful people, and the visit that Hassan organized was heartening. The faculty at Tehran University of Medical Sciences and the Jami Institute of Technology I think forged bonds of friendship.

When the negotiations for the nuclear accord were undertaken at the end of the Obama administration, there was a lot of press for John Kerry and the Iranian foreign minister, Javad Zarif. But underneath all that was a friendship which resulted from Ali Akbar Salehi and his friend from MIT, Ernest Moniz. Ernie Moniz was a young faculty member at MIT and met Salehi when he was a graduate student, and they were the key scientific personnel behind these negotiations. These negotiations, we know, were very successful. But the accord has now been reversed by the U.S., but I think it is a temporary reversal. I think the science will prevail.

Visiting Turkey

As Bill mentioned, he asked me a few months back if I would be interested and willing to visit the Middle East Technical University in Ankara, Turkey. I was not familiar with Middle East Technical University, though I subsequently learned that a couple of American colleagues were graduates from there. It's a wonderful university built on the American Land Grant University pattern at 45 square kilometers of arid countryside, which they built a campus by planting a million trees, re-foresting that part of Turkey, and taking in top undergraduates with a focus on science and technology. It was a wonderful visit. It was heartwarming. I think the faculty and students at the Middle East Technical University would love to be here on your campus today. They have so much enthusiasm, and they would share your enthusiasm.

At the same time, 200 miles to the southeast, Recep Erdogan and the Turkish forces have invaded Syria. This was not something discussed on the campus of the Middle East Technical University. I think the basic viewpoint of those I talked to be that Erdogan is a hard guy, probably a passing phenomenon, and doing something that was then regarded as stupid. Sound familiar?

I think the friendships though that were launched during that visit I hope will lead to new paths forward for the U.S. And as Bill mentioned, the fact that they often have ties to the U.S., my host at that event, Professor Irvin Callay, earned his Ph.D. at Iowa State University. So, I think relations between countries remain difficult, but the friendships that we build with our counterparts in these countries can be a positive force for bringing peace to the world.

Navigating Science Diplomacy within Society

Ronit Prawer
Eastern U.S. Region Director, U.K. Science and Innovation Network

Hello again, everyone. For those whom I didn't meet yesterday, I'm Ronit, and I am the director of the British Government's Science and Innovation Network, in the Eastern U.S. In effect, my team serves as the Science Attachés at consulates all across the U.S. As Bill said in his kind introduction, we work to nurture, broaden, and deepen the relationship between the U.S. and the U.K. on science and technology. In that respect, I am a practicing Science Diplomat, but speaking to you today from my personal perspective with a view to sharing my own experience in this career, rather than as a representative of the British Government.

When Bill gave me the homework question for today (and I always do the homework that Bill assigns), I began to consider the recent political developments which are making it more difficult to sustain scientific cooperation, even for fundamental and for unclassified research. There are two specific emerging phenomena that I wanted to raise, which, if they continue, will have a significant negative impact on science advancing nations toward achieving their "social, economic, and environmental goals," as Bill so eloquently put it.

This first of these phenomena, which we touched on yesterday, is what scientists perceive as a growing rift between science and society, which many have gone so far as to term the "war on science." The second phenomenon, which colleagues alluded to this morning, are shifts in political approaches which will render the enterprise of scientific collaboration much more difficult from a practical perspective. I thought in today's talk I would examine each of these phenomena, in turn.

Science and Society—Are We Growing Apart?

Let's begin with the first of these, the science-society divide. Yesterday, as we were sitting here enjoying our first excellent session of this meeting, Alexandra Ocasio-Cortez questioned Mark Zuckerberg in Congress. She asked him, and I quote: "Could I run a political ad that said that one of my Republican colleagues, completely incorrectly, could I lie and say that my Republican colleagues support the Green New Deal, or would you take that down off Facebook?" And Mark Zuckerberg said, "Congresswoman, we would think about it…" (C-SPAN 2019). The exchange went viral very quickly, seemingly hitting a nerve with the American public. What strikes me about this is that *political misinformation* is a concern, taking center stage in public discourse. But we are not asking the same questions about *scientific* misinformation, and the insidious spread of pseudoscience.

It is telling, from my perspective, that we are not asking Mark Zuckerberg "will you take down the click-bait that tells you that 'these three foods will cause cancer,' or 'if you insert this object into your body, you'll feel better,' or 'these smoothies are all you need to ensure you don't get the flu,'" or all kinds of damaging scientific misinformation? We've seen people seriously endangered by this phenomenon because they are now afraid of science and its emissaries like researchers and doctors. Likewise, we continue to see the planet endangered, as pseudoscience undermines climate research.

I am personally very wary of the current climate, and I think that what we have is a situation where conspiracy theories are drowning out legitimate facts. Conspiracy theories trade on the appeal of certainty, whereas we all know that science is imbued with uncertainty; it's par for the course. So, being able to pedal these theories online has a certain power. People are attracted to surety, these theories get shared, and as we all know social media algorithms keep people in their own bubbles. This is a critical area to tackle if we want to look at where we are driving a wedge between science and society.

And I don't think this is specific to a particular country. It's of course worse in some places, but for the most part the interconnectedness of our online worlds means that this is feeding into the global political developments to which Bill referred earlier. That really matters because it affects the confidence that one has in the scientific community, which, in turn, affects how one votes, which then affects support for science funding, and so forth. It affects our views on more specific questions like policy around EVs or whether or not we open a new coal mine. And so, as a community, we have to get this question of *scientific* misinformation to be seen as *just as insidious* as political misinformation.

And that brings me quite neatly to the second of the issues that I'll briefly touch on.

Practical Impediments to Collaborative Research

I am a diplomat, so let me *diplomatically* state that both the U.S. and the U.K. have had what I would call our adventures in democracy over the past few years. I was hoping I would not have to discuss "the B word" [Brexit], but I think that given the framing of today's panel, I do have to briefly touch upon Brexit.

There was a moment in 2012, which I think many of you will remember if it struck you as it did me: during the Brexit debate in the U.K., one of the British ministers was talking to a political pundit who spoke to him about what the economists were projecting would be the economic impact of Brexit. He said: "You know what? This country has had enough of experts!"

That was, I think, a pivotal moment for me on a very personal level. For the country I serve—a country that has a history of people spilling out of the doors of the Royal Institution to listen to the Christmas lectures; a country with a great and storied history of both science communication and science diplomacy; a country with enormous respect for the scientific method—to have a leader say that expertise was not welcome, was something "other" to the people, was a real wake-up call. In many ways, this is a reflection of the first issue we discussed regarding the rift between science and society. And when we don't repair that trust, we land in situations where we risk endangering international collaborations because science does not have the mandate to actually scale up, and policies guide funding elsewhere, guide public investment elsewhere, etc.

In terms of the Brexit debate, the U.K.'s research base is incredibly international. Something like 30% of academic researchers in the U.K. come from another country. Over 70% of U.K. scientists have trained or worked abroad. So scientific collaboration is absolutely fundamental to British research communities, as it is with almost all successful scientific enterprises. Actually, much of that collaboration is right here in the U.S.: close to 15% of papers published by the U.K. researchers include an American co-author, and we have 27 shared Nobel prizes. The extent to which collaboration and scientific success are inextricably interwoven is completely undeniable. What you will be relieved to hear, though, is that despite the Brexit debate and some of this rhetoric, the U.K.'s commitment to science as a foundation of prosperity has remained unwavering. In fact, while politics rages in the U.K. and the U.S., we have had very successful conversations around scientific research, so it's not all doom and gloom. Let me give you a bit of proof for that because I definitely still see worried faces!

First, contrary to what you might have thought and after many years of flat budgets, the U.K. politicians began actually significantly boosting science spend in the lead up to Brexit. In fact, I think we have probably broader and stronger *political* support for science than we have in a long time. I won't bore you with all the statistics but there have been significant commitments to increase science spending every year since the Brexit vote in 2013. In 2017, the government actually set a ten-year goal of increasing R&D spend as a percentage of GDP from 1.7, which it is now, to 2.4% (Stokstad 2020).

Second, and one of the things that I think is the best-kept secret around Brexit, is what we call the government underwrite. The U.K. government has actually committed to underwriting the funding of any U.K. researcher who is involved in a Horizon 2020 project, regardless of what happens with Brexit.

Third, we just announced a whole new visa scheme. We've talked today about how visas and mobility are a critical issue in this debate around national scientific collaboration. This was something my team was incredibly vocal about, and in response, the U.K. government established a dedicated new visa scheme to help people with specialized skills in STEM come to the U.K. to work, study, and live there. So for me, these things are really impressive examples of how, in spite of some of the political divisions, science in government has remained really well-supported.

To close, I thought I would share a little story about some of the ways the science community has mobilized around these issues. In 2013, I was traveling with some of Israel's elite scientific establishment to the U.K. for what we call the Israel-U.K. Science Day. I'm going to give you a little bit of a behind the scenes of what happened there because for me, it really exemplifies the way the scientific community rallies around precisely the issue of protecting ease of scientific collaboration.

As I mentioned, I have a bit of experience in the Middle East. Israel is, this year, the fifth most innovative country according to the Bloomberg Innovation Index, has one of the highest per capita rates of filed patents, and is one of the most technologically influential cities around the world (Jamrisko et al. 2019). In 2013, the negotiations around the Horizon 2020 were in process. There was a very tense moment in the negotiation process, which came just as I took this group of senior Israeli researchers to London. At that time, the Israeli government began to debate backing out of the European research program because of specific clauses regarding the Israeli settlements (Sherwood 2013).

What happened at that moment in the U.K. was actually fascinating for me. These people who were the heads of academies, leaders of the scientific establishment, real visionaries who were in the U.K. with us for a week, they realized they had no time to lose. They basically dropped everything and went to bat for international collaboration. They went on television, they went on the radios, they said yes to all sorts of interviews; they basically took every moment they had to fight hard in the court of public opinion. They were talking to the public, and they were talking to the Israeli government, and their message was, if I might paraphrase: "It's not about the money involved in H2020. It's about ease of collaboration. It's about our colleagues and our peers in Europe and around the world. Israel might be doing brilliantly in all these metrics related to publications and research and innovation metrics, but we *don't do it alone*. If we don't sign on to this Horizon 2020 agreement, you are going to stifle the way in which we collaborate with our colleagues abroad. We must not have the arrogance to think that our scientific endeavor can be successful without that collaboration." And they were absolutely adamant. They held the government to ransom in some ways, and eventually the Israeli government signed the Horizon 2020 agreement (TOI Staff 2014).

I think we've really seen that exemplified in the U.K. Many scientists have been vocal about what they see to be the benefits of international collaboration and have said to the government: these are the implications for the scientific community. And the government has taken much of that on board and responded using some of the policies that I mentioned earlier—visa policies, underwriting policies, uplifts in science funding. It's really gratifying to see how responsive government has been in listening to the scientific establishment in this way.

Turning away from Israel and Europe and the U.K., I think you also see this in the U.S. to an extent. One of the most amazing moments last year was the March for Science and seeing the scientific establishment rally around the importance of international collaboration.

Let's close with Bill's question around optimism. I think there are a lot of reasons to be optimistic, and those reasons are the people in this room. The reasons are that people are passionate and committed to the importance of widescale, international scientific collaboration as the fundament of scientific progress.

We looked at two issues today: the rift between science and society and then political or practical impediments to international collaboration. It's my opinion that if we don't focus on fixing the former issue, governments will have a diminishing mandate to listen to the scientific community. If they have public that doesn't trust their scientific process and doesn't trust the impartiality of scientists in civic debate, then they are going to have no mandate to fight for science inside policy and political communities.

So, I think there are a lot of reasons for optimism, but there are even more reasons for activism. We have a lot of power to change the way governments operate in terms of international scientific collaboration, but we have to recognize the urgency of public with a decreasing faith in science, and public that is being fed pseudoscience without any repercussions.

Questions and Answers

Question

Deborah Hughes Hallett: First, I would like to just support what Peter said. I taught at METU in Turkey for two years. It was almost thirty years ago. I still know my students, I still hear from them, they still show up unexpectedly to ask me things, so yes it does make a difference. Their professors, who were my colleagues when I was there, are now having quite a difficult time because not only is it difficult to get visas to come here, but you have to get permission to leave to go to a conference, so tolerance and persistence on our part is important.

I'd like to ask a question to Dr. Prawer: Do you think that we (or the community generally) need to think about how science is perceived because of who it's allied with, particularly for getting funding? I am struck by the paper that I just read with my class from DRC Congo on the Ebola outbreak where, in the cities where the disease has killed thousands, 25% of the inhabitants of those cities believe the disease doesn't exist because they believe it was cooked up by the government to oppress them (Vinck et al. 2019, 5).

Answer

Prawer: I think that is such a good point. I think we see this all the time. Again, we're so focused on political misinformation and how you tackle it, but there are such limited consequences for pseudoscience and scientific misinformation. Your example, you think about what happened in Kerala where people believed social media stories about diseases are spreading through chicken (Vijaykumar 2019); think about what happened and continues to happen with vaccine hesitancy.

Let me tell you a story: I was at a dinner party a while ago. I was supposed to be diplomatic and reasonable, but I was sitting next to someone who said, "Oh you're a geneticist. You people use animals in your scientific research. How can you do that?" So, we had *that* conversation. I asked her if she gave her daughter the cervical cancer vaccine, it so happened that she had. Anyway, then we were served our main course and she asked the host if it was organic because she didn't eat GMOs. I asked her why not and she said that *she didn't want to eat DNA*. So, that was the end of my ability to be diplomatic at that moment.

Dinner parties aside, you can really see the implications of the fact that there are no real repercussions for those spreading dangerous scientific falsehoods. You know, Goop and Gwyneth Paltrow were given a minor fine for telling women to do something that is scientifically unsound (Hetherington 2018). But those fines are a drop in the ocean for companies like that. When are we going to hold people accountable for pedaling pseudoscience to the same account that we hold people pedaling political misinformation?

Agre: I like the emphasis you placed on being diplomatic. I think that's sometimes a big challenge. I grew up in Minnesota. My dad was a chemistry professor at a small college and my mother was a farm girl. Their entire married life there was sort of a gap in communication. Now Dad's been gone a long time, and mother's becoming increasingly conservative. We don't talk about politics. But I've noticed that sometimes with the opposition is a firm rejection of science. If you try to talk about evolution in one of these communities you will be soundly rejected, but if you talk about natural selection, treating children with antibiotics to which the bacteria are sensitive, there is no opposition. So, I think that's part of being diplomatic; not alienating people before you can even have a dialogue. I think the dialogue is where positive things emerge.

Colglazier: For the students and post-docs here, you can see that us elders have kind of made a mess of things. So what is your advice, both for working at the scientific level and working at the interface of science and society, where can we actually get the most bang for our buck to do things that will help our country and help the world. You don't have to ask a question, but we're looking for your advice.

Question

Audience Member: I'm not a student, but arguably we should all be students for life, so I'll qualify myself that way. I wanted to invite you all to consider the flip side of what we've heard so effectively from you this morning and yesterday evening, which affirms in many ways the challenges of diplomacy as scientists. Can you maybe share some perspectives that leave us, and especially our future in the room, with some hope?

Answer

Agre: I think a clear advance in science, which is evident here, is the emergence of women in science. During my lifetime there has been a major transformation. Certainly in medicine, where I trained. I entered medical school in 1970. 10% of my class was women. Admissions officers were so astonished because it had never been that high before, and they thought it would never be that high again. Of course, now the majority of medical students are women. The same is true of our graduate schools. There have been excellent women scientists all along, but not in the numbers we see today. I think that they are going to have more than twice as many heroes in the future. A favorite joke I like to tell is about a Nobel Prize winner and his wife driving in the desert. In the future, that will be a Nobel Prize winner and her husband driving in the desert. I think the emergence of women is going to change everything.

Prawer: Here, here! The emergence of women *is* going to change everything. I think the optimism-pessimism debate is the wrong dichotomy. I think it's an activism-fatalism debate. There was a Jonathan Franzen article quite recently in *The New Yorker* that caused a huge amount of debate titled "What If We Stopped Pretending?" (Franzen 2019). He basically was saying "When are we going to accept that we can't undo what we've done? When are we going to accept that climate change is a reality?" I don't think that that is the right approach.

I think the right approach is not fatalism; it's activism. I think Greta Thunberg is right when she says we are in crisis. "I want you to act as you would in a crisis. I want you to act as if the ho use was on fire" (Thunberg 2019). I think that some of that gives science diplomacy the ability to have some real traction.

One of the examples that I'm following very closely at the moment, because I think it's a fascinating example of science diplomacy, is the BBNJ debate at the U.N. For those of you who haven't looked at it, science diplomats brought the issue of ocean biodiversity loss to the floor of the U.N. for the first time. There was certainly a lot of pessimism. We have grotesque islands of plastic forming in the High Seas, a grieving orca carried her calf along the coast for nine harrowing days—that's enough to make anyone pretty pessimistic.

But what is optimistic is that countries have convened at the U.N., and they have two years to form an internationally legally binding instrument to govern our use of the High Seas and to protect biodiversity in areas currently beyond national jurisdiction. I am optimistic about reaching a treaty, and it is a fascinating negotiation to watch. Two years to get the U.N. member states to agree to something legally binding is pretty impressive. I think they're going to do it, and I think it's going to go a long way toward protecting biodiversity in two-thirds of the world's oceans.

There are certainly processes that, when you watch the bureaucracy down here, it's horrendous and exhausting. But when you watch it from up here and you think, "okay we are actually going to end up with something the international community has agreed on," that is my hope. And that doesn't happen unless you're optimistic enough to be an activist.

Colglazier: I like emphasis on avoiding fatalism. What makes me optimistic, and what we can actually do in practical terms. In terms of optimism, I visit many countries across the world. Even with the dysfunction occurring in the U.S., in our politics, in our government, I have always been warmly welcomed as just another American scientist there. They want to continue collaborating, they want to continue working. As Peter said, great respect for American science. They want to be part of this worldwide enterprise with us. I talk and meet students from universities around the world. They are all the same. Many of them are very idealistic, they all want to find out how they can use their expertise to make the world a better place.

Of course, as scientists, we have to remember that the most important thing of all is continuing the scientific enterprise. Most scientists are going to continue working in their field, and that's fine, but also trying to engage and going across the science-society interface, being able to understand what the needs of our societies are and how we can actually contribute. Vaughan Turekian and I are very much involved in the U.N. in the role of science, technology, and innovation and their role in achieving what are called Sustainable Development Goals. The social, economic, and environmental, they cover the waterfront. Many countries are paying very serious attention at the political level to this. The diplomats at the U.N., as well as the political leadership of almost every country, have recognized that they have to upgrade their scientific and technical capabilities and if they are going to make progress on whatever their national goals of their country are, they are going to have to do a much better job getting input and advice from their scientific community, trying to figure out how they can harness science, technology, and innovation to be secure, prosperous, competitive in this very fast-moving world.

Even with all the political challenges we face right now, my optimism comes from the young people that I meet. The scientific enterprise is going to continue at a very fast pace. We have to make sure that it's sustained well. In fact, in many countries, even at the highest levels, even at the diplomatic level, they now see the importance of trying to harness their capabilities in science and technology to achieve whatever the social, economic, and environmental goals of their country.

REFERENCES

American Immigration Council. 2020. "Immigrants in the United States Fact Sheet." August 6, 2020. www.americanimmigrationcouncil.org/research/immigrants-in-the-united-states

C-SPAN. 2019. "Representative Ocasio-Cortez Questions Mark Zuckerberg on Facebook Fact-Checking Policy." House Financial Services Committee Meeting, October 23, 2019. www.c-span.org/video/?c4824531/representative-ocasio-cortez-questions-mark-zuckerberg-facebook-fact-checking-policy

Dionne, Jr., E.J. 2020. "Trump's Betrayal of American Greatness." *The Washington Post*, June 21, 2020. www.washingtonpost.com/opinions/trumps-betrayal-of-american-greatness/2020/06/21/724f26ec-b264-11ea-8f56-63f38c990077_story.html

Franzen, Jonathan. 2019. "What If We Stopped Pretending?" *The New Yorker*. September 8, 2019. www.newyorker.com/culture/cultural-comment/what-if-we-stopped-pretending

Hetherington, James. 2018. "Gwyneth Paltrow's Goop Hit with Big Fine for Telling Women to Put Eggs in Their Vaginas." *Newsweek*, September 6, 2018. www.newsweek.com/gwyneth-paltrows-goop-hit-big-fine-telling-women-put-eggs-their-vaginas-1108779

Jamrisko, Michelle, Lee J. Miller, and Wei Lu. 2019. "These are the World's Most Innovative Countries." *Bloomberg Innovation Index*, January 21, 2019. www.bloomberg.com/news/articles/2019-01-22/germany-nearly-catches-korea-as-innovation-champ-u-s-rebounds

Kerr, Sari Pekkala and Kerr, William R. 2018. "Immigrant Entrepreneurship in America: Evidence from the Survey of Business Owners 2007 & 2012." NBER Working Paper No. w24494, Available at SSRN: https://ssrn.com/abstract=3158934

Machiavelli, Niccolò. 1981. *The Prince*. Harmondsworth, England; New York, NY: Penguin Books.

Pethokoukis, James. 2020. "Trump's Visa Ban Is Economic Suicide." *The Week*, June 23, 2020. https://theweek.com/articles/921644/trumps-visa-ban-economic-suicide

Polak, Paul. 2009. *Out of Poverty: What Works When Traditional Approaches Fail*. San Francisco, CA: Berrett-Koehler Publishers, Inc.

Sherwood, Harriett. 2013. "Israeli Government Split over Joining Prestigious EU Science Programme." *The Guardian*, November 26, 2013. www.theguardian.com/world/2013/nov/26/israel-eu-science-programme-settlement-boycott

Stokstad, Erik. 2020. "UK Cues Up Big Funding Increases for R&D." *Science*, March 11, 2020. www.sciencemag.org/news/2020/03/uk-cues-big-funding-increases-rd

Thunberg, Greta. 2019. "I Want You to Panic: 16-Year-Old Issues Climate Warning at Davos." *Guardian News*, January 25, 2019. Retrieved May 13, 2020 – via YouTube. www.youtube.com/watch?v=RjsLm5PCdVQ&t=2m23s

TOI Staff. 2014. "Israel Signs Key European Scientific Funding Deal." *The Times of Israel*, June 9, 2014. www.timesofisrael.com/israel-signs-key-european-scientific-funding-deal/
Vijaykumar, Santosh. 2019. "Pseudoscience Is Taking over Social Media and Putting Us All at Risk." *Independent*. August 7, 2019. www.independent.co.uk/news/science/pseudoscience-fake-news-social-media-facebook-twitter-misinformation-science-a9034321.html
Vinck, Patrick, P. Pham, K. Bindu, J. Bedford, E. Nilles. 2019. "Institutional trust and misinformation in the response to the 2018–19 Ebola outbreak in North Kivu, DR Congo: a population-based survey." *The Lancet*. 19, 5. https://doi.org/10.1016/S1473-3099(19)30063-5.
Zogby, James J. 2002. *What Arabs Think: Values, Beliefs, and Concerns*. Utica, N.Y.: Zogby International.

TECHNOLOGICAL GOVERNANCE IN LATIN AMERICA: INSTITUTIONAL CHALLENGES TO NAVIGATE EMERGING TECHNOLOGIES (2010–2019)

Martín Pérez Comisso
Ph.D. Student in Human and Social Dimensions of Science and Technology, School for the Future of Innovation in Society, Arizona State University

"*Either we invent or we fail. Take what is good, leave what is bad, imitate with judiciousness, and for what you lack invent*" is a phrase enunciated by Simon Rodríguez (1828 [1990]), mentor of one of the liberators of Latin America, Simon Bolivar. Its revolutionary meaning has been interpreted in various fields of knowledge as a call for action, not only because of its eminent learning and anti-colonialist dimension but in particular because of its reference to innovation and proactive reflexivity toward foreign models.

This call takes on special relevance when we observe the technological trajectories of the region, which has stood out for its dependence on the extraction of natural resources, great socio-economic inequality, and an oscillating history of internal conflicts since the independence of the nations from Spain during the 19th century. Rodríguez defies the governments and institutions of the region to find their own way, something that has been a permanent challenge for all countries in the region. Scholars of science policy Dutrénit and Sutz reflect on the current situation that "After decades of development efforts, Latin America continues to be the most unequal region in the world" (Dutrénit and Sutz 2014).

But in front of any new technological revolution, there is a new opportunity to make things different. It is common in the region for a discourse of opportunity to "lead" a new technology revolution in their own way, and the last decade of the 2010s has not been an exception. But "Latin America is a very heterogeneous region in terms of countries levels of development and maturity of their NSI" (Dutrénit and Sutz 2014), and these differences are key to understanding to navigate regional challenges and opportunities which emerging technologies made for the governance and institutionalization of technologies in Latin America.

In this chapter, I want to argue that Latin America is still not ready to proactively lead or govern emerging technologies, with a focus on the institutional transformations in science and technology institutions during 2010–2019, following the question: Is S&T governance of Latin American countries prepared to regulate emerging technologies?

To approach this question, the chapter is organized into four sections: the first one situates the notion of emerging technology in terms of governance, based on science policy and technology studies literature. Then, I discuss the historical trends of S&T governance that characterize the region with an emphasis on a lack of historical role of emerging technologies in national discourses. The next section provides a comparison of those institutional transformations based on executive, legislative, and funding trends on the national innovation systems (NIS) of six countries in the region: Argentina, Brazil, Chile, Ecuador, Mexico, and Uruguay. Later, I address some illustrative emerging technologies agendas from the discussed countries, to understand the similarities and differences in the models of governance for them. Finally, I summarize the findings, to propose some regional policy recommendations that can be useful for further discussion about the challenge to govern emerging technologies.

A Local Notion of Emerging Technologies

Before introducing the institutional transformation about emerging technologies in Latin America, it is necessary to situate the concept of "emerging technology." In this section, some considerations are introduced to understand the political challenges in this concept, and to study these kinds of technologies at an institutional context, as well as to present some methodological challenges to attend in the incorporation of emerging technologies in Latin American culture (or in any peripheral context).

First, this idea of emerging technology is highly recurrent in innovation studies and science and technology governance, in particular the scientometric study of the emergence of new technologies. We still can find applications of the concept in different scales, classes, and stages of emerging technologies, being used to frame large socio-technical systems as big data techniques (Huda et al. 2018) or synthetic biology (Wellhausen and Oye 2007), as well as very niche techniques like digital PCR (Salipante and Jerome 2020) or nanogenerators (Zi and Wang 2017). The concept of emerging technology is presented to frame the most diverse range of innovations, but in this chapter I am going to focus on its use for a specific set of examples in the Latin American region.

Moreover, the idea of emerging technology has an increasing relevance as a policy term since the framing of the Fourth Industrial Revolution for the World Economic Forum during the last decade (Schwab 2017), because "greater discourse on, and around, emerging technologies offer unparalleled possibilities for stakeholders, to critically engage with the public and policymakers on the trajectory of the technologies, and their deployment in different societies" (Bowman et al. 2017). Governments around the world have incorporated this concept on some level in their policy discussion, and Latin America is no exception.

Certainly, the conceptualization of emerging technology offers a compelling framing conversation about development, futures, and expectations in the local context, and has been used as an ad hoc and versatile political discourse in favor of industrial, economic, and scientific programs and plans. Authors such as Carlota Perez (2014) propose that these technologies articulate a new technological paradigm in the making, which can drive societal, productive, industrial, and economic change on a global scale.

Paradoxically, these definitions of emerging technologies remain ambiguous. For instance, some authors use the concept to reflect a novel field of application or diffusion for an existing technology (Small et al. 2014; Eguchi et al. 2018), as others focus on the potential impact of them in society (Porter et al. 2002), while at the same time unpredictable effect on the implementation of them (Boon and Moors 2008). Media has even more diversity of use for the term (Balistrieri 2013).

Rotolo et al. observed that there is a "lack of consensus over definitions is matched by an 'eclectic' and ad hoc approach to measurement" (Rotolo, Hicks, and Martin 2015). Under this observation, the authors articulate a systematic review on the conceptualization of emerging technology in the literature. They argue that a precise definition must include these five elements: (i) radical novelty, (ii) relatively fast growth, (iii) coherence, (iv) prominent impact, and (v) uncertainty and ambiguity. With these considerations, they propose an operational definition for emerging technology:

> A radically novel and relatively fast-growing technology characterized by a certain degree of coherence persisting over time and with the potential to exert a considerable impact on the socio-economic domain(s) which is observed in terms of the composition of actors, institutions and patterns of interactions among those, along with the associated knowledge production processes. Its most prominent impact, however, lies in the future and so in the emergence phase is still somewhat uncertain and ambiguous.

Rotolo et al. also acknowledge the limitations of their definition, with an emphasis on the methodological practices for the characterization of emerging technologies (mostly scientometric methods). Despite the fact that they seem to share attention for the context, the phenomenon's definition is incomplete, because "[t]he understanding of emerging technologies also depends on the analyst's perspective" (Rotolo, Hicks, and Martin 2016). The authors don't incorporate the notion of situatedness in time and place in their description, and that makes room in their definition for improvement, in particular for regional and national emerging technologies.

The last point is central for the regional governance of emerging technologies. It needs to be situated in national and local contexts to propose effective actions and approximations to the radical, impactful, and uncertain effect of them in a culture. Questions arise like: When does a technology emerge for a peripheral country? Who has the power to establish when technology has begun or end its emergence? What are the factors resisting or supporting the appropriation of these technologies in specific organizations or general users?

No technology is diffused or distributed seamless or equitably in a region, country, or the world. This intrinsic unevenness of emergence needs to be considered in the institutions, capabilities, and agendas for emerging technologies in a region. There is a huge implicit power dynamic about the existence of emergent technologies, that goes beyond the scope of this chapter, but it is relevant to highlight to understand the institutional challenges that any country has from the novel and disruptive incorporation of emerging technologies in their practices and actions.

In the next section, I will introduce some historical elements that characterize innovation systems of Latin America, with particular attention to some institutional commonalities.

Institutional Arrangements in Technological Governance in Latin America

Sadly, for Simon Rodríguez, at the beginning of the 21st century, Latin America is not a region characterized by their inventiveness. Technological governance in Latin America is an area with limited attention of government officials, several times incorporated as a sub-area of education or economy agendas. Hebe Vessuri, an eminent Latin America science scholar, gives us a historical perspective in her book called "*Either we invent or we fail*" (O inventamos o erramos): "The story of science in developing countries is full with examples of attempts for institutionalization following by collapses, with unlimited optimism followed for a pessimist indifference, and a lack of public confidence about the long-term intellectual work" (Vessuri 2007, 94). With NIS highly dependent on global developments, the region plays a minor role in the production and development of new technologies, generating less than 2% of the world's patents.

NIS of Latin American countries reproduce a model of institutions, in which scholars like Lemarchand (2015) have studied the formation and evolution at a national level of "national innovation systems" based on equivalent institutions for research funding, nuclear research, agriculture incentive, and others since the 1960s.

The dictatorial periods of the 70s and 80s intervened in the democratic normality of the countries in the region. Albornoz describes that for science and technology policy, "many governments reconsidered the role of the state, which led to a reduction in its scope of action, abandoning sectors of activity that had hitherto been unquestionably its own to adopt other functions, supposedly linked to regulations and quality control, but which in practice often led to a decline in public funding for many activities, including academic, scientific and technological activities" (Albornoz 2014).

The first major transformation happened in 1990 based on the idea of "technological catch-up" and incentivized for international organizations. Vessuri also warns us about this transformation claiming, "The recognition of the role of government planning of science in the developing countries does not imply the presence of positive outcomes. As a matter of fact, the national experiences have been subject to strong criticism in several respects" (Vessuri 2007).

As an exemplar case, the role of the Interamerican Bank for Development (IBD) since 1990 has been to fund and frame programs aligned with its own understanding of development for the region. Several local scholars had investigated those cases and programs for Argentina, Colombia, or Uruguay, with a focus on the negotiations and outcomes. Among the criticism, they observe a reproduction of the dependency-model that has been recurrent for the science and technology policy in the region (Aguiar, Aristimuño, and Magrini 2015; Aguiar, Davyt, and Nupia 2017; Aristimuño and Lugones 2019). This external influence has also been observed for other intergovernmental institutions, like the United Nations Educational, Scientific and Cultural Organization (UNESCO) and the Organisation for Economic Co-operation and Development (OECD), that have slightly competing paradigms about what is for science and technology in the region.

The formulation, focus, and establishment of S&T programs have been largely configured outside the region. Hurtado argues that "[t]he determining and constitutive obstacle is geopolitical: the place assigned to Latin America in the world capitalist order explains the technological and scientific 'failures' that are manifested in the lack of organizational and institutional capacities for the management of strategic technologies—or 'technologizing technologies,' as Jorge Sabato, strategist of the Argentine nuclear plan, called them—and of useful knowledge for social development and the complexity of the productive fabric" (Hurtado 2019).

Institutional Transformation and Presence of Emerging Technologies in NIS

Currently, most countries of Latin America have NIS with more than 50 years of history; new institutional transformation is ongoing for adaption to current challenges of the Fourth Industrial Revolution, a highly damaged and changing natural environment and social issues emerging at overpopulated cities.

At the beginning of this decade, the common diagnostic about technology governance is summarized by Marco Aurelio Jaso Sánchez: "The absence of critical masses, of experts, the low levels of institutionalization and the vertical design of governmental STI policies made it difficult to disseminate foresight and its horizontal and vertical exercise in different national, sectoral and governmental spheres" (Jaso Sánchez 2015).

Nonetheless, Albornoz observed a different trend. "In the early years of the 21st century [...] science and technology policy is gradually being incorporated into the agenda of the countries of Latin America. Investment in science, technology, and higher education has increased in almost all countries. Most of them started to formulate and implement policies to stimulate innovation" (Albornoz 2014). New institutional models have been proposed and implemented in the region during the last ten years. In Table 2.1, I summarize some examples of institutional transformations for six selected countries: Argentina, Brazil, Chile, Ecuador, Mexico, and Uruguay.

Table 2.1 shows how several institutional arrangements have been developed for the research policy in the region, with a particular emphasis on the transformation and/or integration of research funding agencies in sectoral ministries, with different configurations for each country. Interestingly, the last two decades is the time that most of the selected countries articulate or update their institutionality based on national or federal board regulations of the scientific activity. It should be noted that limited capabilities for the assessment and monitoring of these changes have been observed in the region, with several gaps about the historical trajectories, design of new indicators, and research to describe the consequences of these new arrangements (Baptista 2018).

An interesting factor for a new generation of research policy in the region is the renewal of human capabilities for research during the last decade. With strong incentives for training, publishing indicators show increasing scientific productivity in all selected

TABLE 2.1
Main institutional regulations and institutions for the promotion and governance of technology in selected Latin American countries during the 21st century

	New Institutional organizations (since 2000)	New Regulations/Legislation (since 2000)
Argentina	SecCTI (2018), MinCyT (2007–2017; 2019).	Law 25.467 of Science, Technology & Innovation (2001)
Brazil		Decreto Federal nº 9283, 9574 (2018) law 13243 (2016) and 13674 (2018)
Chile	Minister (2018), national innovation council (2005). National Agency of Research and Development (ANID, ex-CONICYT, 2019)	Ministry Law (2018)
Ecuador	Ministry of Education, Knowledge and, Culture (2015)	Executive Decree No. 1829 (2006) Law of Science, Technology, Innovation and Indigenous Knowledge (2015) Plan Ingenio (2016)
Mexico	Research Council (2017–advisory)	Law of Science (2002)
Uruguay	Governmental Ministry for Innovation – GMI (2005); DICYT (2005)	Plan Ceibal (2007) National Plan of Science and Tech (2019)

Source: author's elaboration.

countries, provoked by a pick on higher education graduated during the last decades. Albornoz also suggests, "The countries of the region were beginning to regain interest in science, but they were doing so under the gaze of the linear model, with a focus on research, having basic science as an implicit normative model" (Albornoz 2014).

Nevertheless, those new institutions and professionals carry another structural problem: there is not adequate public or private investment in science and technology in the region. Figure 2.1 shows a comparison of the percentage of GDP invested in science between 2008 and 2017 for selected countries. Most countries maintain a steady investment, such that all of them but Brazil are less than 1% of GDP.

In addition, a right-wing political transition during the second half of the period decreased and even eliminated investment in several programs, in part because S&T investment is seen as unprofitable in the short term for this political paradigm. An example of this is a disinvestment in philosophy and sociology programs in Brazil during the government of Jair Bolsonaro (Redden 2019). There are several other political elements involved in those decisions that go beyond the scope of this paper, but as a result of a regional shift, the funding in S&T tends to decrease since 2016. For the first time since 2000, the economic resources devoted to science and technology in the region have decreased and has continued to decline in recent years. "Economic constraints in LAC continued in 2017 causing the percentage of spending to fall even further" (Barrere 2019).

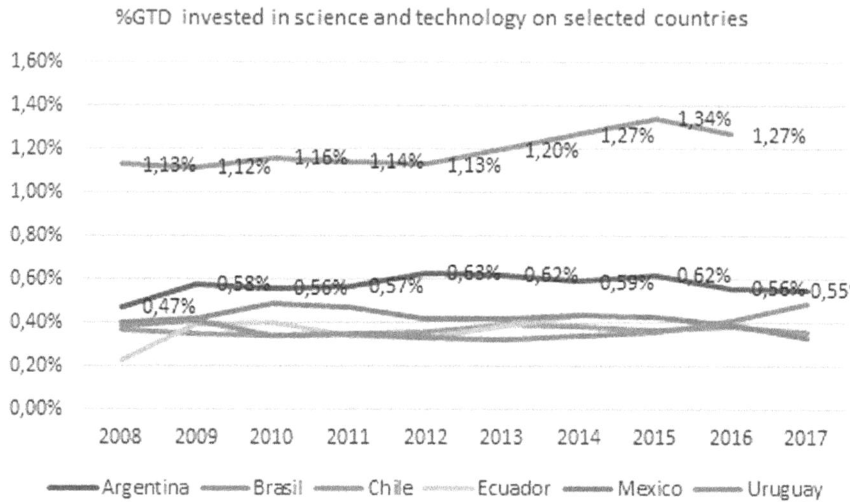

FIGURE 2.1 GTD invested in S&T for selected countries between 2008 and 2017. (Source: RICYT.)

On top of that, Van Noorden, comparing the regional performance of NIS, considers that "given the region's share of the world's population and gross domestic product (GDP), publication rates still fall short of what would be expected. Research quality has not kept pace with rising output, and the continent's research papers still struggle to attract citations from the rest of the world" (Van Noorden 2014).

This tendency is consistent with other indicators, like, for instance, patent applications per capita. In 2015, Latin American countries represented 8% of the global population, but just produced 5% of scientific production and 0.3% of patents (Lemarchand 2015). Even in these patent applications, 81% of them came from international corporations protecting their industrial property on the region's markets. Based on Barrere (2019), Figure 2.1 illustrates the national percentage of annual GDP on S&T investment, compiled for RICYT (Spanish acronym for the Iberoamerican Network of Indicators in Science and Technology).

Under this scenario, the contributions to create, produce, and innovate emerging technologies in Latin America, under the current definitions, institutions, and economic regime, are not auspicious. Under these definitions, the region's profile has become known as a place of little invention and not much progress. Most emerging technologies came from other parts of the world, or even when they do exist, they are not able to become visible in the global conversations until a later stage of development.

In particular, there are difficulties with graphing data of diverse NIS using current study methods around the creation and impact of emerging technologies, particularly due to the large scope of patent producers or publications outside of Latin America. Thus, all these conditions keep the production of emerging technologies as an exogenous matter to the region and favor late discussions and framings of technological policy.

In the next section, I will exemplify how NIS have been arranged (or not) for the regulation and implementation of a set of emerging technologies during the last decade.

Emerging Governance of Technology in Latin America

Emergent technologies are also a continent phenomenon, as I mentioned above. "To make sense of the technology, nations develop regulatory styles based on their particular framings of risk, institutional arrangements, and public participation modes" (Salazar et al. 2019). Each nation and institutional cultures have several moments when an emerging technology began to be appropriated for local users, institutions, and regulations. The last decade has been a series of national plans of strategies to situate emerging technologies in the contexts of the selected countries (Argentina, Brazil, Chile, Ecuador, Mexico, and Uruguay).

In this section, I perform an overview of a series of examples that is not a systematic review but illustrates the kind of institutional reactions observed in literature and media about emerging technologies during the last decade in Latin America. Evidence for this emerging technology regulation was observed in legislative minutes, media reports for these countries, and executive resolutions of research agencies or national ministry, when appropriate.

Table 2.2 summarizes those findings, with a focus on the governance actions oriented to the implementation, promotion, use, and assessment of seven emerging technologies: biotechnology, nanotechnology, artificial intelligence, cybersecurity and fintech, Augmented Reality/Virtual Reality and Internet of Things (coupled for the enhanced layers of digital information, as referred to as AR/VR and IoT), Renewable Energies, and programs for the incentives of technology-based start-ups and entrepreneurship. The focus was to find at least the presence of some kind of regulation of these selected emerging technologies.

There is no common pattern to describe how these countries incorporate these plans and strategies in their NIS: sometimes, there is a local controversy which provokes a window of opportunity for a regulation (like biotechnology in Mexico); other times,

TABLE 2.2
Presence of National Plans, Programs or Committees to advise about selected emerging technologies on selected countries in the region

	Biotech	Nanotech	Artificial Intelligence	Cybersec/ FinTech	AR/VR and IoT	Renewable Energies	Start-ups local entrepreneurs
Argentina	X	X	X	X	X	X	X
Brazil	X	X	/	/	/	X	X
Chile	X	-	/	X	/	X	X
Ecuador	X	/	/	X	X	X	-
Mexico	X	X	X	X	-	X	X
Uruguay	X	X	X	/	X	X	X

Key: / represents ongoing efforts; X represents the existence of a national plan, strategy, or regulation currently under implementation

Source: author's elaboration.

an international or local organization champions the adoption of successful policies made for another country before (like start-up incentives in Brazil) based on a particular case, or also is the result of long process uncoordinated implementation, resulting in a redefinition of obsolete regulation (in the case of renewal energies and its transitions in Chile).

Nonetheless, the literature about technology policy in the region warns about particular difficulties to study emerging technologies. In the late 1980s, Argenti et al. claimed that "[y]ou can't make politics out of what you don't know [...] Therefore, knowing something about the situation is a prerequisite for politics" (Argenti, Filgueira, and Sutz 1988, 163). As shown previously, there are uneven institutional capacities in Latin American countries for the assessment and limitations.

In addition, complex relations between institutional governance and emerging technology agendas have been observed as an interactive element for the success of technological policy in the region. For instance, a study of Mexico's biotechnology industry by Amaro Rosales and Morales Sánchez recognized, "In the literature on innovation that the institutional context co-evolves with technological development and that the trajectory depends on initial socio-economic conditions, there has been little recognition of the fact that in developing countries it is necessary to think about social needs and therefore about innovations that solve these problems" (Amaro Rosales and Morales Sánchez 2016).

Stability of organizations is not related to success or failure of programs; instead, it is the effect of the interaction between agendas, assessment, and local understanding of the own process of technological emergence. In this respect, the institutional transformations detected for selected countries put the focus on the arrangements of NIS, instead of the situatedness, framing, or effectivity of local agendas for emerging technologies. For this reason, the rapid institutional transformations in recent years diminish the ability of NIS to anticipate, evaluate, and maintain complex governance systems, despite the increase in the number of specialists in the countries of the region in the last decade.

Amaro Rosales and Morales Sánchez clearly articulate this situation in their research, which is found frequently in the reviewed sources: the lack of preparation of NIS in Latin America for emerging technologies. They claim:

> In the country, there is an institutional framework that could give certainty to the different agents that make up the sector; however, based on the fieldwork carried out, it is possible to say that the regulatory framework is still weak, so it is not a decisive factor in the conduct of the agents, especially due to their low capacity to generate economic incentives and apply sanctions. What can be observed is that most decrees, standards, and laws, as well as the organizations responsible for promulgating and implementing them, have not played a decisive role in the sector, either in promoting technological development or in regulating it. This situation creates problems in a number of ways, since, although regulatory organizations and laws exist, they do not represent a coherent and orderly framework with common objectives.
>
> (Amaro Rosales and Morales Sanchez 2016)

In contrast, there is a growing body of work outside the governmental apparatus that contributes to situating appropriately emerging technologies in Latin America. Some great examples of this multilayered approach for them can be found in the case of nanotechnology. Foladori et al. (2017) apply an indicator-centered analysis of the policy of nanotechnology in Mexico. In a similar direction, Quevedo (2017) incorporates the use of the actor-network theory to map the assemblage of institutions, regulations, and processes of Brazilian nanotechnology. For their part, Montoya and Rivas (2019) use a comparative approach to assess the direct impact on citizens of AI readiness, with particular attention to automation and data privacy. These experiences, among others, go in the adequate direction for a better understanding and implementation of emerging technologies in the region, and it is a weak point for all innovation systems compared in this paper.

The last section will summarize the main findings and articulate some policy recommendations for the governance of emerging technologies in the region in the following years.

Toward Better Governance of Emerging Technologies in Latin America

Hebe Vessuri has been a champion for the "endogenization of technology." Defined by her mentor Amilcar Herrera, this idea "refers to the process which determines the characteristics that technology should have in Latin America. What is endogenous is the process of definition and not necessarily the technology itself, which can be imported, as long as it is appropriate." She identifies a key role of technology in the material and cultural transformation of societies. For her, "the proposal implies nothing more or less than the autonomy or independence to define in which way a particular country wants to function, with which technology, available or not, it wants to proceed" (Vessuri 2007, 120).

The transformation during the studied period of S&T governance in the region has been intense, from many institutional, regulatory, and funding changes. Even so, this notion of making emerging technologies "endogenous" is a function that, although historically desired, is outside the actions that the region's NIS carry out. It is observed that a discontinuous evolution of NIS in most of these countries, as well as the absence of instruments for the participation and governance of emerging technologies, makes it very difficult to effectively anticipate or regulate imported technologies.

There is an ontological disconnect between emerging technologies and their governance, particularly because of the unpredictable, global scale, rapid growth, and potential impact of these technologies. Not being a production center is a central condition among the lack of institutional tools to anticipate the design, development, and implementation of emerging technologies. Nevertheless, prioritize and incorporate these capabilities will bring "greater awareness of the benefits and risks of such technologies, and their applications, bring with it an array of complex societal issues that will need to be addressed in real-time" (Bowman et al. 2017).

The opportunities that emerging technologies bring to the innovation systems in Latin America are different to previous historic challenges: there is a larger awareness in the region about the value and impacts of emerging technologies, a large number of graduate programs had trained and produced relevant tools and research to understand the local impacts of emerging technologies, and despite the decreasing trend in funding, there are new institutional arrangements that allow experimentation and exploration of new political strategies to approach the production, use, and assessment of emerging technologies.

I suggest three policy recommendations to increase the future capabilities of Latin America in respect of emerging technologies: (i) increase regional cooperation in the agenda-setting of emerging technologies, (ii) assimilate local and regional research capabilities and frameworks to assess emerging technologies, and (iii) continue to fight back hegemonic regimes of properties, to value the actual contribution of the region in the production of technologies.

Milton Ladikas, a German researcher in technology assessment, summarizes the situatedness of emerging technologies as "Since circumstances for technological transformation differ globally, there is an understanding that global cooperation should transcend the existing tools and develop TA approaches that can consider contextual circumstances and aid decision-making in different settings" (Ladikas 2019). A regional technological assessment, like in Europe, is a good strategy to emulate in Latin America for the governance of emerging technologies.

Due to cultural and political similarities among the countries, a multi-governance collaboration centered around emerging technologies is an opportunity to articulate NIS as a responsive ecosystem for governing emerging technologies. Nevertheless, Vessuri observes, "[t]he regional institutional collaboration has been recommended as one of the most effective ways to quickly build a larger scientific and technological infrastructure, but often there has been a total concentration of advanced scientific training in some regional centers. Those centers have a moderate success when the institutions get international support through bilateral agreements with specific governments" (Vessuri 2007, 98).

In this respect, a regional observatory for emerging technologies can be a better alternative for this case. Based on common challenges (as sustainable production, aging, larger urban populations, social inequality), a regional integration will leverage capabilities that countries by themselves are unable to enact in the global conversation of emerging technologies, such anticipation, framing, and assessment. In addition, this kind of institution has the potential to articulate technological diplomacy in the common good.

The second point is about the incorporation of local researchers into the conversation and assessment of emerging technologies. At the moment, an increasing number of research centers around the S&T policy are in the region (Baptista 2018), but with promising research about emerging technologies and related topics.

In the short term to build up these capacities, a clear incentive in this direction can be expressed in specific funding programs to assess and anticipate the emerging technology. Then, an expansion of the number of professionals working in the NIS (and adequate indicators in this regard) could strengthen the system as a whole, with professionals better trained in the skills that the NIS currently lack in the region.

Finally, as I addressed during the paper, there is a collection of local paradigms about science and technology in Latin America that explicitly fight back assumptions about knowledge production and technological understanding. These traditions have the value to contest the meaning of standardized indicators, policy discourses, and agendas with the focus on the local appropriation of knowledge. For instance, there is a question of whether measurement of emerging technologies based solely on scientometrics of patents is accurate, when consideration of the characteristics and limitations of the patent systems for non-central nations is taken into account.

Increasing of attention of co-dependent relations with nations and corporations producing and exporting emerging technologies, that is expressed differently in every case, is another potential outcome for not restricting the theoretical and empirical approximation to emerging technologies. These critical Latin American traditions of science and technology policy did not just inform this chapter, but those diverse epistemologies also offer new ways to deal with emerging technologies due to their uncertain, diverse, and contingent applications. I am convinced that these alternatives will be needed in the coming years, to discover and arise the effects of emerging technologies in peripheral contexts, and in the worlds to emerge within.

REFERENCES

Aguiar, Diego, Aristimuño, Francisco, and Magrini, Nicolás. 2015. The role of the Inter-American Development Bank (IDB) in the reconfiguration of institutions and policies for the promotion of science, technology and innovation in Argentina (1993-1999). *Revista Iberoamericana de Ciencia, Tecnología y Sociedad - CTS* 10 (29): 11-40. Centro de Estudios sobre Ciencia, Desarrollo y Educación Superior: Buenos Aires, Argentina. ISSN: 1668-0030.

Aguiar, Diego, Davyt, Amilcar, and Nupia, Carlos. 2017. International organizations and policy convergence in science, technology and innovation: the Inter-American Development Bank in Argentina, Colombia and Uruguay (1979-2009). *Redes* June 2017. National University of Quilmes. ISSN: 0328-3186.

Albornoz, M. 2014. "Memoria y balance de la RICYT: lecciones aprendidas y desafíos futuros." *Revista Iberoamericana de Ciencia, Tecnología y Sociedad (CTS)* 9 (25): 229–239.

Amaro Rosales, M., and Morales Sánchez, M.A. 2016. "Sistema sectorial de innovación biotecnológica en México: análisis y caracterización de sus principales componentes." *Redes* 22 (42): 137–40. https://repositorio.esocite.la/275/

Argenti, G., Filgueira, C., and Sutz, J. 1988. *Ciencia y Tecnología: Un diagnóstico de oportunidades (Science and Technology: What Can Be Done?)*. Montevideo, UY: CIESU.

Aristimuño, F. J. y Lugones, M. J. 2019. "El BID y las políticas de Ciencia, Tecnología e Innovación en Argentina (1990-2015)." *Ciencia, tecnología y política* 2(3): 031. doi: 10.24215/26183188e031.

Balistrieri, T. 2013. "Emerging Technologies are Changing the IT Job Market." *Wired Magazine*, May 23, 2013. www.wired.com/insights/2013/06/emerging-technologies-changing-it-job-market/

Baptista, B. 2018. "Una aproximación a las capacidades de diseño e implementación de políticas de ciencia, tecnología e innovación en América Latina." *Revista Iberoamericana de Ciencia, Tecnología y Sociedad (CTS)* 13 (38): 85–125.

Barrere, R. 2019. *El Estado de la Ciencia: Principales indicadores de la Ciencia y Tecnología Iberoamericanos/Interamericanos*. Red de Indicadores de Ciencia y Tecnología - Iberoamericana e Interamericana- (RICYT). www.ricyt.org/wp-content/uploads/2019/10/edlc2019.pdf

Boon, W., and Moors, E. 2008. "Exploring emerging technologies using metaphors: a study of orphan drugs and pharmacogenomics." *Social Science & Medicine* 66 (9): 1915–1927.

Bowman, D., Dijkstra, A., Fautz, C., Guivant, J.S., Konrad, K., Shelley-Egan, C., and Woll, S. 2017. "Emerging technologies: politics, problems and the unparalleled possibilities of engagement." In *The Politics and Situatedness of Emerging Technologies*, edited by Bowman, D., Dijkstra, A., Fautz, C., Guivant, J.S., Konrad, K., Shelley-Egan, C., and Woll, S., 1–9. Clifton, VA: IOS Press.

Carrozza, T., and Brieva, S.S. 2018. "Las políticas de CTI y el desarrollo inclusivo y sustentable en la Argentina: ¿constituyendo nuevas institucionalidades?" *Revista Iberoamericana de Ciencia, Tecnología y Sociedad (CTS)* 13 (39): 207–232.

Dutrénit, G., and Sutz, J., eds. 2014. "Introduction." In *National Innovation Systems, Social Inclusion, and Development: The Latin American Experience*, 1–13. Cheltenham, UK: Edward Elgar Publishing.

Eguchi, A., Gronseth, S., Antee, A., Ardito, G., and Christodoulou, A. 2018. "What is an emerging technology? The SITE community's perspective – Online interactive tools." In *Proceedings of Society for Information Technology & Teacher Education International Conference*, edited by E. Langran & J. Borup, 1445–1448. Washington, DC, United States: Association for the Advancement of Computing in Education (AACE).

Foladori, G., Figueroa, E. A., Lau, E. Z., Applebaum, R., Robles-Belmont, E., Vásquez, L. L. V., … & Leos, V. 2017. "La política pública de nanotecnología en México." *Revista Iberoamericana de Ciencia, Tecnología y Sociedad (CTS)* 12 (34).

Huda, M., Maseleno, A., Atmotiyoso, P., Siregar, M., Ahmad, R., Jasmi, K., and Muhamad, N. 2018. "Big data emerging technology: insights into innovative environment for online learning resources." *International Journal of Emerging Technologies in Learning (iJET)* 13 (1): 23–36.

Hurtado, D. 2019. "Proyectos de país en disputa: ¿Qué hacer con la ciencia y la tecnología?" *Ciencia, Tecnología y Política*, 2 (2): 015–015.

Jaso Sánchez, Marco A. 2015. "La prospectiva tecnológica en la planeación de la CTI en Argentina, Colombia, México y Perú. Una revisión desde el institucionalismo ideacional." In *Mirada Iberoamericana a las Políticas de CTI*, 135. www.academia.edu/40199086

Ladikas, M. 2019. "The need for an International Technology Assessment." Blog UNCTAD. United Nations. May 19, 2019. https://unctad.org/en/pages/newsdetails.aspx?OriginalVersionID=2081

Law of Science, Technology, Innovation, and Indigenous Knowledge – Ecuador SENECYT. 2015. "Código Orgánico de la Economía Social de los Conocimientos, Creatividad e Innovación." https://flacso.org.ar/wp-content/uploads/2015/07/Codigo-Ingenios.-MPI.pdf

Law of Science – Mexico DOF 05-06-2002, 2002. "Ley de Ciencia y Tecnología. Diario Oficial de la Republica de Mexico. 5 de Junio de 2002." http://www.diputados.gob.mx/LeyesBiblio/ref/lct.htm

Lemarchand, G.A. 2015. "Chapter 7: Latin America." In *UNESCO Science Report: Towards 2030*, edited by S. Schneegans, 175–209. Paris: UNESCO Publishing.

Montoya, L., and Rivas, P. 2019. "Government AI Readiness Meta-Analysis for Latin America and the Caribbean." In *2019 IEEE International Symposium on Technology and Society (ISTAS*), 1–8. Nov. 15–16, 2019. Medford, MA: IEEE. doi: 10.1109/ISTAS48451.2019.8937869

Oye, Kenneth A., and Wellhausen, Rachel. 2009. "The Intellectual Commons and Property in Synthetic Biology." In *Synthetic Biology*, edited by Markus Schmidt, Alexander Kelle, Agomoni Ganguli-Mitra, and Huib Vriend, 121–140. Dordrecht: Springer.

Perez, C. 2014. "A Green and Socially Equitable Direction for the ICT Paradigm." Globelics Working Paper Series 2014–01. Globelics-Global Network for Economics of Learning, Innovation, and Competence Building Systems, Aalborg University, Department of Business and Management.

Porter, A.L., Roessner, J.D., Jin, X.-Y., and Newman, N.C. 2002. "Measuring national emerging technology capabilities." *Science and Public Policy* 29 (3): 189–200.

Quevedo, J.P.d. 2017. "A rede de atores das proposições de regulação da nanotecnologia no Brasil." *Redes* 23 (44): 77–101.

Redden, E. 2019. "Brazil's Bolsonaro Takes on Philosophy, Sociology." *Insider Higher Education*, April 29, 2019. www.insidehighered.com/quicktakes/2019/04/29/brazils-bolsonaro-takes-philosophy-sociology

Research Council – Mexico Consejo Consultivo de Mexico. 2017. "Inversion privada para la Ciencia, Tecnología e Innovación." http://ccciencias.mx/es/fichas/item/496-inversion-privada.html

Rezende, S. 2010. "Brazil: challenges and achievements." *Issues in Science and Technology* 26, 3. http://issues.org/26-3/rezende

Rodríguez, S. (1990). *Sociedades Americanas* [en 1828]. Caracas: Fundación Biblioteca Ayacucho.

Rotolo, D., Hicks, D., and Martin, B. R. 2015. "What is an emerging technology?" *Research Policy* 44 (10): 1827–1843.

Sagasti, F.R. 1989. "Crisis and challenge: science and technology in the future of Latin America." *Futures* 21 (2): 161–168.

Salazar, M. P., Valenzuela, D., Tironi, M., and Gutiérrez, R. A. 2019. "The ambivalent regulator: the construction of a regulatory style for genetically modified crops in Chile." *Tapuya: Latin American Science, Technology and Society* 2 (1): 199–219.

Salipante, S. J., and Jerome, K. R. 2020. "Digital PCR—an emerging technology with broad applications in microbiology." *Clinical Chemistry* 66 (1): 117–123.

Schwab, K. 2017. *The Fourth Industrial Revolution.* New York: Crown Business.

Small, H., Boyack, K.W., and Klavans, R. 2014. "Identifying emerging topics in science and technology." *Research Policy* 48 (8), 1450–1467. http://dx.doi.org/10.1016/j.respol.2014.02.005

Tobar, L. 2016. "Codigo Ingenios: Nuevo regimen de propriedad intellectual en Ecuador." https://www.pbplaw.com/es/codigo-ingenios-nuevo-regimen-de-propiedad-intelectual-en-ecuador/#_ftn1

See also: Solis Murillo, M. de L. 2018. "Los Principios del Código Ingenios y su Incidencia en las Instituciones de Educación Superior." *Revista Científica Hallazgos* 21, 3. https://revistas.pucese.edu.ec/hallazgos21/article/view/266

Van Noorden, R. 2014. "The impact gap: South America by the numbers." *Nature News* 510 (7504): 202.

Vessuri, H. 2007. *"O Inventamos o Erramos": La Ciencia Como Idea-Fuerza en América Latina.* Bernal: Universidad Nacional de Quilmes Editorial Bernal.

Vessuri, H. 2011. ""O inventamos o erramos." La ciencia como idea-fuerza en América Latina." *Ciencias* 102: 58–59.

Wellhausen, R., & Oye, K. A. 2007. "Intellectual property and the commons in synthetic biology: Strategies to facilitate an emerging technology." Presented at 2007 Atlanta conference on science, technology and innovation policy. October 2007. Atlanta: IEEE.

Zi, Y., & Wang, Z. L. 2017. "Nanogenerators: an emerging technology towards nano energy." *APL Materials* 5 (7): 074103.

SCIENCE IN LEGISLATION: APPLYING SCIENTIFIC PEER REVIEW AND RIGOR TO PUBLIC POLICYMAKING

CJ Pommier
March for Science Southern Arizona

Ansel Andrews
March for Science Southern Arizona

Josh Hoskinson
March for Science Southern Arizona

Henry Werchan
March for Science Southern Arizona

INTRODUCTION

The March for Science continues to be a unique and unprecedented grassroots movement. It stands for the role of science, evidence, and objective reality in our society that resists the barrage of "alternative facts" in our current post-truth era of politics in the U.S. The March for Science, formerly the Scientists' March on Washington, was originally an international series of marches and rallies held on Earth Day in 2017 meant to raise a unified voice in support of science and celebrate its role in shaping modern society. These marches and rallies were primarily in response to the growing anti-science movement in local, state, and federal levels of government resistant to scientific ideas such as evolutionary theory, global climate change, vaccinations, and genetically modified organisms. Policymakers crafted science-related public policy primarily out of political ideology and rhetoric instead of scientific evidence. However, the sparks that ignited the fire of the 2017 March for Science were the actions occurring in quick succession viewed as anti-science by the newly inaugurated administration, which included but were not limited to (1) proposed resumption of construction of the Keystone XL Pipeline, (2) appointments made at the Environmental Protection Agency and the Department of Interior, (3) removal of the U.S. from the Paris Climate Accord, (4) freezing of federal research grants, and (5) issuance of a gag order on climate scientists in the EPA which prevented them from disseminating climate science information to the public. This quick succession of activities taken by the new administration forced scientists, science enthusiasts, those recognizing the importance of science to society, and others to come together to object to these actions. March for Science Tucson was the local satellite chapter for Tucson, Arizona, which saw approximately 2,500 attendees at the 2017 Rally for Science at El Presidio Park in Tucson, Arizona. These international marches and rallies, including the one in Tucson, Arizona, solidified the notion that science is not only critically important to American society but is also incredibly important to its sustainable future and success. Therefore, these international marches and rallies, including the one in Tucson, Arizona, solidified one inalienable point: **scientific research should inform public policy**.

Nearly three years after the original march and rally, March for Science Southern Arizona, formerly March for Science Tucson, continues to advocate for the preservation and expansion of the critical role of science and evidence in our Southern Arizona community. March for Science Southern Arizona seeks to bring science and evidence to the forefront of Arizona's political conversations. For this purpose, we propose the program: *Science in Legislation.*

SCIENCE IN LEGISLATION: A HIGH-LEVEL FRAMEWORK

Ideology and political rhetoric have overtaken policymaking at the local, state, and federal levels of government rather than argumentation and sound, logical reasoning. For effective and responsible governance to occur, evidence from reputable sources must support policies developed at these various government levels. We theorize that adopting the scientific principles of sound logic and evidence-based review improves policymaking at all government levels. Therefore, it must be subject to peer review by experts in the various fields these policies aim to impact. As such, the scientific peer-review process must be applied to legislation to maintain a robust and evidence-based foundation for all legislation at these various levels of government.

Importance of Local and State Policies

Local policies have the most impact on daily life. Administration of local policies influences tangible impacts on quality of life, such as road repair, night lighting in our parks, and public services distribution. However, local advocacy often receives little attention as compared to advocacy at the federal level.

Advocacy at the state and local levels is equally, if not more, important. In many cases, states can act where the federal government cannot (or will not). An example of states acting where the federal government will not is the standard for vehicle emissions in California, which are much more stringent than the federal standards. Even when state legislators will not take actions the voters want, Arizona's veto-proof ballot initiative process allows voters to take direct action. This demonstration of direct democracy is frequently exercised as part of the Arizona Constitution.

Although states may set stricter environmental standards, states may also inhibit local regulation. An example is the Arizona restriction on enacting environmental regulations more stringent than the federal limits, codified in Arizona Revised Statutes Ann. § 49-255.01(B) (Arizona pollutant discharge elimination system). This statute inhibits municipalities from making rules about water quality that would protect state residents from the federal government's absence of rules. For example, perfluorinated and polyfluorinated alkyl substances (PFAS), a class of long-lived chemicals often called "Forever Chemicals," are unregulated. These are commonly used in household items and firefighting foams on airports and military bases to fight fuel fires. Wells in several locations, including near the Tucson International Airport and the Davis-Monthan Airforce Base in Tucson, AZ, have been shut down due to concern over human health from high levels of PFAS found in the water. However, due to the state restriction on water quality regulations, the city cannot take regulatory action.

The state can also prevent action at both the local county and city government levels. One recent example where the state reversed a local ordinance was an ordinance passed by the Bisbee Council in 2013 (O-13-14) that targeted single-use plastic bags through a tax on each bag. The Council intended to promote recycling and reduce litter caused by the bags, which readily snag on the area's thorny trees and cactus. The 2016 Arizona Legislature passed HB2131, which prohibited local restrictions on retailers. Facing a lawsuit from the state Attorney General, which would result in an estimated $2 million loss of funding to the city, the Bisbee Council repealed its law.

Another critical reason to place more attention on local policy is the fact that local policymakers are often more accessible than state ones, which are, in turn, more accessible than the U.S. Senators and Representatives. There are often more opportunities to meet and advocate with state and local policymakers. City Council meetings are much easier to get to after work than Washington, DC. The local representatives travel out of state less frequently and participate in more local events to interact with constituents. Sometimes, they live down the block or around the corner. From the constituents' perspective, the local representative may be less intimidating than a federal one. The voter may be more willing to speak out and advocate for road repair or neighborhood park funding.

Analysis of Presence of Similar Programs Operating in Arizona

There are no known similar programs in Arizona that focus on applying evidence-based science to the legislative review process. Some states have programs that employ scientists as advisors to state policymakers, such as the California Council on Science and Technology. However, there do not seem to be other programs whose primary focus is on applying evidence-based science to support legislation. In Arizona, there is no such program.

Aside from formalized programs at the state level, some national professional societies, such as the American Chemical Society (American Chemical Society 2019; American Chemical Society n.d.) and the American Association for the Advancement of Science (Pain 2014), instruct their members on effective methods to engage legislators. There are also national and international scientific advisory boards such as the National Academy of Sciences and the Scientific Advisory Board. However, there do not appear to be any efforts strictly focused on scientific peer review of legislation.

Given this void, grassroots advocacy groups, such as the Arizona Science Policy Network, attempt to address Arizona's shortfall. Unfortunately, they are far out-funded by other organizations that disguise themselves as evidence-based but in reality place ideology ahead of science and evidence, such as the Center for Arizona Policy that "promotes and defends the foundational values of life, marriage and family, and religious freedom."

March for Science Southern Arizona is concerned with legislation on environmental and wildlife protections, global climate change, natural resources, vaccines and public health, STEM education, genetically modified organisms, science funding, and other natural science topics. Our program, Science in Legislation, would allow a confluence of community members, science professionals, science communicators, public policy experts, and others to apply the scientific peer-review process to science-related legislation. This review process would center on comparing proposed science-related policies with the current scientific literature to determine if the proposed policies are consistent with researchers' findings. Collaborators would compose "synthesis reports" that present their analyses in a nonpartisan manner easily understood by the general public and presented to both the general public and elected representatives. By doing so, we hope to increase the scientific literacy of elected officials and encourage evidence-based policymaking that benefitted from a scientific peer-review process.

Therefore, to create a high-level framework for the Science in Legislation program, we present the generalized methodology below:

1. Review legislation associated with the natural sciences.
2. Assess legislation by consulting current scientific literature on the topic of inquiry.
3. Present the findings in a "synthesis report" for use by policymakers and the general public.
4. Establish a permanent reference source for future legislative sessions.

We aim to focus on science-related policies being proposed in the Arizona State Legislature for our initial iteration of the Science in Legislation program. This focus arises from the general principle that local and state policies have a much larger and more immediate impact on residents. Few organizations are focusing on Arizona-based legislation.

Science Applicability to Arizona Legislation

For this report, we analyzed the 1,416 bills considered in the First Regular Session of the 54th Arizona State Legislature from January to June 2019 (Legiscan 2019). Public records categorize each bill into the categories of Introduced, Engrossed, Failed, Vetoed, and Passed, as reflected in Figure 2.2. Each bill was briefly reviewed and tagged as "science-related" if we determined that peer reviewers could research the subject via scientific literature to provide adequate scientific evidence for or against the bill's underlying assumptions or intended effects. For example, we considered bills related to the procurement of funds or budget allocations as not "science-related," and bills about vaccinations or watershed management were considered "science-related." As shown in Figure 2.2, of the 850 bills that remained as "Introduced" during the session, approximately 128 were deemed "science-related."

An example of a "science-related" bill, and therefore a candidate for our Science in Legislation program, was "HB2472 (SB1116) – Vaccinations; antibody titer testing; notification." This bill states the following:

> *Requires Health Professionals to notify individuals that they have the option to take an 'Antibody Titer Test' to determine whether they have an immunity to a disease before receiving a vaccine for it.*

For HB2472 (SB1116), the underlying assumption is that an antibody titer test proves immunity. Subjecting the bill to scientific rigor would establish the needed body of evidence, both in support and in opposition of this assumption, to critically assess

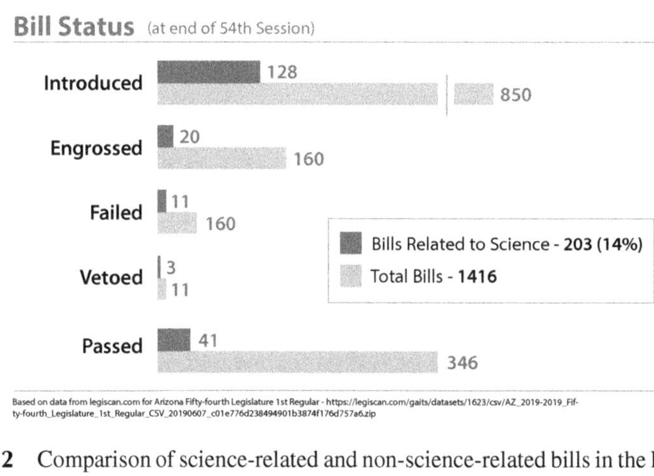

FIGURE 2.2 Comparison of science-related and non-science-related bills in the First Regular Session of the 54th Arizona State Legislature.

this assumption's scientific validity. Additional evidence would also be collected, including impacts on public health and costs.

Analysis of Science Advisory Reports

The synthesis report is a key deliverable for this program. To establish best practices for the science advisory report structure, we evaluated 12 reports released primarily from 2017 to 2019. These reports represent a variety of sectors: the private sector (Arizona Technology Council 2019; Krastios 2019; Thomas and Wessel 2018), government/military (Nemer 2019; Sargent and Shea 2014; Snow and Coleman 2019; Council of Canadian Academies 2019; U.S. Air Force Scientific Advisory Board 2018; U.S. EPA 2019; Tang and Lascari 2017), and academia (Talati and Higgins 2019; Weller and Govani 2018). These reports also ranged in localizations, including state, national, and international levels of government. The reports analyzed ranged from high-level one-page summaries to in-depth analyses spanning more than 100 pages.

Assessment of the major sections of the reports identified several common structural elements, including a summary or letter to the commissioning body, an introduction, information on three to seven current topics, recommendations, information about the report, and a list of experts who participated in crafting the report.

To implement best practices in the synthesis, our organization would not provide expertise. Still, it would coordinate perspectives on the report's topic(s), drawing from relevant local scientific expertise representing private industry, government, and academic sectors.

Communication Plan

Any opportunities to apply our program's findings to policymaking will require an effective communication plan. Unlike many other states, this is further challenged in Arizona because the state legislature is only in session for part of the year (from January to May). However, August–November presents an opportunity when the legislature is not in session, and lawmakers are typically more accessible in their home districts. Consequently, the Science in Legislation program will focus on targeting lawmakers for in-person meetings while out of session. In addition to the in-person meeting, which will present the high-level findings in our synthesis reports, leave-behind summary sheets will be made available for the policymakers' later use. In addition to the elected representatives, our meetings will also target policymakers' staff, who will be more available and will be charged with developing policy position details (Table 2.3).

In addition to engaging the policymaker with a targeted communication plan and strategy, the Science in Legislation communication plan will also target the public through local town hall discussions and social media. Lastly, March for Science Southern Arizona will provide the Science in Legislation findings, including the detailed synthesis reports, through our website. The website will also serve as an enduring resource for these reports, as many bills appear on legislative agendas for multiple years.

TABLE 2.3
Communication plan for science in legislation synthesis reports

Target Audience	Method	Timing
Lawmakers	In-person review of high-level synthesis report findings	Aug-Nov
• Co-sponsors		
• Not Committed	Leave-behind summary sheet	After in-person meeting
Lawmakers' staff	Leave-behind full report	After or in lieu of in-person meeting
Public	Town hall discussion	Periodic throughout year
	Website distribution	As soon as report is published, permanent source

Vision: Where Do We Go from Here?

Our near-term objectives for our Science in Legislation Project are to validate the concept by applying it to one or more legislative bills with scientific and public policy experts' consultation. Upon validation, we intend to formalize and sustain the program to broaden and institutionalize the application of scientific peer review to public policy.

To maximize our opportunity for success and broaden our impact, it is key that we establish collaborative relationships with policy and scientific partners, develop connections with policy experts, and leverage our network to identify scientific experts. A partnership with academia and eventual internship is a natural evolution of this program.

REFERENCES

American Chemical Society (ACS). 2019. "How to Call Your Members of Congress." www.acs.org/content/acs/en/policy/memberadvocacy/advocacy-tools/how-to-call-members-of-congress.html

American Chemical Society (ACS). n.d. "Ways to Advocate for Chemistry." www.acs.org/content/acs/en/policy/memberadvocacy/ways-to-advocate.html

Arizona Technology Council. 2019. *2019 Public Policy Guide*. Retrieved from www.aztechcouncil.org/wp-content/uploads/2018/12/2019-AZTC-Public-Policy_web_single.pdf

Council of Canadian Academies. 2019. *Canada's Top Climate Change Risks: The Expert Panel on Climate Change Risks and Adaptation Potential*. Ottawa, Canada: Council of Canadian Academies. Retrieved from https://cca-reports.ca/wp-content/uploads/2019/07/Report-Canada-top-climate-change-risks.pdf

Krastios, Michael. 2019. *Emerging Technologies and Their Expected Impact on Non-federal Spectrum Demand*. Washington, DC: Executive Office of the President of the United States of America. Retrieved from www.whitehouse.gov/wp-content/uploads/2019/05/Emerging-Technologies-and-Impact-on-Non-Federal-Spectrum-Demand-Report-May-2019.pdf

Legiscan. 2019. *Arizona 2019 Fifty Fourth Legislature 1st Regular* (20190607) [Dataset]. Retrieved from https://legiscan.com/gaits/datasets/1623/csv/AZ_2019-2019_Fifty-fourth_Legislature_1st_Regular_CSV_20190607_c01e776d238494901b3874f176d757a6.zip

Nemer, Mona. 2019. *Annual Report of the Chief Science Advisor 2018*. Office of the Chief Science Advisor of Canada. Cat. No. Iu35/2019-1E-PDF. www.ic.gc.ca/eic/site/063.nsf/vwapj/OCSA_AnnualReport_2018.pdf/$file/OCSA_AnnualReport_2018.pdf.

Pain, Elizabeth. 2014. "How Scientists Can Influence Policy." *Science*. February 16, 2014. Chicago: American Association for the Advancement of Science (AAAS). www.sciencemag.org/careers/2014/02/how-scientists-can-influence-policy

Sargent, Jr., John F., and Shea, Dana A. 2014. *The President's Office of Science and Technology Policy (OSTP): Issues for Congress*. Congressional Research Service (RL34736). Washington, DC. January 13, 2014. Retrieved from https://fas.org/sgp/crs/misc/RL34736.pdf

Snow, John, and Brad Coleman. 2019. *Second NOAA SAB EISWG Report to the US Congress*., as required by the Weather Research and Forecasting Innovation Act of 2017 (P.L. 115-25, 18 April 2017), as amended (P.L. 115-423, 7 January 2019)." https://sab.noaa.gov/sites/SAB/Documents/SAB%20Working%20Groups/Standing%20Working%20Groups/EISWG/Reports/SAB_MtgRpt_Jul2019_2019EISWGRpttoCongress_FINAL07032019.pdf?ver=2019-08-22-114709-943.

Talati, S., and Higgins, P. 2019. "Policy sector perspectives on geoengineering risk and governance." *Journal of Science Policy & Governance* 14, 1.

Tang, X., and Lascari, R. 2017. *20 Years of the Information Technology Agreement*. World Trade Organization. Retrieved from www.wto.org/english/res_e/booksp_e/ita20years_2017_full_e.pdf

Thomas, D., and Wessel, C. 2018. "Emerging Therapeutic Company Investment and Deal Trends." Retrieved from www.bio.org/emerging-therapeutic-company-investment- and-deal-trends.

U.S. Air Force Scientific Advisory Board. 2018. *Maintaining Technology Superiority for the United States Air Force (MTS) Abstract (2018–0558)*. Retrieved from www.scientificadvisoryboard.af.mil/Portals/73/documents/MTS%20Abstract%20PA%20Approved%2030%20Oct%202018.pdf?ver=2018-11-29-093107-043.

U.S. EPA. 2019. "Consultation on Updating EPA Guidelines for Carcinogen and Non-Cancer Risk Assessment" EPA-SAB-19-003. https://yosemite.epa.gov/sab/sabproduct.nsf/LookupWebReportsLastMonthBOARD/357DC7E5C59BA9AD85258438005BA457/$File/EPA-SAB-19-003+.pdf.

Weller, Nicholas, and Govani, Michelle Sullivan. 2018. "Prop 127: Mapping Arizona's Energy Future." *Consortium for Science, Policy and Outcomes*. Retrieved from https://cspo.org/library/prop-127-mapping-arizonas-energy-future/.

3 Foundations of Science Diplomacy

SUMMARY

This chapter contains the transcription of the keynote speaker session at the conference, starting with Prof. Kevin Lansey giving an introduction to the keynote speaker, Dr. Glenn E. Schweitzer, Director of the Program on Central Europe and Eurasia at the National Academy of Sciences. Dr. Schweitzer's keynote discusses historical cases of science diplomacy, looking all the way back to Ivan the Terrible. He explains the development of science diplomacy-type positions in the U.S. foreign policy as the

DOI: 10.1201/9781003220503-3

political terrain shifted post-World War II, and discusses the impacts that he has seen throughout his long career in conducting science diplomacy. This is followed by a Question-and-Answer period with the audience.

Next is the transcription of a keynote lecture called "20 Years of Science Diplomacy" given by Dr. Colglazier at the World Science Forum in November 2019, a few weeks after the conference. Colglazier shares experiences throughout his career, which coincides with the growth in popularity of "science diplomacy" as a term, as well as its development into a focus of study and a global objective. He shares five important elements that make science diplomacy an effective tool, and five areas that could benefit from the use of science diplomacy.

INTRODUCTION TO GLENN SCHWEITZER

Kevin Lansey

Civil and Architectural Engineering and Mechanics, University of Arizona

Glenn Schweitzer is a distinguished practitioner of science diplomacy. I've known Glenn for a number of years now. He started as a foreign service officer working in both Belgrade and then Moscow. He was the first director of the Office of Science and Technology. He shifted over to the EPA after that and became the director of the Office of Toxic Substances and then the Environmental Monitoring Systems. His final government assignment was the responsibility for responding to the nuclear accident at Three Mile Island. Since then, he's moved over to the National Academies. He has been there for 30 years and has been working with Russia and Iran.

I've been in about a dozen days of meetings with Dr. Schweitzer and on a range of topics. We've had meetings on climate change, on infrastructure, on wetlands, and Glenn is the person who will sit back and watch, and you'll wonder if he's really listening to what's going on, but when the talk is over, he will ask the most insightful question that no one in the room has thought about and brings up other topics and ties things together in a really unbelievable way. I have respect for what he can do and I'm really looking forward to his words about the field of science diplomacy.

LEGACY FROM IVAN THE TERRIBLE TO THE NOBEL LAUREATES

Glenn Schweitzer

Director, Program on Central Europe and Eurasia, National Academy of Sciences

I've recently been accused of being a historian of science diplomacy. I'm not trained as a historian at all, but I can look back on a few years. So today, I thought I would reflect on some of the experiences we've had over the years and the lessons learned that could be relevant today.

Let me start with this picture (Figure 3.1). I am sort of an action person. I don't like to sit back in the office to just make recommendations. I like to see change happen. This image is from 2007 in Tehran. We were on one of our usual annual visits there and I was lucky enough to be put in the carriage down in the Central Park of Tehran and have my picture taken.

FIGURE 3.1 Photograph of Glenn E. Schweitzer in a carriage in the Central Park of Tehran, Iran in 2007.

Ivan the Terrible Reaches Out to Europe

The topic of this presentation may have set you back a little bit, but I wanted to talk about several things. First, 450 years ago, Ivan the Terrible decided to send 17 Russian engineers to Europe to learn about how they were building their cities, how they were expanding their transportation networks, how they were organizing a Navy, and so forth. The idea was that these Russian engineers would then come home to St. Petersburg and transfer the lessons to Russia, and Russia would be ready to cooperate in the future. So, it was a two-year expedition. They would go, come back, and then Russia would really enter the real world. There was only one problem: none of them returned.

Thomas Edison, Albert Einstein, and Others Transform America

Let me fast forward 300 years to Thomas Edison who was self-educated and was awarded over 300 patents. But the interesting part for this talk is that over 100 of those patents were also patents in Europe. He really cemented the bonds in many ways between the United States and Europe, and I think he will often be remembered as the person who says, "if I can't sell it, I don't want to invent it."

Then of course, Albert Einstein came along and the famous Einstein-Szilárd letter to President Roosevelt in 1939 about the rapid advances in natural science, and that the nuclear sciences were changing the world. President Roosevelt immediately set up the Manhattan Project, and that, indeed, did change the world (Gosling 1999).

Joe Taylor Inspires Current Generation in Iran

I know very well five Nobel Laureates who went to Iran. Three of them are deceased. One is here with us, Peter Agre, who was among that group who were pioneers in Iran. The fifth one is Joseph Taylor at Princeton, who wasn't able to come up to Tucson for this conference, but he has been remembered in Tehran in a variety of ways, including his bust in the Pardis Technology Park about 50 miles from Tehran where he shares space with other giants of the world in the Scientists' Garden (National Research Council 2010).

In 1989, I had my first book published by a commercial organization, and it was titled *Techno-Diplomacy: US-Soviet Confrontations in Science and Technology*. I just spent several years trying to sort out what the Soviets were up to in Silicon Valley and a publisher in New York City said that this would be a great topic for a book. I started delving into what all this transfer in technology was meaning and I came to the conclusion that it probably wasn't too bad from some points of view, but possibly bad from other points (Schweitzer 1989).

I thought I was participating in techno-diplomacy, because science wasn't the real issue, but technology was the issue. Over the years, I've had the feeling that science diplomacy is relatively simple compared to technology diplomacy. And then, we get into this inevitable hassle of what is meant by science diplomacy? Some people have tried to standardize it. I usually disagree with the definitions people are putting up there. So, I try to stick to my definition and believe that there are three aspects of technology.

First of all, there are conflicting technological issues in the countries you're dealing with. Second, you don't want to do harm. You want to try to do things to improve this situation. And third, you try to move toward peace, not toward war with whatever you're trying to do. Since that time, I've stuck to my title of techno-diplomacy. These terms get mixed up and so if I talk about science policy or technology policy or science and technology (S&T) policy, it is sort of a confused area, but I myself am principally oriented toward technology, perhaps because I was trained at Caltech as an engineer.

Science Diplomacy at Foggy Bottom: The Early Days

Well, the starting point for any discussion of techno-diplomacy or science policy and diplomacy, in my view, is the Department of State. The Department of State named its first science advisor to the Secretary of State in 1951. It was a professor from Caltech. The reason he was named was because the United States had invested an enormous amount of money in the reconstruction of Europe, and the feeling of Washington was "what are we going to get from our money? Maybe we could benefit scientifically as well as economically and so forth from Europe." So one of the charges of this new science advisor to the secretary was to focus on how we can cooperate with Europe. He was given a broad portfolio to address any topic he wanted except that he could not address the topic of atomic energy; that was too sensitive a topic and apparently he didn't have clearances. So he was kept out of the atomic energy business (Koepfli 1983).

After Sputnik was launched in 1957, a second science advisor was appointed, and his job was to see what the United States could do to keep pace with the Soviet Space Program. Several years later in 1965, a bureau for science and technology was established in the Department (Pincus 2014), which was headed by an international affairs expert, but a very good guy. I knew him very well, and at one time I was named as a special assistant to him. The way he obtained his science advice is quite interesting. Once a month, he would have a luncheon in the Secretary's private dining room with ten of the top officials of the Department. He would have a guest scientific speaker to talk on some topic of science. So once a month, the leadership of the Department was getting a good dose of one small piece of the science overall or other elements or developments of the Department.

The United States Information Agency (USIA), which was our old cultural affairs agency, operated truly around the world. It was folded into the Department of State. Perhaps now it's best known for the Fulbright program and others which it takes care of, but it is a powerhouse within the Department of State, and not that many people realize that. My experience is that you can learn plenty from the Cultural Affairs Office, where this operation rests. And if you're interested in science and diplomacy, you don't want to ignore that office.

Our first science attachés were really based at the end of World War II in London. It all started in London in about 1958. And the reason was the same as we have an advisor in Washington: how can the U.S. benefit from the scientific advances in Europe? So, a lot of our science diplomacy efforts began in London (Pielke and Klein 2009).

Then, there were five Science and Technology Advisors, two of whom are at this meeting [Dr. William Colglazier and Dr. Vaughan Turekian] who were appointed in succession. Specialized offices for science and technology were established throughout the Department; there was a special office for Information Technology, a special office for the PEPFAR program, which was the attack on AIDS around the world, a special office for the Arctic, and so forth. At one time under the Obama administration, there was something like 45 special advisers for special topics reporting to the Secretary of State under Obama (Thomas 2017). So there is a centralized office in the Department, but there are a lot of other specialized offices where a lot of people are working (Hourihan 2017). And then of course, there are the dual-use political-military activities that people worry so much about.

Science Diplomacy at the U.S. Embassies and Missions

I think knowing something about the U.S. Embassies abroad and what's going on there is also important because that's where a lot of the science diplomacy activities are funneled through. In 1963, there were around 25 senior professors and senior industrialists who were recruited by the Department to populate about 20 of the embassies around the world. Some embassies got two of these people; most got one (National Archives of the U.S. 1995). So this is where I came into the picture, on an unbeknown path.

The attachés were doing okay except for Moscow. What happened in Moscow? Well, a professor from Princeton was sent over there for two months to see what he could do at the embassy. The only thing he could do at the embassy was cause chaos. This professor was interested in being a professor and not interested in being a diplomat. The embassy came back to the State Department and said, "don't send us any professors. We don't want any professors over here; we want somebody that we can trust." And who can you trust? The only ones they trusted were foreign service officers.

Well, I just happened to be a foreign service officer. I think I was one of the few scientifically trained foreign service officers in the foreign service, and I also knew some Russian. So one day, the phone rang, and they invited me over to the Soviet desk. You want to go to Moscow? Why? Don't you want to go to be the first science attaché there? So I said, "Fine, I'll go." That's how I ended up in Moscow.

"All-Hands" (All-of-Government/All-of-Society) Approaches at Home and Abroad

There was also a special cone setup in the Department to get career and promotion opportunities for the scientifically oriented people. This lasted about three or four years, but there weren't enough scientifically oriented people to have a competitive cone for promotions, so that cone disappeared and was folded into the economics cone. So then the science-trained foreign service officers would have to compete with the economics people, and that was a big deal for the internal people, who were unhappy. If you were not a science person, you were happy, and if you were a science person, then you were unhappy because you longer had a promotion spot.

Most recently (as of 2017), this was the lineup: we have abroad 100 full-time diplomats working for the State Department on science; we have 200 part-time, 100 short-term, and 250 local full-time science diplomats. So we have a lot of science diplomats out there in various categories, but it's pretty hard to find an embassy around the world where there's not somebody there working either full-time, or at least has devoted a substantial part of their time on science. Maybe at some posts with a staff of only two or three, but there aren't many of those. If it's an embassy post with a reasonable number of people, you can be sure to find someone in there who is charged with the responsibility for science. That doesn't mean that the person has enough time to spend on it, but there's somebody out there designated.

The Department also has dozens of people at the United Nations Organizations, such as the WHO (World Health Organization) and UNESCO, around the world with science-trained people. However, with UNESCO, we are merely observers because we're not a member of UNESCO. But in the Organization for American States (OAS) and other organizations around the world, there are also science-trained people.

The last item in this lineup is interesting: we have 58 bilateral science agreements with other countries. A lot of those science agreements are just about dormant, and they maybe have one event a year, if at all. But if you tried to kill one of those inactive agreements, then you're in trouble because they mean a great deal to the other countries. To have a science relationship with the United States is really diplomatically important to them.

But the Department of State is just a small piece of the science diplomacy activities that the U.S. government engages in. There are more organizations than you can imagine! We have people scattered around the world all engaged in some kind of science. Sometimes, they check in with the embassies; sometimes, they don't. But here's a list of the various kinds of organizations based in the United States which have people assigned overseas to do scientific-related duties, within and beyond embassies: USAID, NIH, CDC, NSF, Commerce, USDA, NASA, U.S. Government contractors and grantees, U.N. agencies, regional organizations, development banks, U.S. Chamber of Commerce, U.S. multinational companies, U.S. investors, U.S. philanthropies, humanitarian organizations, and special interest groups. And it's almost impossible to coordinate these, certainly in the big countries, but even in the small countries. So, there are a lot of people out there, all well-intended, some blessed by the U.S. government, some not blessed by the U.S. government, but most—hopefully all of them—are doing a good job. But too often, they are hesitant to go to the embassies because there is a fear that "those guys don't know what they're talking about." But you'd be surprised. It's hard to find an embassy where the people can't tell you something that you don't know.

Impacts of Science Diplomacy on Policy

The impacts of the projects. This to me is the real payoff. Impacts to policy include awareness and advocacy of policy officials' results and findings from projects, changes in approaches and/or practices at national and local levels, commitment of resources to transform findings into implementation, and the memorialization of findings from legislation, regulations, or scalable practices.

First, why would we be concerned with and how do we measure the effectiveness of projects? Some types of impacts include joint publications, validation of research or project approaches, continuation of project activities with either local or international support, international and/or national awards for project achievements, cost savings through abandonment or reorientation of unproductive projects, and highlighting the strengths and weaknesses in research concepts and techniques.

Next, I use as my case-in-point Malaysia, because in Malaysia, it's not a money issue. Quite often, we hear that we can't make an impact because there is no money. But in Malaysia, it is not a money issue; it's an organizational issue; how do you organize your people? And that was what the book on Research Management was attempting to do, which I co-edited (Aziz and Schwietzer 1988). I think it became sort of a handbook for the people in Malaysia, because again, it wasn't focused on finding the money. Here in the United States it's "give me the money and I'll do something." But that wasn't the case in Malaysia, and I think it's important to keep in mind. You can always use the lack of money as an excuse for not doing something constructive.

Successes of U.S. Government

Now let me talk about a few of the successes of the U.S. government which I have had the privilege of witnessing over the years.

First let me start with Korea. After World War II, Korea had some well-trained people. A lot of Koreans had served with the U.S. Army in Korea, and they became technically trained. They knew how to repair your jeeps, they knew how to keep their tanks going, and so forth. So, you had a population which was technologically interested. The Korean government said let's set up the infrastructure for science and see if we can start there. The first thing they did was set up the Korea Institute of Science and Technology (KIST), which is now a distinguished university. And they said, "Okay, now we're ready to run!"

But they did several other things that were important, too. They said, "No, we're *not* going to take foreign direct investment, and we're *not* going to have foreign companies build their facilities in our country. Just the opposite of the Singapore model, we're going to have Koreans build the facilities in Korea" (Sung and Chung 1996).

The other thing they said was that a lot of Koreans need money to buy some equipment from abroad to put in their facilities. There was only one bank in Korea authorized to make loans for people to buy equipment from abroad and bring it in. They wanted to review what people wanted to bring in and if a Korean entrepreneur wanted to bring in some equipment that could conceivably be made in Korea, the loan was denied. So they became experts in reverse engineering (Abelson 1988). The combination of reverse engineering, the dependence on their own people, their educational and their innovation infrastructure, coupled with the incentives for Koreans to come back home, making high salaries and living in good facilities, were very important in moving Korea near the top of the scale of technologically oriented countries in the world (Yang et al. 2020, eaba5894).

Second, in Taiwan. One person in Taiwan made a great difference, a fellow named Bruce Billings. In the 1960s, he was assigned to the embassy in Taipei as a science attaché. Billings came from a computer company, and he knew basically anybody worth knowing in the computer field in the United States. He took it upon himself to bring over these computer types in the United States, one by one, for three years and expose them to the Taiwanese and vice versa, focusing on technical parks and suddenly, out of nowhere, Taiwan became a computer giant (Fuller and Rubinstein 2013). And even today, one of the principal sources of technology for the People's Republic of China is Taiwan. If you look at the contracts between Taiwan and Republic of China, you'll be amazed. So when you say we're blocking imports from China, we're really blocking sales to China from Taiwan.

The third one is the University of Science and Technology in Hong Kong, which opened in 1991. The Hong Kong University of Science and Technology is interesting because half the proceeds of the Hong Kong horse racetrack go to the university. So, they have a chunk of money there to keep the university going, but they didn't know what to do with it. So, we found one of the best university administrators in the United States: a fellow from the University of Pittsburgh. They got him to go over there to the Hong Kong and told him to spend a week and spend as much money as he wants—just as long as he finds good people. "If you have to pay some guy $250,000 because he's first class, then do it." Money was not an object with horseracing going on every Sunday. But they picked very carefully, paid them very well, and they picked good people. And now, the University of Science and Technology in Hong Kong is a place to be reckoned with (World University Rankings 2020).

Finally, with regard to Israel. Some of you may know the story after the founding of the Israeli nation. After Israel was founded, they incurred a debt to the United States of hundreds of millions of dollars. Well, the United States said, "okay, Israel, instead of paying us hundreds of millions of dollars, take that debt and invest it in a foundation in Israel, and that foundation will be devoted to promoting the U.S.-Israeli scientific cooperation." They did that in three fields: science industry, agriculture. Those three foundations have really been powerhouses in lifting the level of technology not only in Israel, but also in the United States and around the world (BSF 2020). So those are real success stories, when somebody thought that it was more important than taking $100 million from Israel to have Israel put it into a foundation which would fund joint projects with the United States.

Successes of the National Academies

Now, what have the National Academies done? Well, the first one I'll mention is the Sakharov issue. I think many of you know about this Russian scientist, Andrei Sakharov. He was the inventor or the primary designer of the 50-million-megawatt bomb. It exploded, and it was the biggest one ever set off. But then, he became a peacenik. He said, "no, no, we've gone too far. We should draw back and try to reach agreements with the United States and others about how to limit arms and eventually disarm and so forth." And that didn't go over too well with the Soviet leaders, particularly when he was expounding this at press conferences at the height of the Cold War. So he was exiled to Gorky in 1980 (Sakharov 1983).

But in 1986 when the Academies were scheduled to have a new agreement with the Soviet Academy of Sciences, what are we going to do to help Sakharov? That was a big issue in Washington. The Academy people were saying, "do something about human rights. He's a victim of human rights violations" because he'd been in Gorky kept on house arrest. We were able to insert in our agreement with the Soviet Academies the phrase "in carrying out this agreement, due consideration will be given to the environment within which it is carried out," and that meant human rights. So that was quite important, and then he was quickly released from house arrest and he was back in Moscow as a "normal person" and an active member of the Academy and so forth (Dowling 2017).

Then, in 1988, the U.S. government, which had relations with East Germany, decided to open a legation in East Berlin. We weren't prepared to open an embassy, but we were prepared to open a legation.

My organization had an exchange program with East Germany—one of the few exchange programs with the West. It just so happened that we were there. I was with a group of about eight American physicists visiting East Germany at the same time the legation was scheduled to open, and they were going to have an opening celebration. When we made our call on *die fähre*, they said, "can you help me get some Germans to come to the opening because we invite them but they never show up." So we said, "well, we're going to go from Berlin south and then come back to Berlin, so if you have it five days from now, I will give you a list of some Germans to invite." And when we came back and had the opening, by golly, all the Germans showed up.

In 1989, our Institute of Medicine met with the Soviet Academy of Medical Sciences to talk about HIV/AIDS. At the time, it was kind of a quiet topic even in the United States, but particularly in the Soviet Union. But that meeting was unbelievable. The Soviets came over to Washington and I showed up because they were interested, and both sides were astounded by what was going on in the other country. As soon as the meeting was over, NIH said to us, "we want you out of this game, we will take over." And the same thing happened when the Soviets went home. Their government said the same thing. The Soviet Ministry of Health said to the Academy of Medical Sciences, "you get out of this game, we'll take over." So, it morphed into an agreement between the NIH and the Ministry of Medical Health. And would you believe it? Thirty years later, right now, my granddaughter is in the midst of that cooperation, which is still going on. She's at Boston University participating in that cooperation of research on HIV/AIDS between the United States and institutes in St. Petersburg.

Next is the Danube Delta. Well, I was always upset because we weren't getting the young people involved in these programs. I don't have a beard, but a lot of my counterparts do, and I wanted to get the "greybeards" out of this and get the young people into it, so we said, "well what are they interested in?" Well, they're interested in the environment. So, we rounded up around ten of these young post-doctoral candidates. We recruited ten young scholars in the United States, and we were in touch with the Romanians who said that they would recruit ten as well. So we sent these ten young people who are interested in environmental issues to Romania, and they're met by the Romanians. Now, keep in mind that this is at the height of Ceauşescu's reign there. So, these aren't exactly people on friendly terms (Tanasescu and Constantinescu 2020, 110324). But they took these 20 young people all out on a barge and they are all floating around the Danube Delta together. They had their measuring equipment, because some of them studied chemistry, and so forth. They were out there for ten days; they slept on the barge, they ate on the barge, they took field trips off the barge, and so forth. When they came back, they were really enthusiastic. I mean, big-time enthusiastic, but they were really ticked off because the Romanian people who didn't like to follow the rules were destroying the Delta! They were dumping garbage in the place, they were shooting the birds, and it was becoming a mess. And the Romanians, within three months, set up a constabulary force with armbands on to go out and to give warnings to anybody who was sort of mucking around with their environment there and that's helped ever since. I think they've now formalized it into a national police force or whatever. But those young people from both sides really, really had some impact. They took this approach in the United States and brought Romanians down in the Louisiana marshes, but they didn't have such good luck in getting people to back off down there. But anyway, it helped.

In 2000, we at the National Academies wrote the reports that had some influence on the legislation that established the position of Science and Technology Advisor to the Secretary of State. And in 2009, there was an election in Iran with Mahmoud Ahmadinejad. It was Ahmadinejad's second term, and you may recall; he came out and won the election, and there were riots on the street. He won the election, but in some areas, he received more votes for himself than there were people living in the section. So, they figured they knew something was wrong (Maloney 2009). So anyway, people were rioting on the streets and the United States sat it out. So, six months after that happened, we arranged a meeting with the Deputy Secretary of State. We went over to his office and told him that we think we can restart these programs again because we think the Iranians will go along with this.

The State Department had cut off funding to all programs with Iran because of the election. But they said, "okay you can try it. If it works, then good luck to you. We're not confident, but we'll give you a chance." So we arranged two meetings with the Iranians, and they went great. The word got out from the State Department, and so people wanted to cooperate with Iran again. The green light was out again, and our programs started off and shot ahead. We're in a little different form right now, but that was really a success story where we had built enough confidence over a decade that we could act responsibly, and our Iranian colleagues the same way. Nobody participating in our programs was hung out to dry. And that was a success story.

Unsuccessful Science Diplomacy Efforts

Let me first speak about the elimination of the *Aedes aegypti* mosquito in Kenya. A fellow, George Craig, at Notre Dame was a mosquito expert. He came to me (I was at the USAID at the time) and said, we have to do something about this malaria program in Africa. I asked him what he wanted to do. His idea was that he could sterilize the male species and then that sterile male species will go around and have relations with all the females, but none will be able to reproduce anything (Hickey and Craig 1966). I asked him if it would work, and he said, "I think so, but I want to try it on an *Aedes aegypti* mosquito (which is the mosquito that causes yellow fever) before we try it on the *Anopheles* mosquito, which is the malaria-spreading mosquito." I said, "okay, George, what do you need from me?" He said that he needed a couple million dollars, an introduction to the government of Kenya, and a mosquito factory.

So, after some discussion, the government of Kenya said, "let's try it on an island which is an uninhabited mile about a half-mile long. It's got a lot of mosquitos but no people, and we think there are *Aedes aegypti* mosquitoes out there." If you knock out the *Aedes aegypti*, you don't disrupt the ecology because 135 other mosquito species will take their place. So the idea was to take these new types of mosquitoes that can't reproduce out there and release them and watch the mosquito population diminish, and if that works, then try it on *Anopheles*.

We needed a mosquito factory. Well, it turned out that during World War II, the U.S. Army had a mosquito factory. We were going to infect the mosquitos in these factories and let them loose on the enemy (Rostker 2013). So I called up the people in Fort Bliss, Arkansas and I asked about this. They said, "oh, we wanted to get rid of those cages!" So, they said, "yeah you can have our cages because we don't do mosquitoes anymore." So, we shipped a mosquito factory over to Kenya and set it up over there. We were using chemicals to treat these male mosquitoes, and it was starting to go along; we spent about a half a million dollars. The people at AID were very skeptical about something like this and thought this was crazy. So they said, "we're going to do a review of your books out there to make sure you're doing alright." They went out and they came back, and they said, "you know, we have to cancel your project."

"Why?" I said, "I thought it was a good project?" They said that it was a good project, but they have to cancel it because we bought a Land Rover instead of a Jeep, and you can't buy British products with AID money. So believe it or not, the project was canceled. I understand that since that time, now that kind of experimentation has been supported by the Gates Foundation, but I'm not currently up to speed.

Dams, Irrigation Systems, Water Recycling, and Sanitation

I have to make a couple of comments about water, because this is really big-time bad, bad news. There are 1,400 new dams and water diversion projects with cross-border impacts currently underway. Fourteen hundred places where they are going to divert water across boundaries. Only 13% of the water basins that cross borders have treaties that include all riparian states. So, some of the people in these basins who are going to be under treaty aren't even members of the pacts of control, and only 25% of the treaties cover the entire basins (Stefano et al. 2017).

Evaluation of Project Successes

We like to evaluate our projects. Well, we try our best. Sometimes, we're successful, sometimes, we're not, but we're willing to give it a shot. And it's hard. We took our USSR project to support scientists in a number of countries and set up questionnaires at two-year intervals to see what difference the project made for their research (National Research Council 2004).

In East Germany, we also did the same thing. We used to give grants of about $100,000–$200,000 to individual scientists, and in doing these evaluations, we found out the impact of all these individual grants. They really tried to squeeze a lot of good things out of those grants. We're thrilled about that. But the message here is that you shouldn't just give the grants; you should follow up and see whether they made a difference or not, and you can't just take for faith they do.

Lessons Learned

Finally, here are some lessons learned I'd like to bring to your attention:

- American approaches may not be the best
- Replace technical assistance with technical cooperation
- Anticipate the ubiquitous tax collector
- Find interested audiences or users before beginning joint efforts
- Technical data may be considered valuable and not given away free-of-charge
- Documentation of achievements, problems, and innovative approaches.

I think they're kind of obvious, so I'm not going to go through each one except the fourth bullet. Before you start a project, figure out who is going to benefit. If you can't identify who is going to benefit, the specific group or specific kind of people, the specific ministry who will benefit, you ought to forget the project.

Also, the next to last bullet is that technical data may be considered valuable. Well, if you went to Russia and visited scientists' homes in the early 1990s, you would see letters they have from their bosses saying congratulations for inventing this and that. These weren't patents; these were regular letters signed by their bosses. But those scientists were so proud of those things and considered them just as valuable. They considered that to be their personal property; you can't mess around with their property. So, the attitude of somebody who doesn't have a patent thinking that he or she should be rewarded for what he's done is very, very widespread and kind of legitimate.

A very important aspect of our work includes reports that are readable. In the photograph (Figure 3.2), you see a report which was translated into Farsi by the University of Tehran and that was the only time we were able to do that. The reports were for Iranians to distribute to the Iranian population the results of Iranian American activities.

I also want to mention a University of Arizona example. We published the proceedings from the "U.S.-Iran Symposium on Wetlands," held March 28–30, 2016 in Irvine, California. That is around 250 pages I think of meetings with the Iranians.

And the last one I'll mention, I kind of like the best of all. I was asked one time to inform some important people what's happening with industries in Russia. I did a ten-page brochure which was done in graphics to show people this kind of thing and that kind of thing. There were very few words with lots of pictures, diagrams, and

FIGURE 3.2 Photographs of the publication U.S.-Iran Engagement in Science, Engineering, and Health (2000–2009): Opportunities, Constraints, and Impacts by Glenn E. Schweitzer in both English (a) and Farsi (b).

charts so that people would pay attention to it. If you write a 20-page report, no one will read it, but people are intrigued by visuals.

Two Tumblers in Space

And finally, my favorite story. This is the one where I think I earned my keep. In 1965, Alexei Leonov, the Russian cosmonaut, took his walk in space—the first human to ever do so. He was out there for about 15 minutes. And I was able to go and see that film in a Russian facility and I was amazed how this guy was walking around. He wasn't walking; he was tumbling. But a friend of mine, Ed White, was an astronaut down in Houston and he was scheduled to make the first walk in space for America three months later (Rothman 2015).

I thought to myself, I wish Ed had a chance to see this boy tumbling. How do I get this film to Ed? Well, my tennis partner in Moscow happened to be the correspondent for NBC. So the next morning, he played tennis with me and I told him I saw this film of Leonov tumbling in space and, you know, Ed White is going to try to do the same thing in three months. So, I asked him, is there some way we can get that film to him? And he said so, "You know, we'd be interested to show that on NBC TV in the States! Do you think you could help me and give me a contact who I might be able to talk to?" I said, "Well, frankly, it's going to cost money," and he said, "It's okay, I can get the money." So, I gave him a name for contact. He contacted this guy I knew at the space agency and then told me, "Glenn, it costs $25,000 but we're going to get that walk-in space in three days," and said, "give me the address of Ed White." So I gave him the address in Houston, and he got the film in three days, NBC showed it on the U.S. television and they sent a copy down in Houston, and Ed White had a chance to see it. And I saw Ed White six months later and he said, "Boy that really helped me, because I didn't tumble as much as Leonov."

Questions and Answers

Question

Wilder: I'm Margaret Wilder from here at the University of Arizona, and I want to thank you for a great talk! I'm a social scientist but I've only heard a little about social so far. I wonder in your experience whether there were any contributions from social scientists to all these different endeavors over the years, whether you've noticed changes, see that as an important part of it, or not an important part of it, etc. How do you kind of define the social sciences within the scientific endeavor, because we see ourselves as scientists, but I'm not hearing too much about that within the discourse here.

Answer

Schweitzer: First, there are different kinds of social scientists. There are political scientists, ethnologists, and other kinds of social scientists. But I did notice a chance with the change of the regime back in the early 1990s and we had specific projects involving the social scientists. The economists were among the first to change their stripes and we saw dramatic changes. And I will say the American economists did a good job of working with them.

The area in which I've been most involved is in ethnology and I've been concerned about violent extremism and the ethnology problems have been serious in Russia over the years. And again, it was up until about the 1990s; it was always "somebody else's fault."

The ethnologists said just leave them alone. They don't want to talk about it. But since that time, the Russians have really embraced the idea of minority groups specifically. I remember in 1994, we started serious work with them. At the time, they said that there are ten hotspots in Russia which were ethnologically plagued, and the hottest spot was Chechnya and the surrounding area, but they were also around the rest of the country. The Russians I think did a good job in letting the social scientists tell their story.

But then they said, "how are they going to solve the problem?" And it took a lot of steps to buy them off. For example, some of these ethnology conclaves were governmental provinces. The Kazans is a good example. The Gazan's people were very uptight about being considered a minority and not having rights. So the Russians gave them rights to all the oil in their oblast (or territory) because they were scared of a revolutionary uprising.

That said, we still work with the Russians and meet with them regularly. In fact, we met with them four times recently on violent extremism and we try very hard to get the social sciences involved. Social scientists in Russia infamously don't get mathematics too well. Social scientists have to know how to use mathematics when they're doing their thing, but they're coming along better. It's less I think a problem of hostility or segregation, but rather that they are just not used to combining databases, and they may not know what a standard deviation is, that sort of thing. But I think it's not because they want to be devious. They just haven't been used to what we would call "good social science."

Question

Audience Member: I noticed that in the groups that you listed for science transmission into foreign countries, you didn't include the Peace Corps. And I'm wondering, has there ever been a role for the Peace Corps in the transmission of science and technology?

Answer

Schweitzer: That's my mistake. The Peace Corps did a great job in recruiting wildlife conservationists about 10 years ago and I was a big supporter of that, but it should have been listed in there. But the trouble is now the Peace Corps changed their ground rules. It used to be you could go in the Peace Corps for two years and get college credit for it—they cut that off a year ago. I think that was terrible. But that's the way it is. I've never met a Peace Corps volunteer who wasn't glad he was in the Peace Corps and I just left it off the list, my apologies.

Question

Peter Agre: When you're playing tennis with your counterparts in these adversarial countries, do you ever let them win?

Answer

Schweitzer: All the time!

20 YEARS OF SCIENCE DIPLOMACY

E. William Colglazier

Editor-in-Chief of Science & Diplomacy; Former Science & Technology Adviser to the U.S. Secretary of State

Note: This Keynote Lecture was originally given by Dr. Colglazier at the World Science Forum in November 2019 and subsequently published in the online journal *Science & Diplomacy* on June 28, 2020.
(Reprinted with permission from E. William Colglazier (ed.), *Science & Diplomacy.*)

The term "science diplomacy" has grown in popularity over the past two decades, first in the scientific community and now in the diplomatic community. In discussing science diplomacy, I am interpreting science quite broadly to include knowledge from the natural sciences, social sciences, engineering sciences, medical sciences, and the humanities.

The current interest of the diplomatic community in science diplomacy is clearly visible in an upcoming conference of foreign ministries and a recent statement by a foreign minister. The Foreign Ministries Science and Technology Advice Network (FMSTAN), created in 2016, will have its next meeting in Vienna in November 2019 with a number of foreign ministries participating. The Swiss Foreign Minister Ignazio Cassis in a recent speech stated:

> "Science Diplomacy is a key instrument for fostering cooperation between states" ... as part of developing a "new vision for Switzerland's foreign policy"...we "want to focus on a new instrument, still little known, but very promising"...an "opportunity for Switzerland to showcase the excellence of its scientific base in support of global dialogue..." (Cassis 2019).

Switzerland has, of course, been practicing science diplomacy for a long time, including by its many talented science counselors and swissnex Network.

Another example of the interest in science diplomacy is illustrated by the Madrid Declaration on Science Diplomacy. This inspiring statement was signed by participants at a conference in 2019 organized by the S4D4C research program (a scholarly investigation supported by the EU) with assistance from the Spanish Foundation for Science and Technology. For the participants, the statement represents a "common vision of science diplomacy in the future."

If we look back more than twenty years, we find science and foreign policy communities talking about "science and international affairs" or "science <u>and</u> diplomacy." This topic grew in importance after World War II, especially with the advent of nuclear weapons. For 75 years, the foreign policy community has paid significant attention to the impact of scientific and technological advances on international affairs.

I was born 75 years ago, and so have lived through this period. I started out as a theoretical physicist, but my career path was influenced by scientific mentors who were great science diplomats. Although never serving in government, they spent much of their time engaging with foreign scientific colleagues to discuss possible solutions to some of the most pressing global problems. Several participated in the Manhattan Project that led to nuclear weapons. After the war, they devoted a significant part of their careers to dialogues with Soviet and Chinese scientists to ensure that nuclear weapons were never used again.

Paul Doty, a distinguished biochemist at Harvard, and Wolfgang (Pief) Panofsky, a distinguished experimental physicist who directed the Stanford Linear Accelerator Center (SLAC), were two of my mentors. Paul created with support of the Ford Foundation the Center for Science and International Affairs at Harvard in the 1970s. Both Paul and Pief engaged with Soviet and Chinese scientists through the Committee on International Security and Arms Control (CISAC) of the U.S. National Academy of Sciences. The CISAC dialogues with Soviet scientists in the 1980s along with other Track II dialogues such as the Pugwash Conferences on Science and World Affairs helped to provide the framework for arms control treaties negotiated during the Cold War (Colglazier 2018). The Track II dialogues among scientists were influential when a diplomatic opportunity emerged from a leadership change in the Soviet Union with Soviet scientists serving as key governmental advisers.

What is the significance of "science and diplomacy" becoming "science diplomacy" over the past two decades? For me, the term science diplomacy indicates an active instrument rather than an area of overlap between two different domains. Vaughan Turekian and Tom Wang helped popularize and define the term when they created the AAAS Center for Science Diplomacy, the online journal www.sciencediplomacy.org, and in co-sponsoring with The Royal Society in 2009 the conference on "New Frontiers in Science Diplomacy."

My concept of science diplomacy for this talk is quite broad and includes (i) scientific knowledge and expertise being used to advance diplomatic and foreign policy goals for national and/or global interest, doing so in positive and constructive ways often accomplished through international scientific engagements, and (ii) diplomacy being used to advance the national and international scientific enterprises important to scientists everywhere. The diplomatic community emphasized in the UN 2030 Agenda the importance of strengthening the science-policy interface and harnessing science, technology, and innovation (STI) for making progress on the 17 Sustainable Development Goals through creation of the Technology Facilitation Mechanism (TFM). The diplomatic community has also recognized that facilitating international scientific collaboration is essential to maximize the benefits for science advancing diplomacy. Nevertheless, increasing concerns about scientific advances that contribute to dual-use technologies has created additional challenges (InterAcademy Partnership 2019). Preserving international scientific cooperation in fundamental basic research that is published in the open literature is a salient topic for scientists in their future science diplomacy dialogues.

My focus for the rest of this talk is on scientific knowledge and expertise, including expertise in technologies and innovation, being used as a tool for advancing diplomatic goals. First, we can ask who are the practitioners of science diplomacy. They can be inside and outside government, at the international, national, and subnational levels. They can include institutions, individuals, initiatives, networks, and mechanisms. At the international level, UNESCO, the Comprehensive Test Ban Treaty Organization (CTBTO), the International Institute for Applied Systems Analysis (IIASA), the WHO, and the International Science Council (ISC) are examples of institutions. The World Science Forum, the Intergovernmental Panel on Climate Change (IPCC), TFM, FMSTAN, the InterAcademy Partnership (IAP), the International Network for Government Science Advice (INGSA), Science 20 (S20), and the Transnational Red Sea Research Center can be categorized as initiatives, networks, or mechanisms. At the national level, there are foreign ministries, scientific academies, professional societies, NGOs, and research universities that have undertaken science diplomacy initiatives. For individuals serving as science diplomats, we can point to Frederico Major who opened this World Science Forum as well as to many distinguished participants at this and past Forums. Today, there are many young scientists and diplomats taking up the challenge.

What are the elements that make science diplomacy a powerful tool? I would include these five:

1. Every country is today focused on building its capacity in science, technology, and innovation for its future prosperity, security, and competitiveness.
2. Diplomats and foreign policy specialists are especially concerned about the negative implications and disruptive potential of rapidly advancing new technologies.
3. Scientists can communicate easily with foreign colleagues regardless of different ethnic backgrounds, cultures, and national governments, which can be especially valuable when governments are estranged.
4. Advances in science, technology, and innovation are an expanding resource of knowledge that can create new pathways and opportunities that leap over diplomatic hurdles and help achieve agreements between countries, as happened with the Montreal Protocol.
5. Spreading the values that come from doing science (e.g. integrity, evidence, objectivity, peer review, transparency, and valuing excellence) contributes to rational, open, tolerant societies, and the use of scientific evidence in societal decisions.

Then, we can ask what are examples of science diplomacy successes. I would include among the most prominent the nuclear arms control agreements during the Cold War, the IPCC and the Paris Climate Agreement, and the Iran nuclear agreement. Yet, there have been recent setbacks for all three, demonstrating yet again that politics is a more powerful force than science at least in the short run. A few other science diplomacy initiatives with initial progress ultimately stumbled or failed when overwhelmed by political events, such as with Libya, Syria, and North Korea (Kuperman 2015; Stoepler et al. 2017; Stone and Campbell 2018). I take an optimistic long view in believing that the political setbacks are temporary, and that science will help us to achieve our diplomatic goals for some of these cases. As one of my scientific mentors, Richard Feynman, said in his appendix to the Challenger Commission report, politicians discounting science do so at their peril as "Nature cannot be fooled."

Looking to the future, here are five areas where I think there is a great need for science diplomacy and where science can help to provide new pathways for progress:

1. Controlling new technologies of war, which can be used by nation states and terrorists, and advancing arms control treaties to reduce the dangers and proliferation of these weapons (e.g. nuclear weapons and delivery systems, autonomous weapons utilizing artificial intelligence, hypersonic weapons creating the fear of first strike, cyber weapons and offensive information warfare, and biological weapons).
2. Providing foresight and facilitating dialogue on the implications of rapid technological developments that can be disruptive (good and bad) to societies in order to maximize the benefits and minimize the negative consequences and threats from technological advances (e.g. AI, gene editing, synthetic biology, robotics, big data, blockchain, and social media).
3. Maintaining a channel of communication between nations that have estranged relations and conflicts and a potential for a new Cold War (e.g., Russia, China, Iran, and North Korea).
4. Accelerating progress on the global goals, especially the global environmental goals (climate, oceans, biodiversity), to safeguard our planet and help countries understand and commit to their share of global responsibilities.
5. STI capacity building in developing and emerging economies (of which the current UN TFM initiative on "STI for SDGs roadmaps" may become an important tool for a number of countries).

In order to make progress, I believe it is important for scientific institutions to engage with stakeholders at all levels of society to understand the needs and aspirations of the public, and to develop roadmaps for how science diplomacy can be used to achieve their institutional goals and contribute to the progress on the 17 SDGs. Our science communities can make a great contribution to our planet by helping to create more knowledge-based societies. The World Science Forum has been and will continue to be an important contributor to this aspirational goal.

REFERENCES

Abelson, Philip H. 1988. "Research and development in South Korea." *Science* 240 (4848): 9. https://doi.org/10.1126/science.240.4848.9.

Aziz, S.A.K. Abdul, and Schweitzer, Glenn E. (eds.). 1988. *Research Management: Proceedings of the Seminar on Research Management*. Kuala Lumpur, Malaysia: Rubber Research Institute of Malaysia.

BSF (U.S.-Israel Binational Science Foundation). 2020. "About the BSF." *Collaboration between Israeli and U.S. Researchers*. www.bsf.org.il/about/

Cassis, Ignacio. 2019. "Science Diplomacy as an Innovative Tool in Our International Relations." Opinion Article in *La Temps*. November 4, 2019. www.letemps.ch/opinions/diplomatie-scientifique-un-outil-novateur-suisse

Colglazier, E.W. 2018. "War and peace in the nuclear age." *Science & Diplomacy*. January 19, 2018. www.sciencediplomacy.org/editorial/2018/war-and-peace-in-nuclear-age

Couch, Emily. 2019. "In Russia, Women's Rights Are Still Not Human Rights." *International Observatory for Human Rights*. December 9, 2019. https://observatoryihr.org/blog/in-russia-womens-rights-are-still-not-human-rights/

Dowling, Stephen. 2017. "The Monster Atomic Bomb That Was Too Big to Use." *BBC Future*. August 16, 2017. www.bbc.com/future/article/20170816-the-monster-atomic-bomb-that-was-too-big-to-use

Fuller, Douglas and Rubinstein, Murray, eds. 2013. *Technology Transfer between the US, China and Taiwan: Moving Knowledge*. Routledge Research on Taiwan Series. New York, NY: Routledge.

Gosling, F. G. 1999. *The Manhattan Project: Making the Atomic Bomb* (DOE/MA-0001). Washington: History Division, Department of Energy, January 1999.

Hickey, William A., and George B. Craig, Jr. 1966. "Genetic distortion of sex ratio in a mosquito, Aedes aegypti." *Genetics* 53 (6): 1177–1196. December 1966. www.ncbi.nlm.nih.gov/pmc/articles/PMC1211089/

Hourihan, Matt. 2017. "Science and Technology Funding Under Obama: A Look Back." American Association for the Advancement of Science. January 18, 2017. www.aaas.org/news/science-and-technology-funding-under-obama-look-back

InterAcademy Partnership. 2019. "Improving Scientific Input to Global Policymaking with a Focus on the UN Sustainable Development Goals." May 2019 Report. www.interacademies.org/node/51066

Koepfli, Joseph B. 1983. Interview by Elizabeth Hodes. Santa Barbara, California, October 25, November 1, 8, and 15, 1983. Oral History Project, California Institute of Technology Archives. http://resolver.caltech.edu/CaltechOH:OH_Koepfli_J

Kuperman, Alan J. 2015. "Obama's Libya debacle: how a well-meaning intervention ended in failure." *Foreign Affairs* 94 (2): 66–77. www.jstor.org/stable/24483483

Maloney, Suzanne. 2009. "Iran and Mahmoud Ahmadinejad's Second Term." *Brookings Press*. August 3, 2009. www.brookings.edu/opinions/iran-and-mahmoud-ahmadinejads-second-term/

National Archives of the United States. 1995. *Records of the Foreign Service Posts of the Department of State: 1788–1990*. Web version based on Guide to Federal Records in the National Archives of the United States. Compiled by Robert B. Matchette et al. Washington, DC: National Archives and Records Administration. www.archives.gov/research/guide-fed-records/groups/084.html

National Research Council. 2004. *Scientists, Engineers, and Track-Two Diplomacy: A Half-Century of U.S.-Russian Interacademy Cooperation*. Washington, DC: The National Academies Press. https://doi.org/10.17226/10888

National Research Council. 2010. *U.S.-Iran Engagement in Science, Engineering, and Health (2000–2009): Opportunities, Constraints, and Impacts*. Washington, DC: The National Academies Press. https://doi.org/10.17226/12921.

Pielke, Roger, and Klein, Roberta. 2009. "The rise and fall of the Science Advisor to the President of the United States." *Minerva* 47, 7–29. https://doi.org/10.1007/s11024-009-9117-3

Pincus, Erica. 2014. "The Science and Technology Adviser to the U.S. Secretary of State: the history and evolution of the role." *Science and Diplomacy*. December 22, 2014. www.sciencediplomacy.org/article/2014/science-and-technology-adviser-us-secretary-state

Rostker, Bernard. 2013. *Providing for the Casualties of War: The American experience through World War II*. Santa Monica, CA: National Defense Research Institute, RAND

Corporation. www.rand.org/content/dam/rand/pubs/monographs/MG1100/MG1164/RAND_MG1164.pdf

Rothman, L. 2015. "What the First American to Walk in Space Had to Say About the Experience." *Time Magazine*. June 3. https://time.com/3898131/edward-white-walk-space/

Sakharov, Andrei. 1983. *Andrei Sakharov From Exile*. Texts provided by the International League for Human Rights. New York, October 1983. From the American Institute of Physics' Center for History of Physics exhibit "Andrei Sakharov: Soviet Physics, Nuclear Weapons, and Human Rights." https://history.aip.org/history/exhibits/sakharov/essay2.htm

Schweitzer, G.E. 1989. *Techno-Diplomacy: US-Soviet Confrontations in Science and Technology*. Boston, MA: Springer. https://doi.org/10.1007/978-1-4899-6046-7.

Stefano, L., Petersen-Perlman, J., Sproles, E., Eynard, J., and Wolf, A. 2017. "Assessment of transboundary river basins for potential hydro-political tensions." *Global Environmental Change* 45: 35–46. https://doi.org/10.1016/j.gloenvcha.2017.04.008

Stoepler, Teresa M., Creswell, Joel E., Anthis, Nicholas J., Chaverneff, Florence, and Parkhurts, Rebecca R. 2017. "The role of science diplomacy in international crises: Syria as a case study." *Science & Diplomacy*. March 9, 2017. www.sciencediplomacy.org/article/2017/role-science-diplomacy-in-international-crises-syria-case-study

Stone, Richard, and Campbell, Cathleen A. 2018. "A new era of forging connections and trust with North Korea's scientists." *Science & Diplomacy*. October 4, 2018. www.sciencediplomacy.org/perspective/2018/connections-and-trust-north-korea-scientists

Sung, Nack-Chung, and Chung, Yong Seung. 1996. "S & T in South Korea." *Science* 274 (5293): 1591. https://doi.org/10.1126/science.274.5293.1591

Tanasescu, Mihnea, and Stefan Constantinescu. 2020. "The human ecology of the Danube Delta: a historical and cartographic perspective." *Journal of Environmental Management* 262. https://doi.org/10.1016/j.jenvman.2020.110324.

Thomas, Will. 2017. "President Obama Leaves Broad Legacy for Science and Technology Policy." *American Institute of Physics*. January 24, 2017. www.aip.org/fyi/2017/president-obama-leaves-broad-legacy-science-and-technology-policy

World University Rankings. 2020. "The Hong Kong University of Science and Technology." *Times Higher Education*. www.timeshighereducation.com/world-university-rankings/hong-kong-university-science-and-technology

Yang, Guang-Zhong, Quek, Tong Boon, Stranigioli, Stefano, Ding, Han, Sun, Dong, and Yuh, Junku. 2020. "Forging global cooperation and collaboration." *Science Robotics* 5 (38). https://doi.org/10.1126/scirobotics.aba5894.

4 Transition to Sustainability

DOI: 10.1201/9781003220503-4

SUMMARY

This chapter begins with a foreword by the Dean of the College of Engineering at The University of Arizona, Dr. David Hahn, illustrating the importance of transitioning our focus to a sustainability model. Dean Hahn then introduces the keynote lecturer, Dr. Zhishen Wu, who presented his paper via teleconference. Dr. Wu explains the philosophies and methodologies of achieving sustainable infrastructure, and then delves into the challenges and predicaments in developing sustainable materials and "smart" infrastructure by utilizing advancing technologies.

Next are three articles that were presented at the conference and subsequently reviewed by a scientific committee to be included in the proceedings: "Human Macroecology, Energy Use Scaling, and the Sustainability of Cities: A look at Latin America" by Joseph R. Burger and Vanessa P. Weinberger, "Framework of Urban Ecological Sustainability Indicators in Drylands" by Parastoo Parivar, David Quanrud, and Hassan Vafai, and "A Strategic Framework for Creating Resilient, Ecological and Sustainable Infrastructure Systems – An Overview" by Ying Zhao, Mohammad Noori, Wael Altabey, and Zhishen Wu.

INTRODUCTION TO ZHISHEN WU

David W. Hahn
Craig M. Berge Dean, College of Engineering, The University of Arizona

It is my pleasure, on behalf of The University of Arizona, to welcome you to this keynote lecture. My name is David Hahn. I am the new Dean of the College of Engineering. I joined The University of Arizona just recently in July, so it's an absolute pleasure to be here to welcome you to this event. I am even more pleased to introduce you to Professor Zhishen Wu, who will give a keynote speech on sustainability enhancement of infrastructures with smart and resilient materials.

Professor Wu is a very distinguished colleague of ours. He is a professor at Southeast University (SEU) in China - one of the top 20 research universities and among the oldest in China (one of the top three), where he shares his expertise. He also has an appointment as a professor at Ibaraki University in Japan.

Professor Wu's research expertise is vast and includes structure engineering, advanced composite materials, and smart senser technologies and structural health monitoring. He is the author or co-author of over 600 technical papers, has over 15,000 citations, has graduated more than 100 graduate students, and also has published 12 books and holds more than 100 patents, so quite an illustrious career.

You can see numerous accolades from his own colleagues, including the Research Prize by the Japan Society of Civil Engineering, the Technology Award by the Japan

Society for Composite Materials, the 2009 Structural Health Monitoring Person of the Year award by the International Journal of Structural Health Monitoring, IIFC(International Institute for FRP in Construction) medal in 2016, and the very prestigious National Prize for Progress in Science and Technology from China.

In the spirit of this conference and diplomacy specifically, he is a member of numerous international societies, boards, and journals, and is a fellow of the Engineering Academy of Japan and other four professional societies. He is now serving on the president of International Society for Structural Health Monitoring of Intelligent Infrastructures (ISHMII) and the president of the International Society for Basalt Fibers and Composites (ISBFC). Let us please welcome, joining us from China, Professor Wu.

SUSTAINABILITY ENHANCEMENT OF INFRASTRUCTURES WITH SMART AND RESILIENT MATERIALS

Zhishen Wu

Professor, Southeast University, China

BACKGROUND

Key Issues and Requirements of Infrastructures

The demands for efficient and effective infrastructure maintenance in Japan have received strong attention in recent times because of several factors such as the imminent risk of natural disaster, accumulation of aging infrastructure, increasing public demand for high-quality infrastructure, and limited government expenditure on maintenance. Traditional post-seismic inspection methods of infrastructures, though provide knowledge about the vulnerability of structural systems and mechanisms of failure, are often unreliable for effective maintenance-based monitoring against major earthquakes. For example, in the 2011 Great East Japan Earthquake and tsunami which alone killed more than 10,000 people and damaged major infrastructure system, the financial loss was estimated at US$ 235 billion. Many buildings and bridges suffered severe damages due to tsunami or strong ground motion effects. These catastrophic events confirmed the inability of the current state of knowledge in evaluating the ultimate capacity of structures to withstand damage under various types of seismic activities. To achieve this goal, it is essential to investigate the response of infrastructures, and to develop robust methodologies that would facilitate the reliable prediction of infrastructures' performance in real-time continuously or periodically. However, we are still looking for scientific and technical solutions to keep our infrastructures safe in disasters.

Another serious issue in Japan is the massive cost of maintenance. The peak of infrastructure development in Japan took place between 1955 and 1975. During this period, rapid economic growth accelerated government spending on infrastructure and as a result Japan's infrastructure stocks have increased significantly. Due to land situations in Japan, a great percentage of bridges and tunnels make up the transportation systems. For transportation infrastructure systems, there are approximately 650,000 bridges with a total length of about 1.2 million kilometers constructed in

370,000 square kilometers. Japan has over 140,000 existing bridges and it is estimated that the number of aged bridges will constitute one-half of all the road bridges by the end of 2030. Of these, most of the railway bridges were built prior to 1970, and over 50% of roadway bridges were built prior to 1980. Bridge extension is about 72% of the total length of the Tōhoku/Joetsu Shinkansen (Japanese high-speed railway line), and the tunnel extension reaches 23% of the same length. Deterioration and aging are also great issues for many highway bridges in Japan, and many of these bridges are in need of extensive maintenance. Therefore, we must come up with creative solutions for these massive maintenance costs.

Development Trend of Engineering Structures

The theme of the development trend of future engineering structures is higher, longer, and lighter, and the life cycle cost (LCC) of structures should be effectively reduced by performance-based design and advanced management methodologies of structures. Thus, major engineering structures must obtain high performance and longevity, which is essential for the mitigation of natural disasters and other disasters. High performance and longevity mean that not only engineering structures have a high degree of safety and performance, but also the sustainability in the whole life cycle in serious environments and under extreme loads is drastically decreased in order to realize the objective of full LCC minimization (less maintenance costs than traditional structures of equal service life, or equivalent maintenance costs with longer service life). Therefore, higher and longer structures are expected to be realized through the application of novel lightweight materials, while meeting growing demand with limited resources.

Serious Environmental Issues

In recent years, increasing number of serious environmental issues has occurred due to various factors such as depletion of metal resource, increase emission of CO_2, and waste of forest resources. Thus, new construction materials are required to solve the problems of resource crisis, high carbon emission, and sustainable development.

Philosophies and Methodologies of Infrastructure Sustainability

Sustainability

As mentioned in previous studies, sustainability is a model that addresses simultaneously today's needs and the impacts on future generations. It is characterized by a holistic view and brings together three dimensions: ecology (environment), economy, and society (Bocchini et al. 2013, 16). As shown in Figure 4.1, in terms of sustainability, the construction work can be categorized into four levels and several aspects. For instance, the design of a sustainable structure requires a general definition as a role model in the first place, considering inter- and intra-generational fairness. Then, through ecology, economy, and society dimensions, various indicators within resources, bio-diversity, and other aspects can be featured and categorized. Finally, the consideration is deepened into each individual indicator such as animal protection and noise emissions.

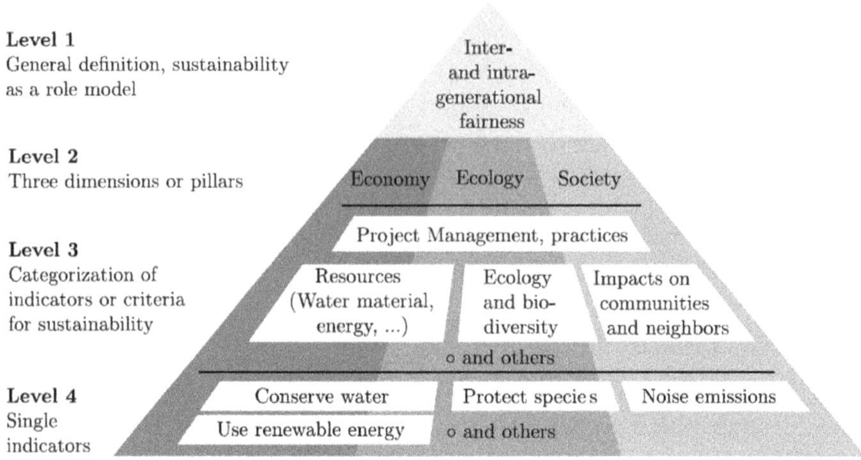

FIGURE 4.1 The hierarchy of sustainability (Bocchini et al. 2013).

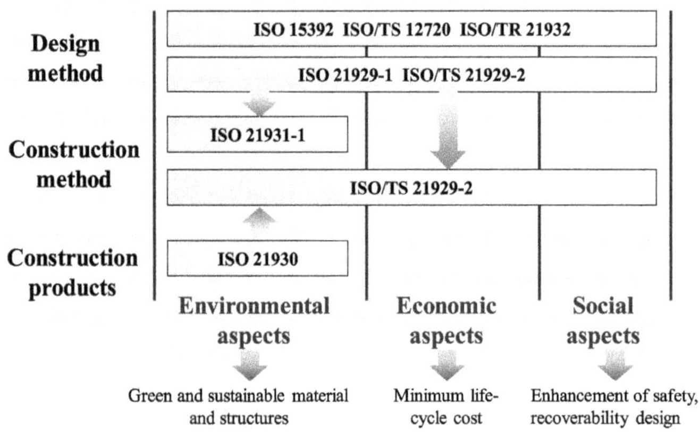

FIGURE 4.2 Sustainable design of structures in ISO standards.

Accordingly, several ISO standards have specified the sustainable design of structures, as shown in Figure 4.2 (ISO 15392:2019, 21930:2017). The environmental, economic, and social aspects claim higher requirements on the design method, construction method, and construction products. Thus, considering these aspects, green and sustainable materials and structures are required to minimize the LCC, enhance the structural safety, and conduct a recoverability design.

Categories of Sustainable Infrastructure

The enhancement of sustainability design of structures mainly includes economic/durable, safety, green/healthy, in which many philosophies were proposed to fulfill the requirement, as shown in Figure 4.3. Within the economic and durable issue,

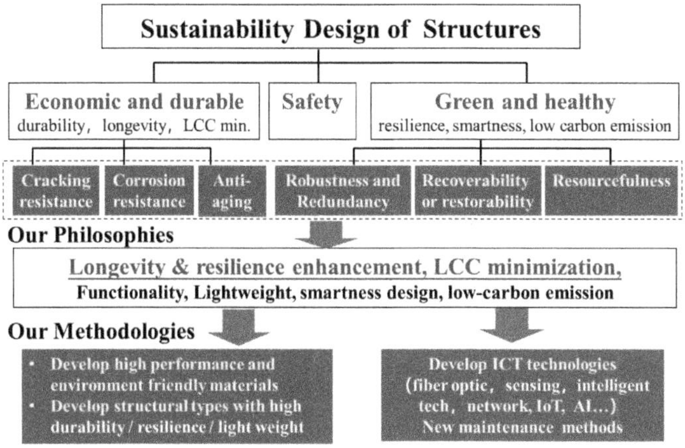

FIGURE 4.3 Sustainability enhancement of infrastructures.

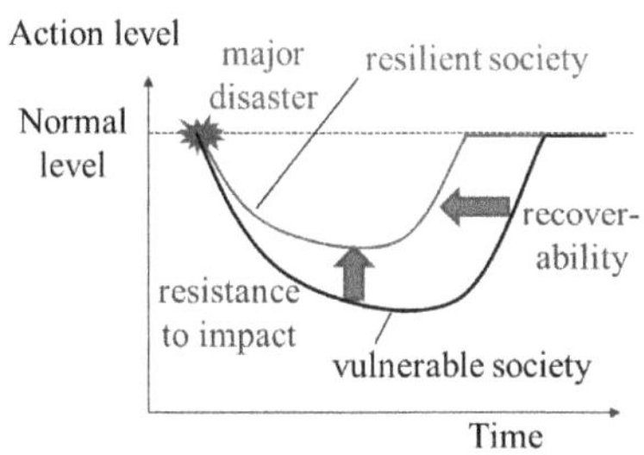

FIGURE 4.4 Vulnerable society changed into resilient society.

emphasis is laid on cracking resistance, corrosion resistance, and anti-aging investigation. In green and healthy design of structures, it is essential to evaluate the robustness, redundancy, recoverability, and resourcefulness of structures. In this study, the longevity design and resilience enhancement of structures, as well as the minimization of LCC, are highlighted. To achieve these philosophies, high-performance and environmentally friendly materials are needed, as well as novel types of structures with high durability, resilience, and light weight. The use of ICT technologies, such as fiber optic, sensing, and artificial intelligence (AI), is also beneficial for ensuring a sustainable and smart structure.

Longevity Design of Infrastructure

Resilience means the ability of a system to withstand an unusual perturbation and to recover efficiently from the damage induced by such perturbation (Francis and Bekera 2014; Holling 1973). It aims to protect human life, minimize damage, protect economic society, and recover quickly (Figure 4.4).

Realizing resilience of engineering systems and structures needs to study the basic principle, take a theoretical approach, and use key advanced technologies of civil

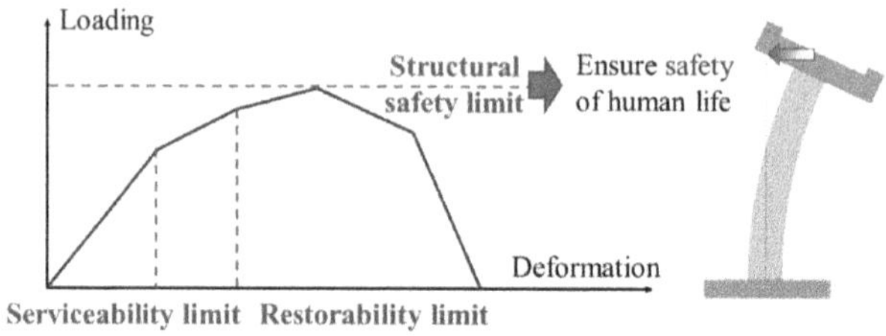

FIGURE 4.5 Three-level limit state in Japanese standard. (Adapted from Japan Cabinet Secretariat 2018.)

engineering systems and engineering structures. A critical analytical definition of resilience is as follows:

$$R = \frac{\int_{t_0}^{t0+t_h} Q(t)\,dt}{t_h} \tag{1}$$

where R is the resilience index, t_h is the time horizon investigated by the analysis, and $Q(t)$ is the area underneath the recovery path; the denominator represents the value of resilience if the event did not occur or had no effects on functionality.

Japan's basic plan for enhancing national resilience is to create a strong and resilient country (Japan Cabinet Secretariat 2018). The basic policies include protecting human lives, avoiding fatal damage and maintaining important functions, minimizing damage and swift recovery and reconstruction, for example measures to prevent fires in densely populated areas, enhancement of disaster resilience of transportation and logistics facilities, and so on. A three-level limit state in Japanese standard was proposed, as shown in Figure 4.5. The most important is the structural safety limit; this state is used to ensure the safety of human life. It associates with collapse, or with other similar forms of structural failure caused by large deformation, large displacement, vibration, etc. To ensure the functionality of the structure, a serviceability limit is adopted, which corresponds to conditions beyond the specified service requirements for a structure or structural element. The last one is restorability limit, which corresponds to conditions beyond which the repair of structure is not possible with the current available repair technologies, with reasonable cost and within a reasonable time period.

Infrastructure Resilience

Longevity design of infrastructure is necessary due to the problems of short life and high maintenance cost during the service life. For the longevity design of infrastructure, the philosophy is to slightly increase the initial cost by advanced material or design, to achieve LCC minimization, along with free or minor maintenance. The goal is to achieve a longer service life of structures with the highest economic benefits. Methods include the design of advanced materials, long-term structural design of key

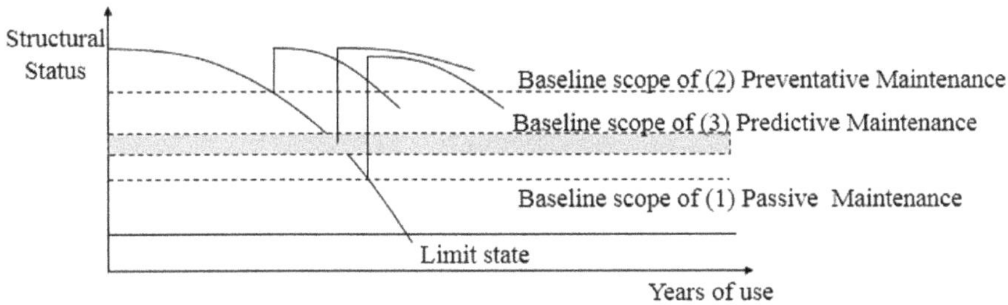

FIGURE 4.6 Threshold value for different structural maintenance implementation.

members, and an effective maintenance strategy. For instance, by taking a long-life bridge design, after a careful design of key members and cable systems, the bridge life can be increased from 25 years to 120 years, with half of the LCC.

To minimize the maintenance cost of the bridge, the original component can be substituted by girders with longevity, durable pavement, and rubber supports. Durable pavement can prevent the traffic break due to repairmen, while rubber supports can extend the replacement period to improve seismic behaviors. Expansion joints are also a good choice, which are long-lasting and easy to replace. The adoption of anticorrosive coating and a longevity bridge deck design also benefits the maintenance cost.

Maintenance managements include passive maintenance, preventative maintenance, and predictive maintenance. Passive maintenance is usually adopted after major damages of the structure; preventative maintenance refers to the regular maintenance to prevent accidents; predictive maintenance, which is based on advanced monitoring technology and analytical technology, is proposed to accurately detect early structural damages. The maintenance management type can significantly extend the lifespan and minimize the LCC of the structures. Figure 4.6 presents the threshold value for the implementation of different structural maintenance strategies.

For a structure, the initial cost, managing cost, and subsequent maintenance cost together constitute the full LCC. Figure 4.7 shows the composition of full LCCs and the benefits of different types of maintenance managements. It is obvious from Figure 4.7 that the managing costs account for a large proportion of the full LCC. The full LCCs can be reduced with the use of new technology of long life, preventative maintenance, and maximize benefits/cost. Figure 4.7 also shows that the benefits vary in different maintenance managements. Longevity is realized through the new technology of long life. Low cost is realized through preventative maintenance. Furthermore, predictive maintenance can maximize the benefits/cost-savings (or minimize LCC).

Challenges in Developing Sustainable Materials and Structures

Green, Resilient, and High-Performance Materials

The combination of steel and concrete is known as a major technological revolution in the history of civil engineering structure development. Reinforcement is isolated

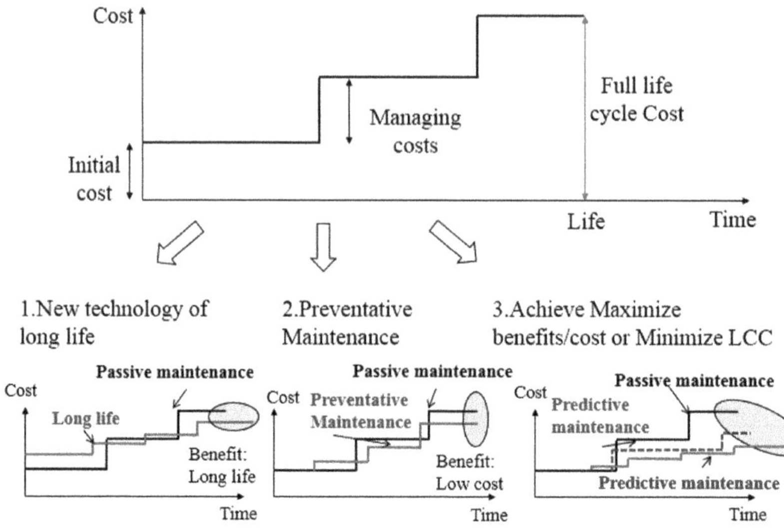

FIGURE 4.7 The benefit of different Maintenance Managements.

from the outside world using concrete, while the corrosive medium is prevented using concrete. Therefore, reinforced concrete is widely used in building structures.

However, under harsh environment, reinforced concrete can lead to steel bar corrosion and expansion, resulting in a decrease in the bonding force between steel bar and concrete, a large number of cracks on the concrete surface, and a decline in the bearing capacity of the structure, eventually leading to the buckling and destruction of the building (Torresacosta and Diazcruz 2020).

In the process of producing steel bar and cement, harmful gases will be released, the environment will get polluted, CO_2 will be released in large quantities, and the Greenhouse Effect will be seen. Meanwhile, in the construction process, a large amount of construction waste will also be generated. The widespread use of reinforced concrete materials will be limited by these factors. Therefore, there is an urgent need to improve the performance of traditional materials and develop new green materials (Hao and Li 2019).

Corrosion Resistance Steel

In the face of increasingly severe corrosion of steel bars, corrosion-resistant steel has been well developed, as shown in Figure 4.8 (Miura et al. 1983). The depth loss after corrosion of ordinary steel, weathering steel, and high-Ni steel was tested for 3, 6, 12, and 24 months. It can be seen from Figure 4.8 that the weathering steel shows better corrosion resistance than ordinary steel; however, it can still be corroded 13 mm after 24 months. The corrosion resistance of high-Ni steel is the best among them as it possesses excellent corrosion resistance; the deep corrosion depth at 24 months is only 4.8 mm, which is 17%–19% compared to ordinary steel. Currently, corrosion-resistant steels have been adopted in many advanced countries to extend service life of constructions.

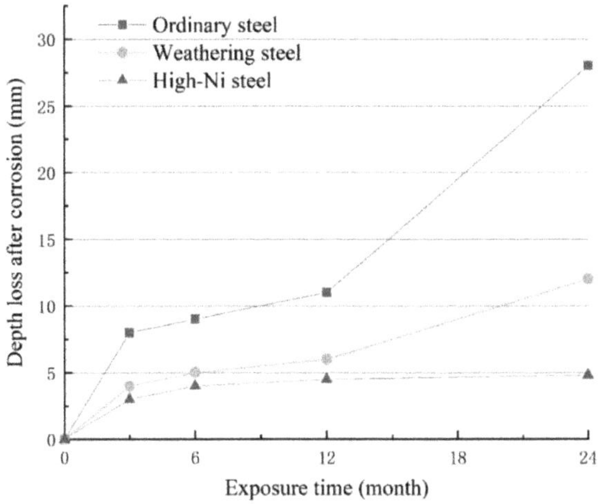

FIGURE 4.8 Steel corrosion depth changes with time (Miura et al. 1983).

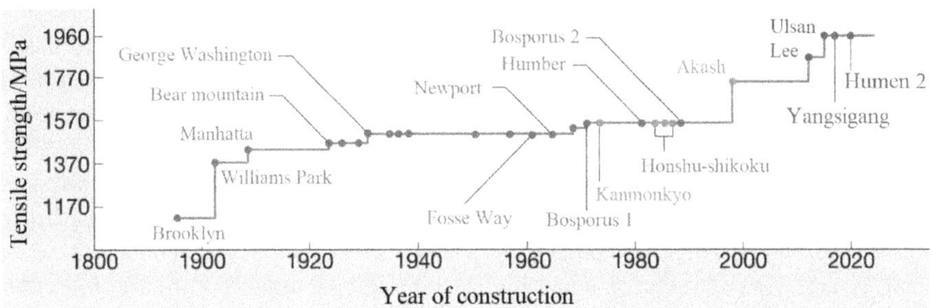

FIGURE 4.9 The development of ultrahigh-strength steel cable tensile strength. (Adapted from Lee, Kim, and Choi 2013.)

Ultrahigh-Strength Steel cables

As shown in Figure 4.9, to enhance the spanning ability and fatigue life of bridges, ultrahigh-strength steel cables are under development (Lee, Kim, and Choi 2013). The world's first modern suspension bridge is the Brooklyn Bridge built in the U.S. in 1883, and its tensile strength is only 1200 MPa. In 1998, Japan built the world's first Akashi Strait Bridge with a tensile strength of 1,770 MPa. In 2014, the tensile strength of Ulsan Bridge built in South Korea reached 1,960 MPa. In 2018, the Yangsigang Bridge and the Humen 2 Bridge constructed in China also reached the tensile strength of 1,960 MPa. In the last 200 years, the tensile strength has almost doubled as shown in Figure 4.9, from around 1000 MPa to 2000 MPa, which will benefit super long-span bridges with better fatigue performance in the future. The bridge is bound to develop in the direction of large span and light weight, and the performance requirements of steel cables used in bridges will continue to increase. Issues such as plasticity, torsion, and bending still need to be solved before using them in wide areas.

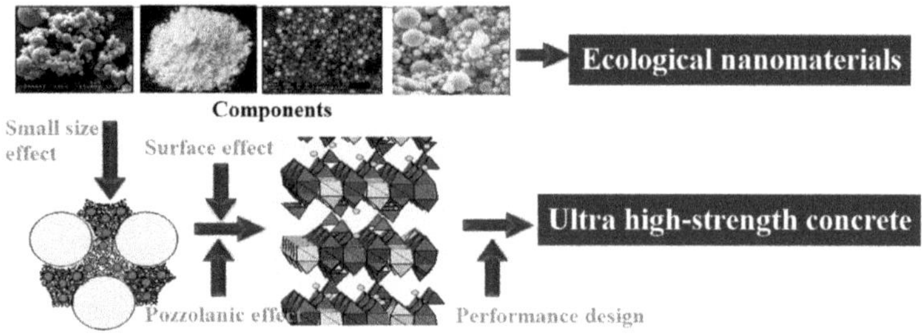

FIGURE 4.10 Ultrahigh-strength and high ductility concrete design process. (Adapted from Ranade et al. 2013a.)

Ultrahigh-Strength and High Ductility Concrete

In view of the shortcomings of traditional concrete that possess low strength and heavy weight, ultrahigh-strength and high ductility concrete was introduced in order to reduce the weight of the structure and improve the spanning capacity (see Figure 4.10) (Ranade et al. 2013a).

Fly ash, kaolin, vitrified microbeads, and silica fume and other biological nanomaterials are incorporated into cement. Their small size helps them to easily combine with the cement surface which leads to pozzolanic effect. Ultrahigh-strength and high ductility concrete is finally made through performance design. The compressive strength of ultrahigh-strength and high ductility concrete is greater than 150 MPa. Ultrahigh-strength and high ductility concrete uses a new type of cement-based engineering materials, with super high strength, high toughness, high permeability, high corrosion resistance, and other excellent properties; Such type of concrete can be used for large-span structures, super high-rise buildings, tunnels, military protection, special engineering, and other fields (Ding et al. 2018). Ultrahigh-strength and high ductility concrete can optimize the structure size, increase the span, and increase the structural bearing capacity, durability, and life cycle. Thus, the requirements of lightweight, high rise, large span, and high durability of civil engineering structures can be achieved using ultrahigh-strength and high ductility concrete (Li and Wang 2013).

Durability Enhancement of Concrete

Under harsh environment, the durability of the concrete structure is mainly affected by three reasons: CO_2, Cl^-, and SO_4^{2-} (as shown in Figure 4.11(a)). First, alkaline substances such as $Ca(OH)_2$ and CSH gel produced by hydration in concrete will react with CO_2 in the air, causing corrosion of the steel bars (Sun and Lu 2015). Second, in the ocean and other environments, the reinforced concrete is continuously invaded by Cl^- in the seawater; therefore the corrosion of steel bars is accelerated, and the durability of the structure is reduced (Tang et al. 2012). Third, when the concrete is under an environment immersed in a sulfate solution, SO_4^{2-} will react with the cement slurry, and the resulting product will have an expansion stress. When the stress is greater than the tensile stress inside the concrete, the concrete is damaged, and the durability of concrete will decrease.

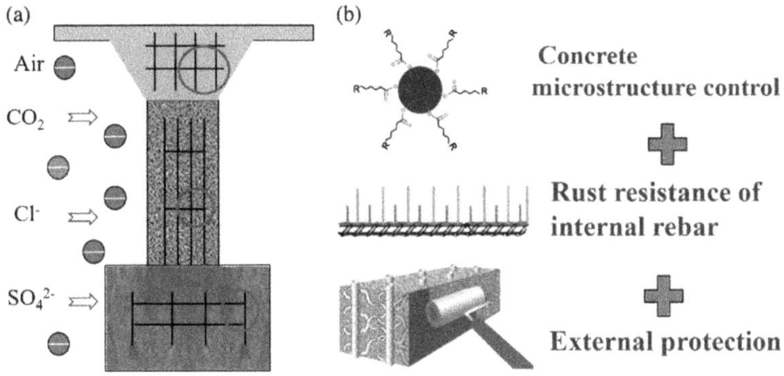

FIGURE 4.11 Influencing factors (a) and protective measures (b).

To solve this problem, integrated three-level strategies, including concrete microstructure control, rust resistance of rebar, and external protection, were developed to enhance the durability of concrete to meet longevity requirement (as shown in Figure 4.11(b)).

Concrete Microstructure Control

Due to the small particle size of nanomaterials such as nano-SiO_2 and nano-Al_2O_3, the pores between cement hydration product particles can be filled better, the porosity can be reduced, the density can be increased, and the cross-sectional structure can be effectively improved, thereby improving the durability of the concrete (Ranade et al. 2013b).

Rust Resistance of Rebar

At present, there are four main measures to prevent steel corrosion (Al-Tayyib et al. 1990): (1) changing the material of the steel bar and using stainless steel bar can greatly reduce the corrosion rate and prolong the service life of the structure; (2) the surface of the steel bar is coated with epoxy coating and galvanized to make it have excellent corrosion resistance; (3) the rust inhibitor is incorporated into the concrete, the corrosion rate of the steel bar is effectively reduced, and at the same time, a passivation film is formed on the surface of the steel bar, and the steel bar is protected; (4) the electrochemical protection method is used to inhibit the electrochemical corrosion of the interface between the steel bar and the concrete.

External Protection

Concrete external protection is one of the measures to improve the durability of the concrete in recent years (Gan et al. 1994). The surface of the concrete is sprayed, painted with anti-corrosion and anti-permeation coating, which forms a hydrophobic protective layer inside and on the surface of the concrete, prevents the development of micro-cracks in the concrete, and isolates the external air, water, and salt ions. This method can effectively suppress the reduction of the bearing capacity of the concrete

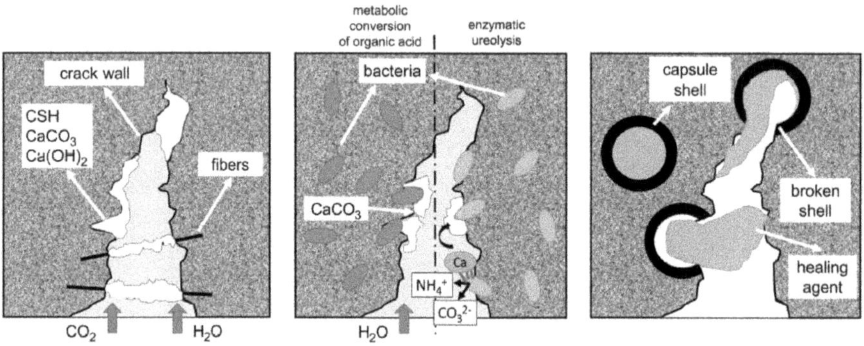

FIGURE 4.12 Self-healing mechanisms: (a) Autogenous, (b) Autonomous bacteria-based, (c) Autonomous capsule-based. (Adapted from Rajczakowska et al. 2019.)

structure, and improve the safety and reliability of the structural service, and extend the service life of the concrete structure.

Self-Healing Concrete

Self-healing concrete provides a methodology of sustainable structures, which can realize an intelligent self-healing ability of concrete after cracking damage without any external intervention (Wu, Johannesson, and Geiker 2012). Owing to low tensile strength and brittleness of ordinary concrete, cracking behavior occurs to concrete structures easily and frequently, which would reduce structural service life and compromise their safety. Self-healing concrete can enhance service life of concrete and minimize maintenance cost by materials and labor work.

Self-healing concrete is usually divided into autogenous self-healing concrete and autonomous healing concrete based on the different working mechanisms. Autogenous self-healing depends on the original components of the concrete material, which swells continuously because of water absorption or relies on continued hydration of cement to fill cracks (see Figure 4.12(a)) (Çopuroğlu et al. 2013). However autonomous self-healing is based on the works of some unconventional engineered additions, including vascular, bacteria, or others. The cracks are healed through bacteria generating limestone (see Figure 4.12 (b)). If a crack were to develop, the capsule would break and then release the healing agent to repair the crack (see Figure 4.12 (c)) (Cailleux and Pollet 2009).

(a) Autogenous (b) Autonomous bacteria-based (c) Autonomous capsule-based

However, self-healing concrete often exhibits characteristics of low initial strength and post-repair strength due to the capsules and the bacteria (Arce et al. 2017). At the same time, cost is another issue that cannot be ignored; some research indicates that the price of self-repairing concrete is higher than twice that of ordinary concrete (Arce et al. 2017; Wang et al. 2015). Thus, those issues must be solved before self-healing concrete gets popularized in engineering applications.

Sustainable Concrete

Sustainable concrete is considered a relatively new green structural material because of its low-level carbon emissions and ability to utilize industrial waste, which comprises three ingredients without cement. The first one is aluminosilicate, including fly ash, ground granulated blast furnace slag, or metakaolin. The second one is coarse and fine aggregate. The last one is a mixture containing sodium hydroxide, water, and sodium silicate or silica fume.

It is confirmed that Ordinary Portland Cement (OPC) produced nearly 10% of the global carbon dioxide emissions (Hasanbeigi, Menke and Price 2010). Some researchers indicate that sustainable concrete would reduce more than 20% CO_2 emissions compared to OPC, which is of great significance for reducing global CO_2 emissions (Assi, Deaver, and Ziehl 2018; Bondar 2013). In addition, sustainable concrete has almost no shrinkage cracks, owing to its few shrinkages compared to normal concrete (Assi, Deaver, and Ziehl 2018). What is more, it possesses better mechanical properties; the compressive strength even exceeds 100 MPa by adjusting the concrete composition (Dinakar, Sahoo, and Sriram 2013). These advantages are the component characteristics of sustainable concrete. As far as material characteristics are concerned, sustainable concrete shows a good constructability because of its low shrinkage, high flow ability, and low calorific value.

However, further studies about the cost of sustainable concrete and restrictions on the production of sodium hydroxide and sodium silicate should be carried out in the future.

Fiber Reinforced Polymer (FRP)

FRP is used as an advanced material with designable, high strength, light weight, and superior corrosion and fatigue resistance properties, which makes it a competitive alternative for steel and an ideal material for strengthening and enhancing the structure (Wu, Wang, and Wu 2012). Carbon FRP composites were initially used for durability enhancement in Japan, the U.S., and Europe since the 1990s, but have not been overwhelmingly successful due to their costs (Wu et al. 2014). Carbon FRP achieved great successful applications through seismic strengthening stimulated by the Great Hanshin Earthquake in 1995 (Wu et al. 2014). After that, different applications of FRP in new construction were developed. At present, FRP has achieved successful application for structural repairing and retrofitting; however, in new construction, problems such as low material efficiency, lack of key application technologies, and high cost of conventional carbon and glass FRPs limit their wide acceptance (Bakis et al. 2002). As a solution, a new type of FRP, i.e. basalt FRP, is developed to meet high integrated cost performance.

Basalt Fibers: Green and High Performance/Cost

Basalt fiber is an environmental friendly and recycling material which is melted and drawn from natural volcanic rocks under 1500 degree (Cao, Wang, and Wu 2011; Wu, Wan, and Wu 2012). Basalt fiber is a green material compared to carbon fiber. For

FIGURE 4.13 The "alchemy" of basalt fiber production.

instance, if we compare the production process of basalt fibers with carbon fibers, it can be seen that the steps required to produce basalt fibers are much less than that of carbon fibers with no toxic gas emission and very low carbon emission (Wu, Wan, and Wu 2012; Wu et al. 2014). Meanwhile, the energy consumption of basalt fibers is just 1/16 compared to that of carbon fibers (Figure 4.13).

At present, we have developed high-performance basalt fibers with high strength, temperature, and alkaline resistance (Cao, Wang, and Wu 2011; Wang et al. 2014; Zhao et al. 2016). The stability of the fiber mass production technology is solved to ensure the wide applications of basalt fibers. Compared with conventional fibers, basalt fibers can achieve 30% higher mechanical properties than glass fibers with a similar cost, 70% strength, with a seventh cost compared to carbon fibers, and competitive high temperature behaviors (Ding et al. 2018; Lee, Kim, and Choi 2013). The strength, stiffness, and durability of basalt fibers also reach up to the levels of aramid fibers. Thus, basalt fibers have a large potential to provide the highest cost performance. In the next 5–10 years, CFRP will enjoy an output of 0.15-0.2M tons, while the value of GFRP is in the range from 6 to 7M tons, and from 0.5 to 1M tons for BFRP. In the future, developing high-performance BF and C/BF for engineering, and selecting different fibers based on structural demands to achieve the highest cost performance are utmost necessary. To enhance the long-term fatigue behavior of construction, the superior fatigue and creep behaviors of BFRP play an important role, which shows much higher performance than GFRP, and even approaching that of CFRP, as shown in Figure 4.14 (Wang et al. 2014; Zhao et al. 2016). With the help of pre-tensioning, the creep behavior of BFRP can be enhanced. These behaviors benefit and increase the service life of long-span structures by more than 100 years and realize a lighter-weight structure.

In addition, BFRP has a strong resistance to alkaline, acid, and salt environment. For a concrete structure application, such as the bridges nearby a strait or river, it can enhance durability from the current 10 to 30 years to a life of 100–300 years (Shi et al. 2020; Zhao et al. 2016). Hybridization of FRP is a good method to achieve high integrated mechanical behavior to fulfill different requirements of engineering, thereby showing the feature of designability of FRP (Wu et al. 2010). The hybrid FRP could realize high strength, stiffness, ductility, and low cost at the same time. Differing from conventional thermal-setting resin in FRP, thermal-plastic resin will

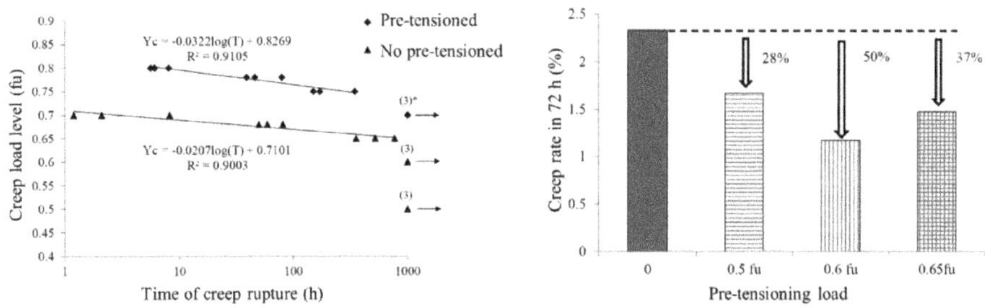

FIGURE 4.14 Creep behavior and creep enhancement of BFRP tendon.

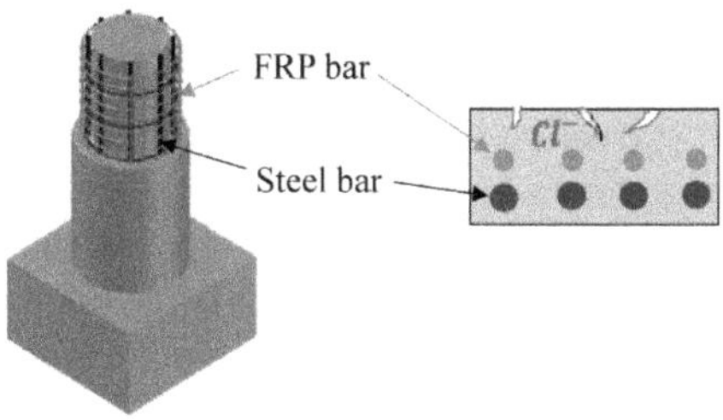

FIGURE 4.15 Hybrid FRP-steel reinforcement.

have a more promising future due to the ease of process for application, superior fatigue and corrosion resistance, and most importantly, it can be used to benefit the sustainable application of materials.

Longevity and Resilience

Reinforced concrete structures have been widely applied in infrastructures. However, several main deficiencies in conventional RC structures are summarized below from practical engineering. First, RC members are generally working at service load with cracks, accelerating the penetration of oxygen, moisture, and chloride ions, and the corrosion of the internal steel reinforcements, which causes the deterioration of the structural performance of RC members. Second, a large number of RC structures cannot be repaired due to their considerably large residual deformation subjected to severe earthquakes, which leads to enormous expenses of rebuilding. In addition, for RC columns under axial forces, the yielding of steel stirrups may result in the failure of constraints on the longitudinal steel reinforcements, causing the buckling of longitudinal steel reinforcements. Furthermore, a high percentage of reinforcement, especially at a beam-column joint, has a negative effect on the constructability of concrete and may threat the construction quality of the RC structure.

Two types of structures are developed to improve the durability, resilience, and constructability of conventional RC structures mentioned above. The first type is realized by hybrid FRP-steel reinforcement, as shown in Figure 4.15. Non-corrosive FRP bars are placed in the concrete cover of steel reinforcements to restrict crack

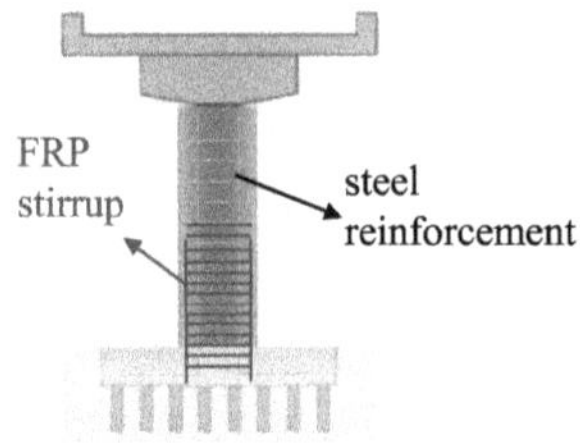

FIGURE 4.16 FRP stirrup.

propagation and improve load-carrying capacity. Due to the contribution of FRP, FRP-steel RC members demonstrate a second-order stiffness after yielding, which can effectively reduce residual deformation when unloading and thus resilient structures can be realized. The second type of structure is characterized by replacing steel stirrups with FRP stirrups in columns, as shown in Figure 4.16. Compared to steel stirrups, FRP stirrups own higher tensile strength and linear elasticity up to failure without yield plateau, and can adequately constrain the longitudinal steel reinforcements and prevent the buckling of longitudinal reinforcements.

Hybrid FRP-steel RC beams exhibit significantly smaller crack width compared with conventional RC beams under same deflections. The durability of hybrid FRP-steel RC beams is verified by tests, which depicts the performance degradation of hybrid FRP-steel RC beams and conventional RC beams in simulated seawater. It can be found that after conditioning for six months, the load-deflection relationship of the hybrid FRP-steel RC beam is nearly identical with that of the original hybrid FRP-steel RC beam, whereas the flexural performance, including stiffness, load-carrying capacity, and deformability of conventional RC beams degrade after conditioning.

Severe problems of structural safety and durability existed in Nanjing Yangtze River Bridge, constructed in 1968. After long-term fatigue loading from vehicles and erosion from river water, various types of damage occurred in different parts of the bridge, contributing to the deterioration of serviceability. Cracking, concrete spalling, and exposure of steel reinforcements were observed in the arch ribs, as can be seen from Figure 4.17. In 2016, a major large-scale maintenance was conducted on the bridge, which adopted BFRP grids and bars (Ibrahim et al. 2016). The BFRP grids were placed in the concrete cover to restrict cracks, as shown in Figures 4.18 and 4.19. Also, BFRP bars were used to resist cracking and their great anticorrosive property was exerted in the post-cast zone on the deck.

According to a Japanese survey, 89.3% of respondents recommended that reasonable repair time of damage-controllable structures should be within one week. The concept of resilience was mentioned in some seismic design codes, whereas systematic specific implementations were not proposed. Thus, the design philosophy of damage-controllable structure is proposed, as shown in Figure 4.20. Similar to conventional RC structures, damage-controllable structures maintain an elastic stage or normal service stage under small earthquakes, in which no damage can be observed and structures can be restored to their original positions when unloading. However, when suffering moderate or rare earthquakes, steel reinforcements yield and damage-controllable structures show an obvious second-order stiffness, which

FIGURE 4.17 Damage on the arch rib.

FIGURE 4.18 BFRP grid used in the arch rib.

is negligible for conventional RC structures. Such second-order stiffness realizes smaller residual deformation and repairability after unloading, while large permanent deformation may occur in RC structures. Furthermore, a reliable slippage between FRP reinforcements and concrete can contribute to a high ductility, as shown in the stage phase of Section 3 in Figure 4.20.

Resilience can be realized by hybrid FRP-steel reinforcement. According to experimental results, the BFRP-steel RC column exhibits smaller residual deformation and larger post-yield stiffness compared to the steel-RC column. Another way to obtain resilience is through the adoption of the steel-FRP composite bar (SFCB),

FIGURE 4.19 Maintenance completed.

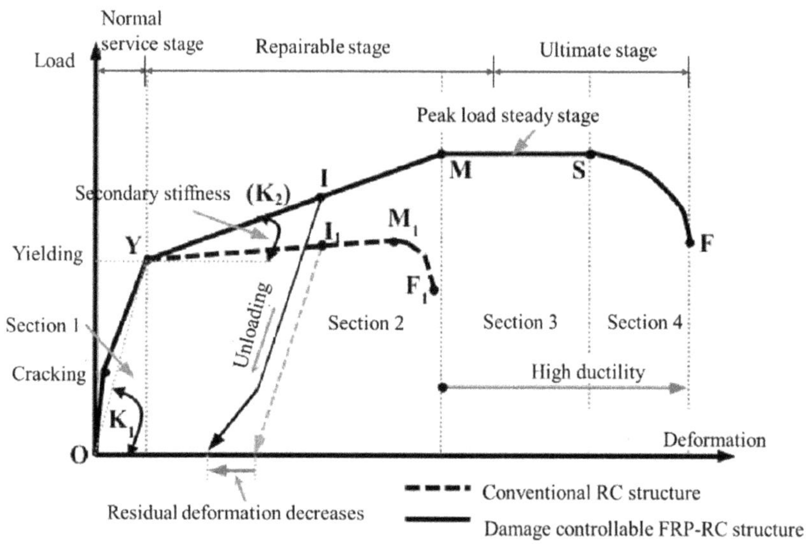

FIGURE 4.20 Load-displacement behavior of damage-controllable structure.

which is compounded by an inner ribbed steel bar and outer longitudinal FRP in a pultrusion process (Sun et al. 2014). SFCB has a second-order stiffness after yielding. Through horizontal cyclic loading tests, it can be observed that the SFCB column has stable post-yield stiffness, and its unloading residual deformation is much smaller than that of the RC column under the same loading displacement.

Apart from the replacement of longitudinal steel reinforcements, FRPs can also be used to replace steel stirrups. Steel stirrups are widely used in concrete columns. However, when subjected to a seismic action, plasticity might occur in the stirrups and their confinement of longitudinal reinforcements decreases seriously, leading to a buckling of longitudinal reinforcements. Due to the higher tensile strength and linear elasticity up to failure without yield plateau, FRP stirrups can be used to replace steel stirrups to effectively confine the longitudinal reinforcements. An integrated FRP stirrup is produced by mold pressing, and it can provide a better confinement on concrete and longitudinal reinforcements, compared with conventionally discontinuous FRP stirrups. The confinement effect of integrated FRP stirrups, arranged in the plastic hinge zone of a column, is verified by a finite element model. The skeleton curves indicate that the ductility improves with the increase of the modulus of FRP and the ductility can be improved up to 35% using a higher modulus of FRP. In addition, the numerical analysis also indicates that compared with steel stirrups, integrated FRP stirrups can effectively resist the buckling of columns.

Light-Weight Structures

Limitations of Long-Span Bridge

For the sustainable development of long-span bridges, limitations lie in the large self-weight of steel cables which limits the spanning ability and decreases the short service life to about 30 years which is lower than the design life of 100 years. As shown in Figure 4.21, the ultra-high strength steel cables are proposed to overcome these problems, which can relieve sag increasing spanning but not enhance longevity. At the same time, considering the superior performance of carbon fibers, CFRP cables are expected to achieve light weight and improve service life. However, it suffers from the limitation of the defects of the materials itself in terms of cost, brittleness, and vibration resistance; also the weak lateral resistance in the anchoring system differs from steel. Moreover, the existing design issues cannot reflect the high performance of the CFRP bridge.

High-Performance BFRP and Hybrid FRP Cables Developed

To meet the above needs, high performance cables are developed, which include BFRP, hybrid Basalt/Carbon, Basalt, and steel wires (see Figure 4.22) (Wang and Wu 2010; Yang, Wang, and Wu 2016). High-performance cables show superior short- and long-term properties closer to that of CFRP cables such as high strength

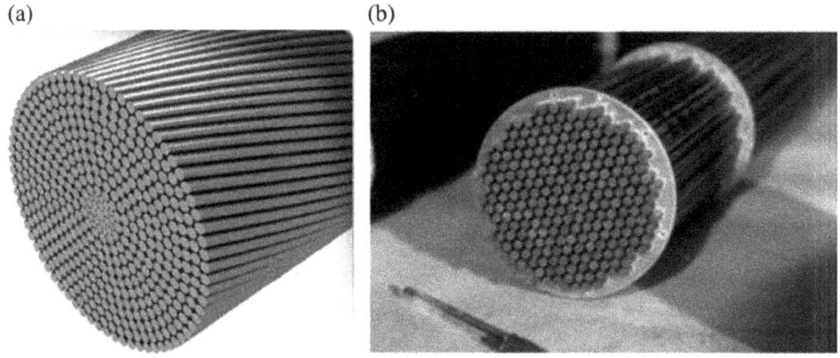

FIGURE 4.21 Ultra-high strength steel cable (a) and CFRP cable (b).

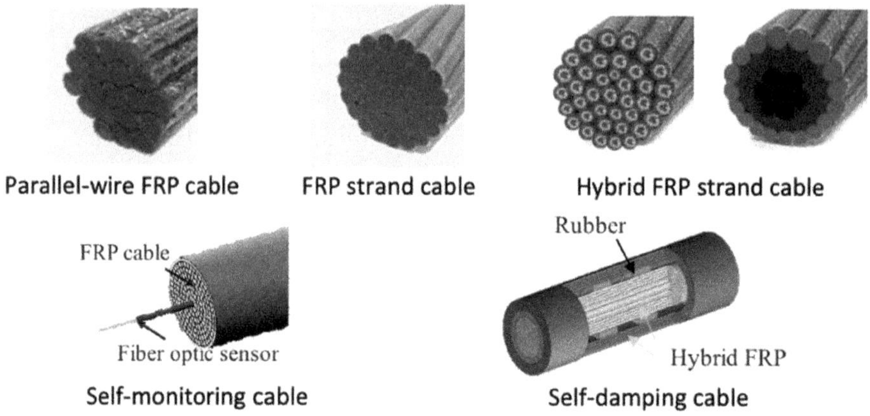

FIGURE 4.22 High-performance cables.

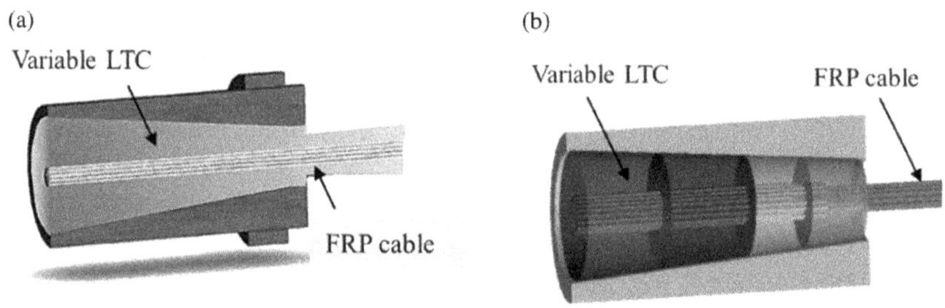

FIGURE 4.23 Variable-stiffness LTC. (a) Variable LTC by fiber winding, and (b) grouting variable LTC.

(1300–2000 MPa), strong fatigue resistance (45%–70% f_u), high creep rupture stress (52%–60% f_u), and low relaxation (<2.5%). Moreover, hybrid FRP cables can realize self-monitoring through fiber optic sensors, and also achieve self-damping function by adding rubber to the cables. Furthermore, as shown in Figure 4.23, the effective large-tonnage anchor systems for hybrid FRP cables are based on variable-stiffness LTCs that have been developed to show the potential strength of cables. Different amounts of optical fibers are wound onto the cable to gradually vary the stress applied to the cables to avoid stress concentration at the anchoring end. In the same way, stress distribution can be achieved by sectioning grouting at the anchoring end.

Realize Long-Span Bridges with Most Suitable Cables

For the structural design of cable-stayed bridges with long-span FRP, only one method of equal stiffness is used. Although the design method ensures the mechanical properties of the FRP cable-stayed bridges, the overall cost of the FRP cable-stayed bridges is limited and the economic performance is not improved obviously. The FRP cable-stayed bridge system is proposed based on the known basic

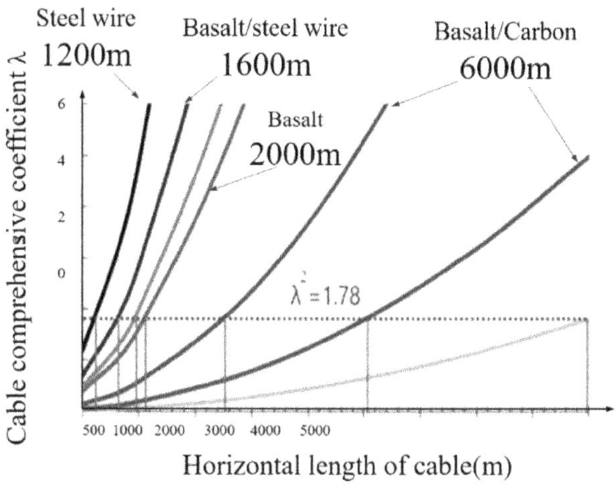

FIGURE 4.24 The applicable main span of cables with different materials.

properties of different FRP materials and the applicable span concept of the cables. By giving full play to the mechanical properties and economic properties of cable-stayed cables with different FRP materials, the optimal design of cable-stayed cables with large span is realized. According to the equivalent elastic modulus ratio and equivalent stiffness ratio of FRP cables with different materials, the applicable spans of different cables are shown in Figure 4.24. The applicable span of FRP cables with different materials can be determined. The applicable main span of the steel cables is 1200 m, while the applicable main span of the hybrid B/SFRP cables is 1600 m. Meanwhile, the applicable main span of the BFRP cables is 2000 m. With hybrid FRP cables, structures can realize a substantial increase in span up to 6000 m for cable-stayed bridges, and increase in the life span of cables from 30 years (steel) to 100–300 years.

The developed high-performance FRP cable systems and technology are used in the design of the Changtai Bridge crossing Yangzi River with a main span of 1176 m, which will be the world's longest combined highway and railway cable-stayed bridge by 2022. First of all, we need to understand the basic properties of the various different materials that make up FRP cables in order to meet the performance requirements of cables. First, carbon fiber cables are the preferred materials for the preparation of stable cables because of their excellent properties such as high strength, high elastic modulus, fatigue creep resistance, low thermal expansion coefficient, and corrosion resistance. However, the ultimate elongation of carbon fibers is low, and there is a possibility of brittle failure under extreme load conditions. Second, basalt fiber cables have outstanding comprehensive properties such as high strength, high elongation, corrosion resistance, fatigue creep resistance, low price, thermal expansion coefficient is only half of steel, and the stress loss under temperature is small; it is a good alternative material for the development of stable cables. Third, hybrid cables combine the properties of carbon fiber cables and basalt cables. Hybrid cables with C/BFRP=4:1 are hybrid design schemes that focus on strength and stiffness control,

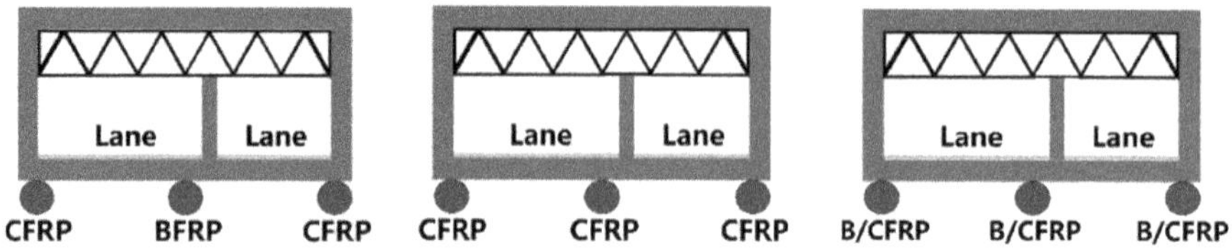

FIGURE 4.25 Design scheme of FRP cables.

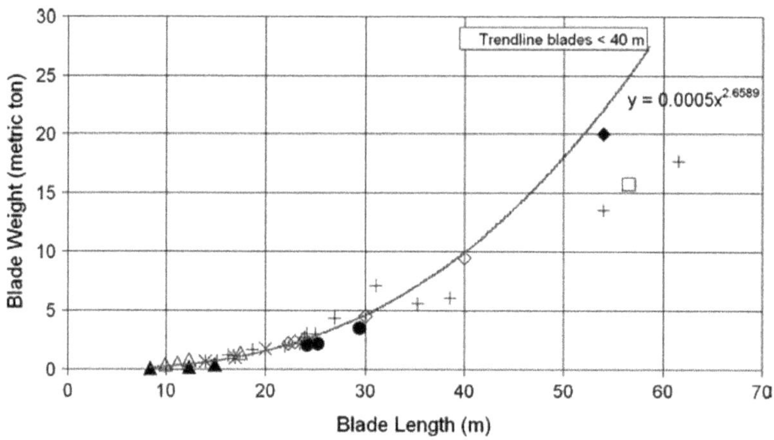

FIGURE 4.26 Development in rotor blade weight versus length. Symbols indicate different manufacturers and processing technologies (Brøndsted, Lilholt, and Lystrup 2005).

while hybrid cables with C/BFRP=1:4 are hybrid design schemes that focus on ductile failure. Based on the above analysis, three possible FRP cable design schemes have been proposed in Figure 4.25.

Large Wind Turbine Blade

The expansion of the renewable energy sector is required to reduce the dependence on fossil fuels. The EU offshore wind energy capacity is expected to grow by 21% annually. Thus, as shown in Figure 4.26, the major trend is to build large wind turbine blades to reduce the leverage cost of energy (Mishnaevsky et al. 2017). The typical sectional configuration of the blades is shown in Figure 4.27.

The blades with composite materials can reach high stiffness, are light weight, and long-time fatigue. Glass/carbon hybrid FRP composites are the most commonly used materials. The use of glass/carbon hybrid composites is a good option that comes up with a balance of cost and performance. To date, the 100m-scale blades have been produced with carbon FRP to match with the megawatt-scale wind turbines. Meanwhile, more types of materials, such as aramid, basalt, and other natural fibers, have emerged as alternatives. More hybridization choices using these materials are available to satisfy different engineering requirements.

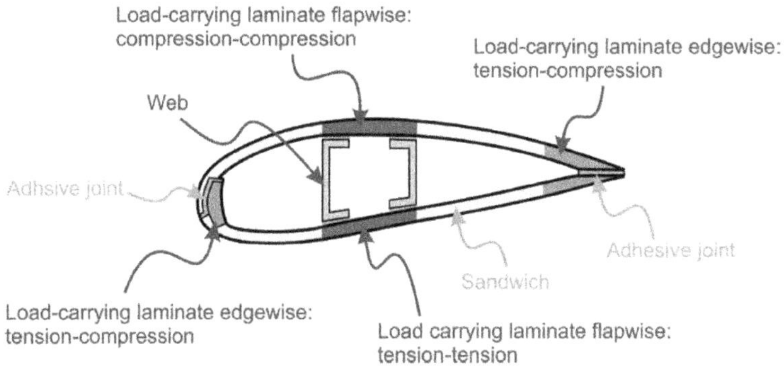

FIGURE 4.27 Typical schema of the blade section (Mishnaevsky et al. 2017).

Recently, an interesting solution was developed to reuse the component of the waste large turbine blades as building materials (Oliveux, Dandy, and Leeke 2015; Bank et al. 2019). Bank et al. preliminarily analyzed the behaviors of building structures made from parts of large wind turbine blades. The relevant codes need to be proposed in the future.

Challenges in Developing Smart Infrastructures

Challenges of Smart Infrastructure and Smart Materials

Generally, the civil infrastructure systems, such as buildings, tunnels, bridges, pipelines, tall chimneys, and dams, are the main and most expensive investment in most countries. However, these systems are deteriorating at an alarming rate due to the forces of nature such as earthquake, adverse environmental conduction and flood, as well as man-made actions such as terroristic attack, excessive or improper usage, and lack of maintenance and repair. Until recent years, sophisticated and precise sensors and actuators are mainly used in the fields of aerospace structures, machines, robots, armament, and so on; and most of today's infrastructure systems are in a "deaf, dumb and blind" state. The introduction of high-precision sensing technology, intelligent materials, and innovative design approaches in civil systems has been deferred due to high costs, heavy reliance on traditional construction and maintenance practices, and the fairly conservative nature of design codes. Recently, some sensing technology and advanced materials have turned to civil engineering structures. The rapid development of new materials, structural sensing and control, data acquisition and processing technologies may result in a new generation of structural systems, which are characterized by health monitoring, diagnosis, damage detection, and self-cure functions. North American, Asian, and European countries have begun the research and development of high-precision sensing technologies, intelligent materials, and smart structural systems with self-diagnostic and auto-adaptive functions, partly aimed at removing, rehabilitating, or strengthening the rather obsolete infrastructure systems with additional structural health monitoring (SHM) and fusing functions, as well as developing new smart structural systems.

The development of SHM technology, and smart materials and structural systems has attracted extensive attention from civil engineering, aerospace, and mechanical industries. Around the world, every year many international conferences and symposiums have been held on the topics of SHM, smart material and structure, rehabilitation and strengthening of civil infrastructures such as the International Symposium on FRP Reinforcement for Concrete Structures that is held every two years from 1993, Structural Health Monitoring that is held every two years from 1997 at Stanford, USA, International Conference on FRP Composites in Civil Engineering held in 2001 in Hong Kong, China, International Workshop on Structural Health Monitoring of Innovative Civil Engineering Structures every two years from 2002 in Canada, International Conference on Structural Health Monitoring and Intelligent Infrastructure every two years from 2003, International Workshop on Advanced Smart Materials and Smart Structures Technology, and many other regional seminars, workshops, and symposiums.

Smart materials are considered as a countermeasure to be applied in SHM technology or to devise more sensitive methods to inspect structures in real time for signs of abnormality, with a purpose of improving the safety and durability of composites. In 1989, based on the combination material, a concept of smart material with sensing and control functions was proposed in Japan that can sense and response to environmental changes.

Generally, smart materials can be divided into two types: passive smart materials with only sensing and response functions to environmental changes, and active smart materials with sensing, response, and control as well as adaptation to environmental changes. Factually, maintenance of engineering structures based on SHM techniques is similar to the health maintenance of human beings. First, discomforts such as pain and itch are perceived via a human nervous system throughout the whole body. Then, the discomforts are primarily diagnosed and located, and decisions are made about whether and when to see a doctor. Finally, more precise diagnosis is made in special hospitals, and appropriate cure measure is taken. The maintenance of practical structures based on SHM techniques also consists of steps such as structural sensing with distributed sensors similar to the "nervous" systems of human beings, diagnosis based on the sensing data that is completed in the "brain" control center, taking appropriate countermeasures that may be carried out automatically by actuators or manually by the "muscle" system infrastructures, and an active and adaptive structural system that act as the "skeleton" to support the whole structure. The fundamental concept of "brain," "muscle," "skeleton," and "nervous" systems for smart infrastructures is schematically depicted in Figure 4.28.

Challenges of Structural Health Monitoring and Damage Identification Technologies

Over the past two decades, SHM systems have been devised, implemented worldwide, and continue to evolve, improve, and monitor the structural performance and operational conditions of various types of engineering structures under their in-service life. The overall purpose of an SHM system is to monitor the in-situ behavior of a structure accurately and efficiently to determine its health or physical condition. A successful SHM strategy is one that is able to promptly and accurately examine the status of the integrity of a structure using embedded or attached sensors, and

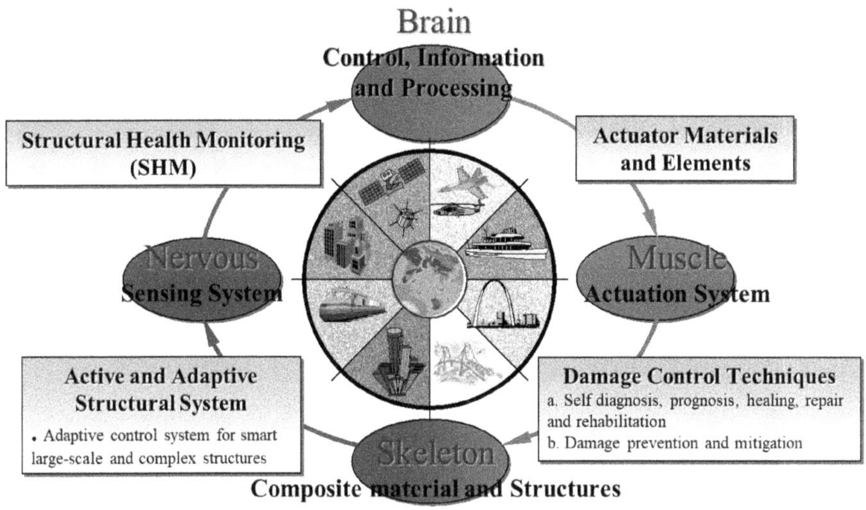

FIGURE 4.28 Brain, muscle, skeleton, and nervous systems for smart infrastructures.

intelligently utilize the data to assess the state of the structure. A SHM system typically comprises two essential components: (1) a sensor network, and a system of data acquisition, data processing, communication, management, and storage; and (2) a data analysis and interpretation, system identification, local and global diagnostic, damage detection, and residual life prediction system.

While the field of SHM has witnessed significant advancements over the past two decades and a wide range of SHM methodologies have been introduced and implemented, there still remains challenges to be addressed in both areas of sensing and signal processing, diagnostics, and analysis methods that hinder the wide application of SHM in engineering practices. For instance, it has been well recognized that the existing SHM systems still have difficulty to identify some structural damages, especially minor damages, such as early corrosion. This is due to the fact that there are two bottlenecks in currently used intelligent sensing techniques. These include: (1) structural damages cannot be effectively analyzed since the current sensing technology has shortcomings in dealing with massive data; and (2) the sensor monitoring device has inadequate comprehensive performance: single function or single performance, and poor durability.

Area-Wise Distributed Sensing and Monitoring: Concept, Theory, and Solutions

The objectives and the purpose of structural monitoring have to be established according to the requirements set forth by the asset management authorities, emanating from the expectations from public, the owner, and the users. The basic requirements for SHM can be summarized as follows: (1) it can accurately and efficiently monitor various health indicators, including strain, deflection, rotation angle, modal parameters, damage, and load; (2) it can identify various types of damage and accurately grasp the "mental state" of the structure; (3) it can predict the structural

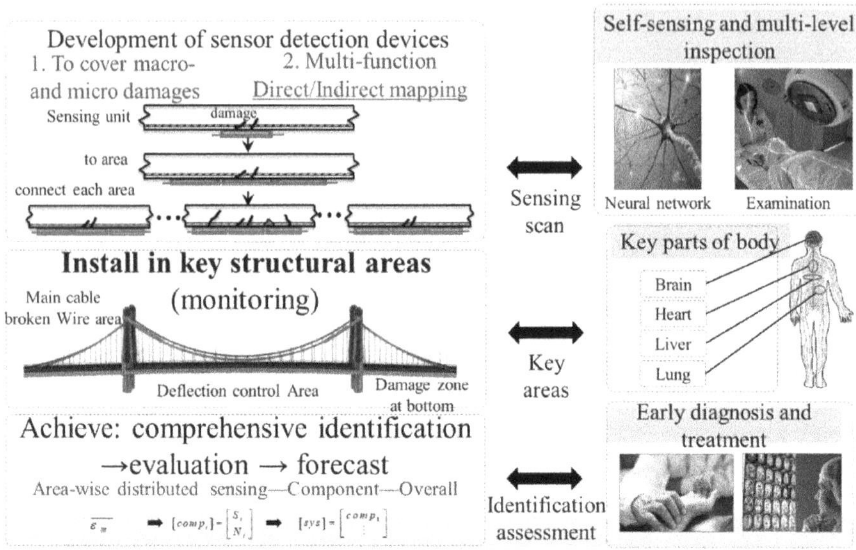

FIGURE 4.29 Concept of area-wise distributed sensing methodology.

life expectancy. In order to address the aforementioned requirements and challenges in the monitoring of civil engineering structures, the concept of area-wise distributed sensing and monitoring for a structure or structural groups is presented in this paper.

In many respects, SHM is analogous to monitoring the health of human body (see Figure 4.29). Both human body and engineering structures have key components. The nervous system in human body comprises a central system, or the command center, consisting of the brain and the spinal cord. In addition, there is a peripheral nervous system which includes a network of nerves that connects the rest of the body to the central system. This complex sensing system collects information from inside the body and from the environment outside the body, such as heat, taste, and noise, processes the collected information, and then dispatches instructions to the rest of the body, facilitating an appropriate response. For smart structures, the arrangements and the number of sensing units are limited and, comparatively speaking, will never be able to reach the magnitude or the effective distribution, or facilitating the response, that would mimic the human body and its nervous system. Although the human body's nervous system is complex, however, human body has a few major organs such as heart, liver, lung, kidney, and others whose health status indicates whether the state of the body is safe, functions properly, and conducts all normal human activities and expected behaviors. It is always a priority to monitor and maintain the health of these vital organs whose health and safety assures the overall health of human body. Similar to human body, civil engineering structures are complex; however, they also have key components and critical areas whose health and safety assures the safety of the structures. Therefore, monitoring these critical areas is similar to monitoring the major organs in a human body, in order to assure the safety and health of a structure. These areas which can easily suffer structural damage, and play an important role in structural performance, are known as critical areas. If distributed sensing systems are placed in these critical areas (area-wise distributed sensing), through dynamic and static, macro and micro monitoring, and multi-level analysis, structural damage and potential failure can be detected with more accuracy, and hence, this will result in early warning. Subsequently, a comprehensive identification of structural state parameters, damage state and loads, and direct evaluation and prediction of structural performance can be achieved.

As discussed above, it is apparent that new sensing technologies need to be developed for the successful implementation of the aforementioned area-wise distributed monitoring which has many unique characteristics. On one hand, the concept of area-wise distributed monitoring requires that the required sensing tool has the capability of macro/micro damage coverage. Traditional strain gauges deployed in an area of a structure may output much different strain values because they are sensitive to local information such as cracks. The proposed distributed strain sensing techniques put forward a promising, novel manner to catch the comprehensive structural information in virtue of the "distributed" sensors' placement and high sensitivity to local damages. However, when the strain response is very small or environmental noise is heavily disturbing, it may be difficult to obtain the effective measurements. Furthermore, as it is well known that ambient vibration testing which can work under the normal operational state of the structures plays a more important role than the traditional controlled experiment in the current large-scale civil SHM, the developed sensors are increasingly expected to improve the ability to enhance the measuring sensitivity and obtain the small ambient vibrating responses. Therefore, we have dedicated special effort toward developing a feasible distributed long-gauge strain sensing system for practical adaptation to civil structural engineering (see Figure 4.30) (Adewuyi and Wu 2010; Hong et al. 2015; Li and Wu 2007, 2008, 2010; Sheng et al. 2010; Wu and Li 2007).

One important advantage of the long-gauge strain sensing is that the damage can automatically be located by the length of the sensor gauge. Therefore, the identification of micro- and macro-damages can be accomplished using a modal macro-strain vector (MMSV). By adapting to the dynamic strain distribution measurements from

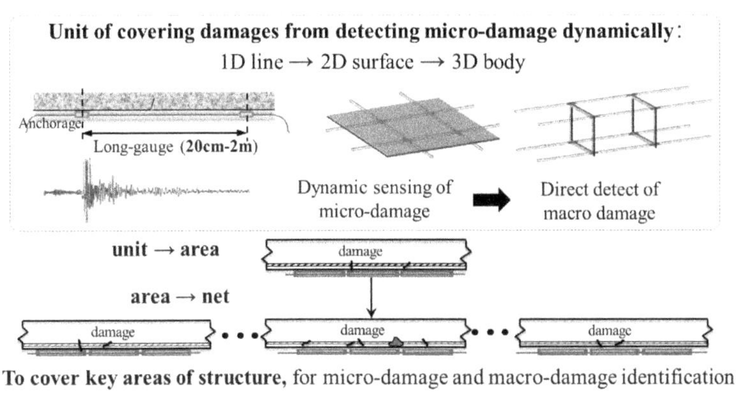

FIGURE 4.30 Vibration-based damage identifications of micro- and macro-damages based on long-gauge strain sensing.

long-gauge strain sensing, the structural dynamic properties can be measured more accurately, reliably, and conveniently. Specifically, strain mode shapes are analogous to the curvature mode shapes. Moreover, the modal curvature can be calculated directly without the central difference method which introduces numerical errors that amplifies the inherent errors in the conventional mode shapes. Consequently, these techniques can provide the required accuracy and reliability needed for complex system identifications of civil infrastructure such as suspension bridges, cable-stayed bridges, and large-span spatial structures.

There are many similarities between structural health monitoring and human health monitoring. To ensure our health state we need to go to the hospital to take kinds of sophisticated examinations and to diagnose whether there are cancer and other diseases in the liver, the lung and other key organs of our bodies. Earlier cancer cells are detected, sooner treatment can be started. Same as the accurate and comprehensive examinations in human health management, the first objective of a good structural health management needs to provide a solution for detecting early deterioration in each key area of the structure. However, in spite of the exhaustive research efforts devoted to SHM methodologies, only a few of SHM methods have been fully validated for applicability on large-scale civil structures. For example, natural frequencies are easy to measure with relative accuracy, but are too "global" to detect local damage. Likewise, mode shapes and their derivatives are spatially specific and more sensitive to local damage, but are very prone to measurement or numerical noise. They cannot provide the required accuracy to identify early damages such as scour, corrosion, torsion, and support failures. As illustrated in Figure 4.31, structural damage, such as broken steel wires, is like the development of cancer, which is slowly increasing in the early stage. Only when the damage intensifies to a certain degree, the degradation of structural performance will be accelerated in the mid- and late-stage. According to existing specifications, cables need to be replaced when the broken wire rate exceeds 2%. However, according to calculation and simulation, it is found that for 2% of broken wires, the frequency change generally does not exceed 0.8%.

To overcome this situation, a combination of long-gauge strain sensing and macro-strain mode analysis is very essential for excellent damage identification. Based on area-wise distributed sensing methodology using long-gauge strain sensors as

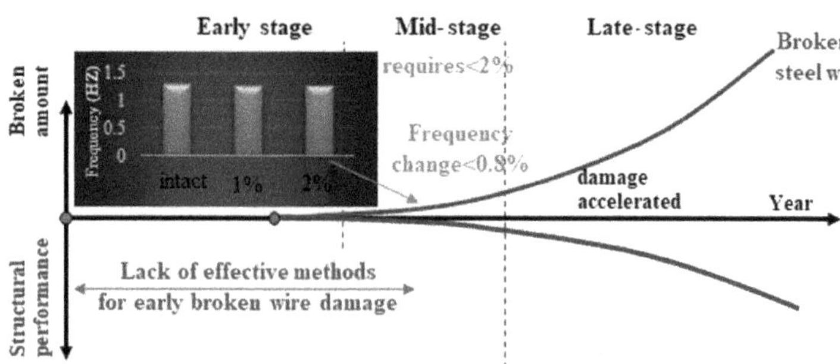

FIGURE 4.31 Relation of damage degree and structural performance degradation in different stages.

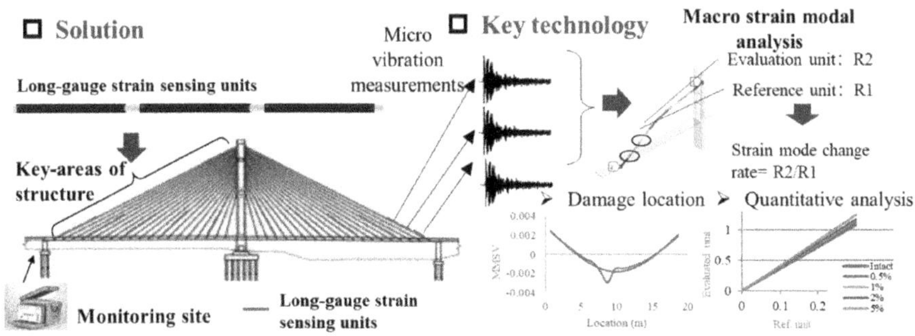

FIGURE 4.32 Solution illustration of broken wire monitoring.

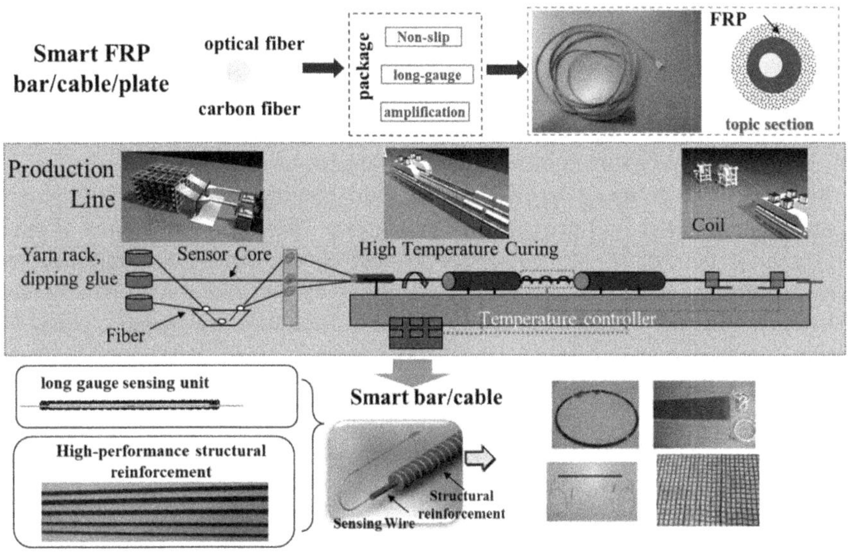

FIGURE 4.33 Production of smart FRP bar/cable/plate/sheet.

illustrated in Figure 4.32, MMSV methods based on macro-strain modal analysis are highly sensitive indicators of small realistic damages and can locate multiple damages quite accurately.

By designing a gauge length usually between tens of centimeters and several meters, the sensing materials can be optical fibers or carbon fibers, the long-gauge strain sensors can monitor average strain in a relatively large area and are more likely to detect early damages (Fouad et al. 2017; Huang and Wu 2012). A practical model of smart FRP composite consists of host FRP composite, sensors, actuators, and a controller. The sensors and actuators are embedded into FRP composites. The abnormality such as the breakage of fibers, delamination between layers, impact loading, and matrix cracks can be detected with the use of sensors. Then, damages are characterized, mechanical condition is evaluated, and command is given by the processor. Finally, proper counter action is taken with the use of actuators. The sensors and actuators embedded into FRP composites should have the following characteristics: (1) sensors and actuators should have very small size in comparison with that of FRP composites for the ease of embedment; (2) they should act harmoniously with FRP composites under environmental changes; (3) the embedment has not obvious influence on the properties of FRP composites. Through our technological breakthrough of the above characteristics (see Figure 4.33), now we have mastered the ability to continuously produce different types of smart FRP materials. Consequently, these smart materials provide more solutions with high operability and simple construction for structures such as roads, landslides, and tunnels.

Case Examples

The sensing technique and strategy thoroughly described above, an area-wise distributed monitoring method, for structural identification and performance evaluation, can be integrated with an SHM system to reveal hidden flaws in a structural state. Moreover, it allows for an SHM system to be designed for infrastructural groups, not only a single structure, and provides a viable method to weave sensing entity into a smart cyber-physical network to manage infrastructure at a collective level and to help study regional factors influencing structural performance.

Infrastructural groups can be classified into continuous and discrete groups. Continuous infrastructural groups, such as highway networks, are defined as structural entities bonded by strong structural connections. On the contrary, discrete infrastructural groups such as architectural groups do not share a strong structural connection and each entity is independent, except that they are spatially close. For continuous infrastructural groups, the SHM system is divided into units and unit stations along the strong structural connection. Each unit relates to a structural entity, while each unit station manages the monitoring of several units. The topology of unit-station-monitoring center adopts a bus structure or a tree structure. In the case of discrete infrastructural groups, each structure corresponds to one unit and all units respond to the monitoring center directly. The topology of unit-monitoring center is a star structure.

Deformation Distribution of a Long-Span Rigid Frame Bridge

Ambient vibration test and analysis of a long-span rigid-frame bridge through the long-gauge fiber optic sensor is investigated, in which calculation of the deformation distribution and identification of the structural modal parameters from the measured long-gauge strains are performed. The bridge is a pre-stressed concrete continuous rigid-frame box girder bridge with the span length of 268 m (see Figure 4.34(a)). The cross-section of the girder is a single box with a single chamber. Cracks and overlarge deflections are two typical problems for general rigid-frame bridges. The FE analysis results indicate that the upper plates of the main girder at both ends of the middle span are the areas easy to have cracks and the bottom plate at the center of the middle span is the area easy to have overlarge deflections. Thus those areas were determined as critical areas and the long-gauge FBG sensor networks were installed on those areas as shown in Figure 4.34(b), in which totally 52 long-gauge FBG sensors are used. The sampling rate was set to be 200 Hz.

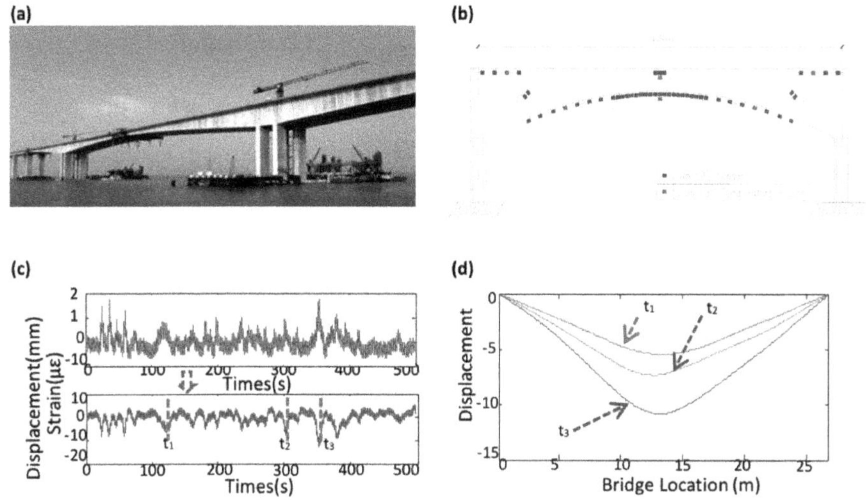

FIGURE 4.34 Area-wise distributed monitoring of the rigid-frame bridge: (a) Bridge photo, (b) sensor layout, (c) measured long-gauge strain and identified deflection. (d) Deflection distribution.

The deformation distributions of the middle span were calculated from the measured long-gauge strains through the method described previously. Figure 4.34(c) illustrates a typical long-gauge strain and the corresponding calculated deformation near the central node of the middle span during ambient vibrations. It is seen that the largest strain and displacement are 17.45 με and 12.69 mm, respectively. The deformation and the strain of the span are small because the studied rigid-frame bridge is stiff. Picking up the displacement values along the middle span at the same time, the shape of the bridge deformation can be plotted. Figure 4.34(d) plots the bridge deformation shapes at time t_1, t_2, and t_3 marked in Figure 4.34(c) It is seen that the maximum values of the span are 5.498 mm, 7.366 mm, and 10.92 mm at t_1, t_2, and t_3, respectively. The deformation distributions of the studied bridge outputted from the measured long-gauge strains provide value information for bridge engineers and owners to understand the bridge performance.

Other than deformation distribution calculation, structural modal parameters including frequencies, damping, strain mode shapes, and displacement mode shapes can be identified from long-gauge strain measurements. Picking up the peak values of strain FRPs corresponding to the first frequency at 0.66 Hz, the long-gauge strain mode shape was identified, in which the FE modal analysis results are also plotted as a reference. The merit of using the long-gauge strain for modal analysis is that not only strain mode shapes but also displacement mode shapes are identified by using the modified conjugate beam method as shown in Figure 4.35.

Seismic Monitoring and Damage Detection of a Reinforced Concrete Continuous Railway Bridge

The monitoring of a reinforced concrete railway bridge was considered as an engineering example to illustrate the damage identification method via damage fingerprint.

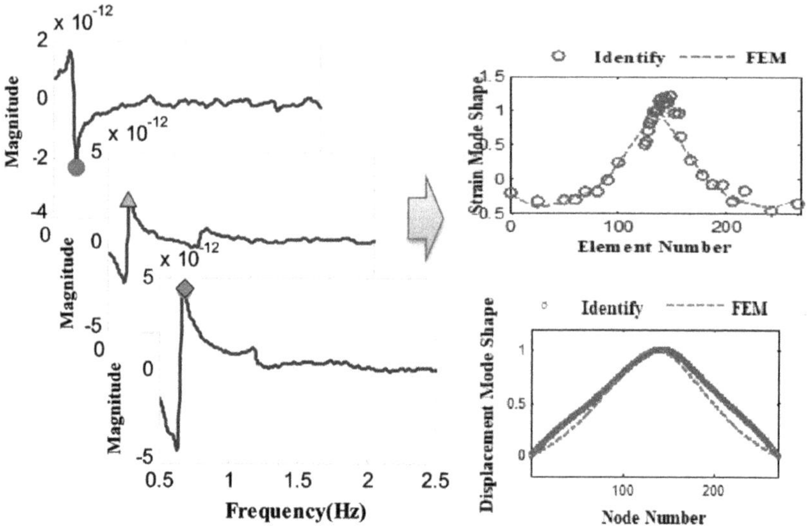

FIGURE 4.35 Modal identification results of the rigid-frame bridge.

Compared with highway bridges, this reinforced concrete continuous railway bridge has less flexibility and small strain. In order to assess the structural condition, the area-wise distributed monitoring was installed in four key areas on the bridge and the piers were covered by long-gauge FBG sensors. The axial and the tangential strains of the beam and the column were measured to infer maxima strain changing, deflection, natural frequency, damping ratio, and modal strain vectors. Comparing with the above-mentioned damage indexes, their results reveal that the modal strain vector from long-gauge distributed strain measurements is more efficient to identify minor structural damages. With the modal strain vectors, it was found that some micro-cracks occurred during an earthquake with a seismic intensity of 4, while the other traditional structural indexes have not obviously changed. The identified structural status before the earthquake is shown in Figure 4.36(a).

The identified structural status during the earthquake is shown in Figure 4.36(b–e). It is clear that the slope of the fitting line – namely the amplitude ratio of strain unit 6 to strain unit 1 – has changed, which means the observed earthquake caused damage to the structure. These cracks were confirmed by the in-situ inspection. Moreover, the later monitoring results showed that the structural responses of the bridge were almost the same as before. It means the crack is closed after the earthquake due to the self-weight and/or internal stress, and it may only be identified with the structural responses during large earthquakes. The successful discovery of this tiny damage proves the effectiveness of the method even to tiny damages.

A Long-Term Monitoring System of a Multi-Span Road Bridge

A road bridge is a topical structure in SHM of civil infrastructures, so an area-wise distributed monitoring system of a multi-span road bridge is presented herein as a

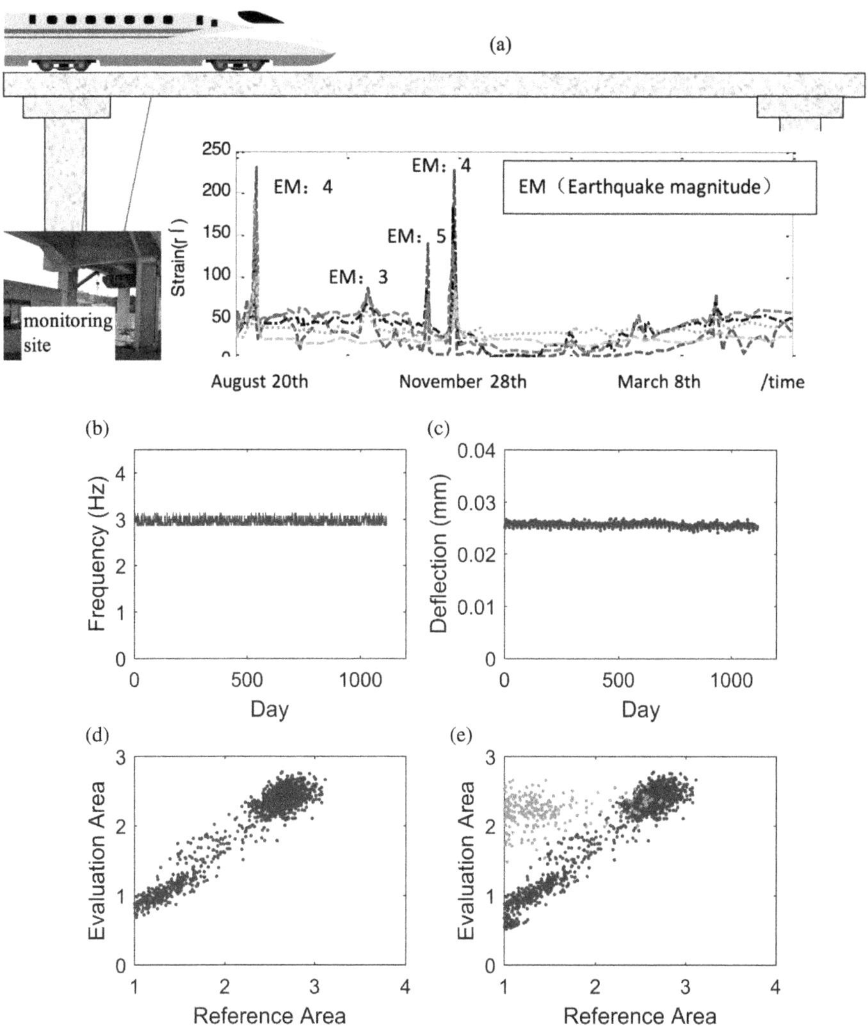

FIGURE 4.36 Damage identification of a railway bridge: (a) results of seismic monitoring; (b) results of natural frequency; (c) results of deflection; (d) modal macro-strain vectors before earthquake; and (e) modal macro-strain vectors during earthquake.

case study to explain how to examine vehicle-induced vibrational responses and topical traffic features based on SHM data (Huang and Wu 2017). The bridge object is a multi-span prestressed box-girder road bridge with a total length of 300 m and four spans of 64.2–85 m. It was constructed in 1972, and is a key national route in Japan which approximately carries 15,000 vehicles per day. During the periodic inspection in years 2006 and 2009, multiple sites of concrete deterioration and steel corroions had been detected. In some girder bridges it was found that at least 12 steel rebars were already broken. In 2011, an external steel strand used strengthing was implemented to

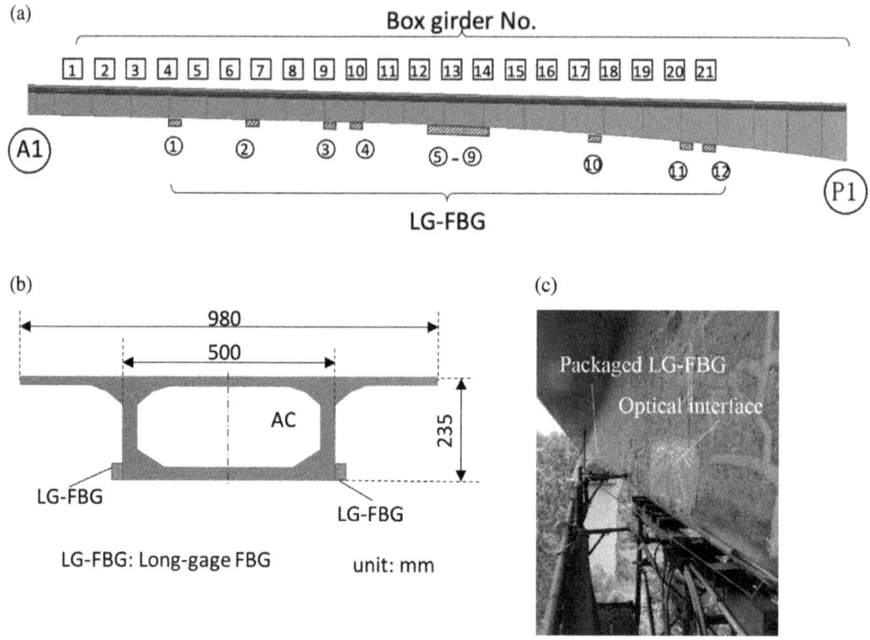

FIGURE 4.37 Area-wise distributed monitoring system using long-gauge FBG strain measurements: (a) side view of sensing locations; (b) section view of sensing locations; (c) photograph of packaged sensors.

prevent bridge collapse. In this study, a long-gauge FBG-sensors-based area sensing system was installed in the bridge before the strengthening project. The main purpose of the sensing system was to examine and identify the structural performance changes before and after the strengthening, and to continuously monitor the service performance. Four key monitoring areas from the full bridge body were picked; the placement of sensor locations in one of the areas is shown in Figure 4.37(a). Both sides of the girder bridge include 12 sets of 1-m-length sensor units, as shown in Figure 4.37(b) and (c).

To address the effect of the implemented strengthening project, results of the measured strain distribution before and after strengthening under the same static inspection test (six 20-ton trucks) were compared. It shows that the measured strain distribution after strengthening are notably larger than the results before strengthening, as shown in Figure 4.38(a). Moreover, the percentage ratios of the measured results under the same vehicle loading from 2011 to 2017 are shown in Figure 4.38(b). The major damage type was noted to be steel corrosion in the pre-stressed concrete girder, and the strengthening project was designed mainly to prevent bridge collapse. However, the damage degree of the steel tendons in the longitudinal and transverse directions was not uniform. Based on the measured strain distributions, it can be concluded that the increased strain is caused by a stiffness reduction which occurred after the strengthening project.

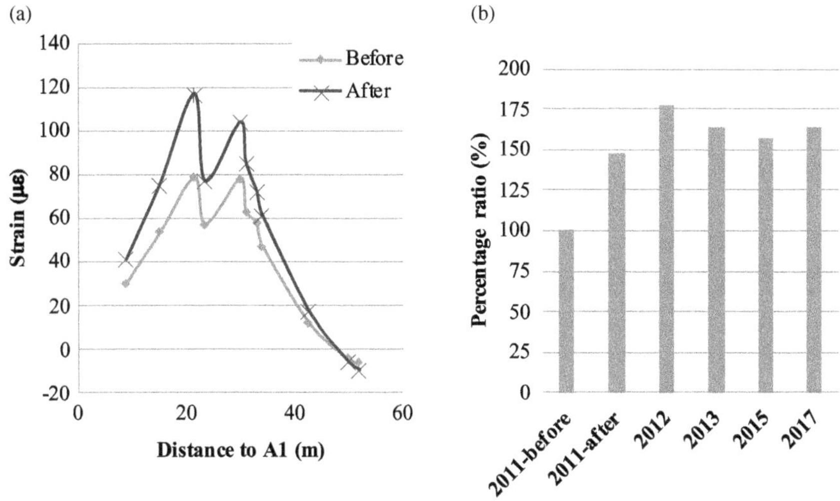

FIGURE 4.38 Strain distrbution under static inspection test: (a) comparison of before and after strengthing; and (b) percentage ratio of strain changes.

Application of Structural Group Monitoring

Urban roads and rail transit structural groups are faced with problems such as high safety requirements, low level of information technology, low ability of emergency warning and emergency control, terrible service performance deterioration, and huge maintenance costs. The safety and health monitoring of these groups can be realized using the areas distributed sensing methodology based on the following considerations: (1) a full urban road and rail traffic structural group can be categorized into multiple key sub-groups; (2) then, key monitoring objects in every key sub-group, such as bridges, tunnels, and buildings, can be determined based on the importance of each structure; (3) finally, key distributed sensing of these import structures will be carried out using structural and fragility analyses. Comprehensive identification and direct evaluation of structural groups can be realized through data analysis and effective monitoring of the entire structural group. That is to say, the area-wise distributed sensing system performs a comprehensive monitoring of the important parts, apart from those with potential risk damage in discrete structural groups combined with the scatter monitoring of other parts. Health monitoring data center for long-span bridges in the Yangtze River area is shown in Figure 4.39; various sub-networks are connected to the data center with IOT technologies. Thus, early safety/health monitoring and emergency warning of each bridge in sub-networks can be established based on area-wise distributed sensing methodology. Furthermore, with this information, comprehensive structural performance evaluation and deterioration analysis and residue life prediction based on big data processing and deep learning methods can be carried out.

Conclusions and Prospects

Infrastructure sustainability with economic/durable and green/health design guidelines can be realized through LCC minimization, longevity, low carbon emission, resilience,

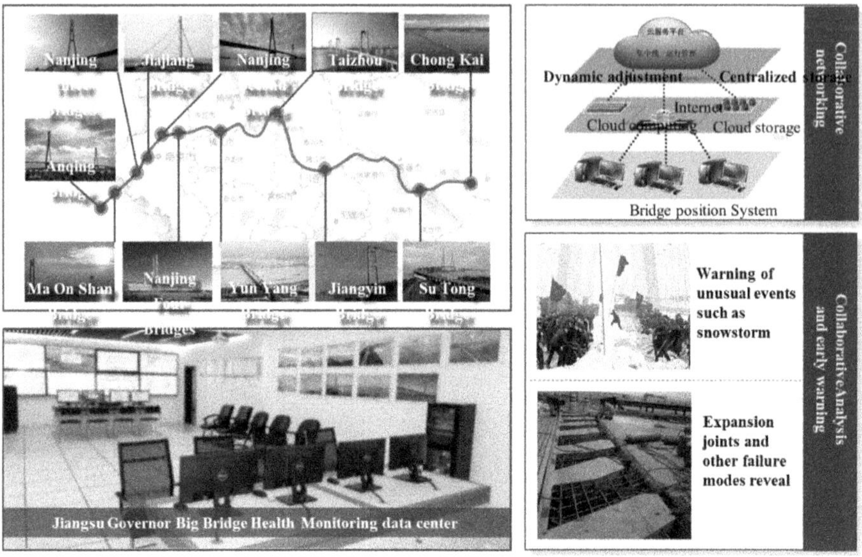

FIGURE 4.39 Health monitoring data center of structural group (11 long-span, 6-kilometer-class bridges).

smart designs with material innovation and ICT. Structural recoverability should be one key index of structural resilience which is an index of structural sustainability.

Focusing on the limitations of conventional construction, the advanced BFRP composites were developed to realize the high performance and longevity of structures. Based on structural performance needs and material properties, the short- and long-term behaviors of FRPs should be enhanced in an integrated manner. Concrete structures reinforced with integrated FRP composites can be enhanced for load carrying capacity, stiffness, ductility, recoverability, and durability. Lightweight long-span structures with advanced FRP cables possess superior static and dynamic behaviors. An intelligent structure with smart FRPs is a promising direction in achieving the potential advantages of FRP and multi-functional structures.

Based on the experience and lessons from developed countries, it is imperative to reduce the maintenance and management costs by establishing an intelligent infrastructure and an asset management system, in order to realize the high safety and longevity of the bridges. One of the keys to prolonging the safety of the infrastructure, such as bridges, is to solve the monitoring precision of damages. Compared with the passive maintenance and prevention maintenance, the effectiveness of predictive maintenance has been achieved recently in many countries. At the same time, based on the automatic detection and health monitoring technology, predictive maintenance management has been gradually recognized in all aspects.

On the other hand, a state-of-the-art methodology called structural area-wise distributed monitoring using a long-gauge sensing technique was introduced. The concept of structural area-wise distributed sensing, which makes up for the inefficiency of the local and macroscopic sensing of the structure, and the area-wise distributed strain data can be effectively used for the early detection of damage and the comprehensive assessment of the structural state. The structural area-wise distributed monitoring methodology has been successfully verified by many practical applications, while it plays an important role in identifying damages at the earliest stage to establish proper maintenance and/or retrofit plans on time.

REFERENCES

Adewuyi, A.P., and Wu, Z.S. 2010. "Modal macro-strain flexibility methods for damage localization in flexural structures using long-gage FBG sensors." *Structural Control and Health Monitoring* 18 (3): 341–360.

Al-Tayyib, A. J., Shamim Khan, M., Allam, I. M., and Al-Mana, A. I. 1990. "Corrosion behavior of pre-rusted rebars after placement in concrete." *Cement and Concrete Research* 20 (6): 955–960.

Arce, G.A., Hassan, M.M., Mohammad, L.N., and Rupnow, T. 2017. "Characterization of self-healing processes induced by calcium nitrate microcapsules in cement mortar." *Journal of Materials in Civil Engineering* 29 (1): 04016189.

Assi, L.N., Deaver, E.E., and Ziehl, P. 2018. "Effect of source and particle size distribution on the mechanical and microstructural properties of fly ash-based geopolymer concrete." *Construction and Building Materials* 167: 372–380.

Bakis, C.E., Bank, L.C., Brown, V.L., Cosenza, E., Davalos, J.F., Lesko, J.J., Machida, A., Rizkalla, S.H., and Triantafillou, T.C. 2002. "Fiber-reinforced polymer composites for construction – state-of-the-art review." *Journal of Composites for Construction* 6 (2): 73–87.

Bank, Lawrence C., Arias, F.R., Gentry, R., Al-Haddad, T., Tasistro-Hart, B., and Chen, J.F. 2019. "Structural analysis of FRP parts from waste wind turbine blades for building reuse applications." In *Advances in Engineering Materials, Structures and Systems: Innovations, Mechanics and Applications*, edited by Alphonse Zingoni, 1520–1524. London: CRC Press.

Bocchini, Paolo, Frangopol, Dan M., Ummenhofer, Thomas, and Zinke, Tom. 2013. "Resilience and sustainability of civil infrastructure: toward a unified approach." *Journal of Infrastructure Systems* 20 (June 2014): 2. https://doi.org/10.1061/(ASCE)IS.1943-555X.0000177.

Bondar, D. 2013. "Geo-polymer concrete as a new type of sustainable construction materials." In *Proceedings of the Third International Conference on Sustainable Construction Materials and Technologies (ICSCMT)*, edited by Peter Claisse, Esmaiel Ganjian and Tarun Naik. Coventry University. http://www.claisse.info/2013%20papers/data/e012.pdf

Brøndsted, Povl, Lilholt, H., and Lystrup, A. 2005. "Composite materials for wind power turbine blades." *Annual Review of Materials Research* 35 (1): 505–538.

Cailleux, E., and Pollet, V. 2009. "Investigations on the development of self-healing properties in protective coatings for concrete and repair mortars." In *Proceedings of the 2nd International Conference on Self-Healing Materials*. Chicago, IL, USA.

Cao, S., Wang, X., and Wu, Z. 2011. "Evaluation and prediction of temperature-dependent tensile strength of unidirectional carbon fiber-reinforced polymer composites." *Journal of Reinforced Plastics and Composites* 30 (9): 799–807.

Çopuroğlu, O., Schlangen, E., Nishiwaki, T., Van Tittelboom, K., Snoeck, D., De Belie, N., and de Rooij, M.R. 2013. "Experimental techniques used to verify healing." In *Self-Healing Phenomena in Cement-Based Materials*, edited by M.R. de Rooij, K. Van Tittelboom, M. De Belie, and E. Schlangen, pp. 19–63. Dordrecht: Springer.

Dinakar, P., Sahoo, P.K., and Sriram, G. 2013. "Effect of metakaolin content on the properties of high strength concrete." *International Journal of Concrete Structures and Materials* 7 (3): 215–223.

Ding, Y., Yu, J., Yu, K., and Xu, S. 2018. "Basic mechanical properties of ultra-high ductility cementitious composites: from 40 MPa to 120 MPa." *Composite Structures* 185: 634–645.

Fouad, N., Saifeldeen, M.A., Huang, H., and Wu, Z.S. 2017. "Corrosion monitoring of flexural reinforced concrete members under service loads using distributed long-gauge carbon fiber sensors." *Structural Health Monitoring* 17 (2): 379–394.

Francis, Royce A., and Bekera, Behailu B. 2014. "A metric and frameworks for resilience analysis of engineered and infrastructure systems." *Reliability Engineering & System Safety* 121 (January 2014): 90–103

Gan, F., Sun, Z.-W., Sabde, G. and Chin, D.-T. 1994. "Cathodic protection to mitigate external corrosion of underground steel pipe beneath disbonded coating." *Corrosion* 50 (10): 804–816. doi.org/10.5006/1.3293470.

Hao, H., and Li, J. 2019. "Sustainable high-performance resilient structures." *Engineering* 5 (2): 197–198.

Hasanbeigi, A., Menke, C., and Price, L. 2010. "The CO_2 abatement cost curve for the Thailand cement industry." *Journal of Cleaner Production* 18 (15): 1509–1518.

Holling, C.S. 1973. "Resilience and stability of ecological systems." *Annual Review of Ecology and Systematics* 4: 1–23.

Hong, W., Zhang, J., Wu, G., and Wu, Z. 2015. "Comprehensive comparison of macro-strain mode and displacement mode based on different sensing technologies." *Mechanical Systems and Signal Processing* 50–51: 563–579.

Huang, H., and Wu, Z.S. 2012. "Static and dynamic measurement of low-level strains with carbon fibers." *Sensors and Actuators A: Physical* 183 (August 2012): 140–147.

Huang, H., and Wu, Z.S. 2017. "Monitoring and structural analysis of a rehabilitated box girder bridge based on long-gauge strain sensors." *Structural Health Monitoring* 17 (3): 586–597.

Ibrahim, A.M.A., Wu, Z., Fahmy, M.F.M., and Kamal, D. 2016. "Experimental study on cyclic response of concrete bridge columns reinforced by steel and basalt FRP reinforcements." *Journal of Composites for Construction* 20 (3): 04015062.

ISO 15392. 2019. *"Sustainability in Buildings and Civil Engineering Works – General Principles."* International Organization for Standardization (ISO). www.iso.org/standard/69947.html

ISO 21930. 2017. *"Sustainability in Buildings and Civil Engineering Works – Core Rules for Environmental Product Declarations of Construction Products and Services."* International Organization for Standardization (ISO). www.iso.org/standard/61694.html

Japan Cabinet Secretariat. 2018. *Fundamental Plan for National Resilience: For Building a Strong and Flexible Country.* www.cas.go.jp/jp/seisaku/kokudo_kyoujinka/en/fundamental_plan.pdf

Lee, C., Kim, K., and Choi, S. 2013. Application of ultra-high performance concrete to pedestrian cable-stayed bridges. *Journal of Engineering Science and Technology* 8 (3): 296–305

Li, P., and Wang, Y. 2013. "Effect of corrosion on mechanical properties of Rebar HRB400." *Advanced Materials Research* (807–809): 2735–2738.

Li, S.Z., and Wu, Z.S. 2007. "Development of distributed long-gage fiber optic sensing system for structural health monitoring." *Structural Health Monitoring* 6 (2):133–143.

Li, S.Z., and Wu, Z.S. 2008. "Modal analysis on macro-strain measurements from long-gauge fiber optic sensors." *Journal of Intelligent Material Systems and Structures* 19 (8): 937–946.

Li, S.Z., and Wu, Z.S. 2010. "Parametric estimation for RC flexural members based on distributed long-gauge fiber optic sensors." *Journal of Structural Engineering* 136 (2): 144.
Mishnaevsky, Leon, Branner, Kim; Petersen, Helga N., Beauson, Justine, McGugan, Malcolm, and Sørensen, Bent F. 2017. "Materials for wind turbine blades: an overview" *Materials* 10 (11): 1285.
Miura, N., Saito, T., Kanamaru, T., Shindo, Y., and Kitazawa, Y. 1983. "Development of new corrosion-resistant steel sheets for automobiles." *ISIJ International* 23 (11): 913–922.
Oliveux, Geraldine, Dandy, L.O., and Leeke, G.A. 2015. "Current status of recycling of fibre reinforced polymers: review of technologies, reuse and resulting properties." *Progress in Materials Science* 72: 61–99.
Rajczakowska, M., Habermehl-Cwirzen, K., Hedlund, H., and Cwirzen, A. 2019. "Autogenous self-healing: a better solution for concrete." *Journal of Materials in Civil Engineering* 31 (9): 03119001.
Ranade, R., Li, V., Stults, M.D., Heard, W.F., and Rushing, T.S. 2013(a). "Composite properties of high-strength, high-ductility concrete." *ACI Materials Journal* 110 (4): 413–422.
Ranade, R., Li, V., Stults, M.D., Rushing, T.S., Roth, J., and Heard, W.F. 2013(b). "Micromechanics of high-strength, high-ductility concrete." *ACI Materials Journal* 110 (4): 375–384.
Sheng, S., Wu, Z.S., Yang, C.Q., et al. 2010. "An improved conjugated beam method for deformation monitoring with a distributed sensitive fiber optic sensor." *Structural Health Monitoring* 9 (4): 361–378.
Shi, J.W., Cao, W., Lei, C., and Li, A. 2020. "Durability of wet lay-up BFRP single-lap joints subjected to freeze–thaw cycling." *Construction and Building Materials* 238: 117664.
Shi, Y. 2016. "Durability study and theoretical analysis on concrete beams reinforced with BFRP with steel bars exposed to marine environments." Master Dissertation, Southeast University.
Sun, J., and Lu, L. 2015. "Coupled effect of axially distributed load and carbonization on permeability of concrete." *Construction and Building Materials* 79: 9–13.
Sun, Z., Wu, G., Wu, Z., and Zhang, J. 2014. "Nonlinear behavior and simulation of concrete columns reinforced by steel-FRP composite bars." *Journal of Bridge Engineering* 19 (2): 220–234.
Tang, F., Chen, G., Brow, R., Volz, J., and Koenigstein, M. 2012. "Corrosion resistance and mechanism of steel rebar coated with three types of enamel." *Corrosion Science* 59: 157–168.
Torresacosta, A., and Diazcruz, L.A. 2020. "Concrete durability enhancement from nopal (opuntia ficus-indica) additions." *Construction and Building Materials* 243: 118170.
Wang, D., Shi, C., Xiao, J., Zhengyu, H., and Fang, Z. 2015. "A review on ultra high performance concrete: Part II. Hydration, microstructure and properties." *Construction and Building Materials* 96: 368–377. DOI:10.1016/j.conbuildmat.2015.08.095.
Wang, X., Shi, J.Z., Liu, J.X., Yang, L., and Wu, Z.S. 2014. "Creep behavior of basalt fiber reinforced polymer tendons for prestressing application." *Materials & Design* 59: 558–564.
Wang, X., and Wu, Z.S. 2010. "Evaluation of FRP and hybrid FRP cables for super long-span cable-stayed bridges." *Composite Structures* 92 (10): 2582–2590.
Wu, M., Johannesson, B., and Geiker, M. 2012. "A review: self-healing in cementitious materials and engineered cementitious composite as a self-healing material." *Construction and Building Materials* 28 (1): 571–583.
Wu, Z.S., and Li, S.Z. 2007. "Two-level damage detection strategy based on modal parameters from distributed dynamic macro-strain measurements." *Journal of Intelligent Material Systems and Structures* 18 (7): 667–676.
Wu, Z., Wang, X., Iwashita, K., Sasaki, T., and Hamaguchi, Y. 2010. "Tensile fatigue behaviour of FRP and hybrid FRP sheets." *Composites Part B: Engineering* 41 (5): 396–402.

Wu, Z., Wang, X., and Wu, G. 2012. "Advancement of structural safety and sustainability with basalt fiber reinforced polymers." In *Proceedings of CICE 2012 6th International Conference on FRP Composites in Civil Engineering,* June 2012. Rome, Italy.
Wu, Z., Wang, X., Zhao, X., and Noori, M. 2014. "State-of-the-art review of FRP composites for major construction with high performance and longevity." *International Journal of Sustainable Materials and Structural Systems* 1 (3): 201–231.
Yang, Y., Wang, X., Wu, Z.S., and Peng, C.H. 2016. "Damping properties of FRP cables for long-span cable-stayed bridges." *Materials and Structures* 49 (7): 2701–2713.
Zhao, X., Wang, X., Wu, Z.S., and Zhu, Z.G. 2016. "Fatigue behavior and failure mechanism of basalt FRP composites under long-term cyclic loads." *International Journal of Fatigue* 88: 58–67.

HUMAN MACROECOLOGY, ENERGY USE SCALING, AND THE SUSTAINABILITY OF CITIES: A LOOK AT LATIN AMERICA

Joseph Robert Burger
Department of Ecology and Evolutionary Biology & Arizona Institutes for Resilience, The University of Arizona

Vanessa P. Weinberger
Center for Genomics, Ecology & Environment (GEMA), Universidad Mayor

INTRODUCTION

Humans are undoubtedly now the major dominant force on the planet (Ellis et al. 2016; Steffen et al. 2011; 2015). In just ~12,000 years, a blink of an eye in evolutionary time, humans have emerged from being *just* another social mammal to becoming the major driver of physical and biological change on Earth (Crutzen 2002). Pressing socio-environmental issues, such as increasing resource scarcity, climate change, social instability, biodiversity loss, pandemics, and food insecurity, require holistic approaches that transcend the physical, biological, social, medical, and engineering sciences with practical recommendations for policy. Here, we synthesize and expand a transdisciplinary human macroecological approach to quantitatively characterize the emergence and future sustainability of modern hyper-dense cities with a special look at Latin America. Our approach uses fundamental biophysical and ecological constraints as macro-indicators (Brown et al. 2014; Burger et al. 2012, 2017; Day and Hall 2016; Hall and Klitgaard 2018; Pacheco et al. 2016, 2018; Schramski et al. 2015) of strong sustainability rather than solely relying on economic indicators of human capital (Ayres et al. 2001), with great potential for policy development. We previously studied global patterns and underlying biophysical processes driving urbanization, energy use, and sustainability (Burger et al. 2017, 2019). Here we take a closer look at the Americas for several reasons.

First, the Americas have experienced two distinct waves of human colonization, during the late Pleistocene (~15,000 years ago) and a second colonization by Europeans in the last ~500 years (Diamond 1997). The first colonization occurred in a narrow time window with limited subsequent migrations, especially into South America, until European colonization (Goldberg et al. 2016; Moore 2014; Rothhammer and Dillehay 2009). European colonizers diminished Native Americans and reduced their populations both directly through genocide and indirectly through the introduction of novel diseases contributing to the loss of distinct culture and traditional livelihoods (Diamond 1997). These rapid, punctual, and recent colonization events to the Americas are considered a unique opportunity for studying the population settlement history of modern humans (Goldberg et al. 2016), and as a simpler contrast to the more complex settlement histories in other parts of the world. Second, the Americas span nearly the full range of terrestrial, marine, and freshwater environments across latitudes and biomes providing an interesting system for comparative geography. For example, the striking similarities in the latitudinal gradient in human population, income, biodiversity, and disease (Bonds et al. 2012; Burger and Fristoe 2018; Tallavaara et al. 2018) beg explanation and offer insights into challenges and policies required for socio-economic development. Third, Latin America is rich in natural resources; yet economic growth has been slower compared to the United States and Canada and much more variable among Latin American countries (ECLAC 2016; Hidalgo et al. 2007). Our macroscopic approach, using spatio-temporal patterns and processes, with a central focus on the role of energy in urbanization attempts to shed light on these issues and the policies that can be taken for regional and global sustainability.

Human macroecology uses theory and data to study human systems across scales from hunter-gatherers to modern urban dwellers (Burger and Fristoe 2018; Burger et al. 2012, 2017; Burnside et al. 2012; Hamilton et al. 2012; Tallavaara et al. 2018). Our approach is “macro” in that it uses biophysical theory rooted in fundamental physical and environmental constraints to motivate hypothesis testing through the compilation and analysis of large datasets (Brown et al. 2011; Burger et al. 2017; Burnside et al. 2012; Hamilton et al. 2012). Doing so attempts to link patterns with processes from local to global scales, and spanning temporal scales from milliseconds required to transmit information in today’s world wide web, to the fossil fuel resources stored on deep geological time that power the current global economy (Burger et al. 2012). It’s “Ecological” in that it uses the fundamental currencies of ecology – energy, materials, and information – and the laws that govern the flows of these commodities between us and our social and physical environments (Burger et al. 2012; Day et al. 2009; Hall and Balogh 2019; Hall and Klitgaard 2018; Odum 1973). We demonstrate how these fundamental principles and general macroecological rules that govern biological and ecological systems can be used broadly as a baseline to understand unique traits of the human species, the sustainability implications behind their current interconnected societies, and to anticipate the challenges and possible policies for achieving future sustainability in a post-fossil fuel world.

Metabolic Scaling Theory

In recent decades, a general theory of energy or metabolic scaling has emerged as a useful transdisciplinary tool to unify across scales and complex systems (Brown et al. 2011; West 2017). General scaling rules spanning the diversity of life can be useful benchmarks to compare and contrast the human species to the rest of life (Burnside et al. 2012; Burger et al. 2017; Hamilton et al. 2012; Moses and Brown 2003). The scaling of metabolic energy, in particular, is a useful currency to quantify the trajectory of the human species from being *just* another social mammal to becoming the Earth's most prolific and environmentally impactful species. Across mammals, individual metabolic rate or energy use (E_i) scales predictably with body size (M),

$$E_i \propto M^{3/4}, \tag{2}$$

where large mammals such as elephants and whales have much higher metabolic requirements than do mice (Kleiber 1932). The ¾ scaling exponent reveals an economy of scale spanning ~8 orders of magnitude variation in body size in mammals, from mice to whales, where large body sizes require less energy per mass. Human biological metabolism is no exception. We have a metabolic rate of ~100 watts at resting and ~120 watts during active exercise or "field" metabolic rate, precisely where Kleiber's law would predict for a mammal of our size (Burger et al. 2012). This corresponds with the 2,000 kcals per day required to maintain healthy diets (Burger et al. 2017; Hammond et al. 2015; West 2017). Hunter-gatherers power their entire societies using the biological metabolism of about 120 watts per capita that they obtain from food calories through hunting, fishing, and foraging.

Unique to modern humans is our ability to harness extra-metabolic energy in the form of fossil fuels and renewables. This increases our per capita energy use from 120 watts of biological metabolism to more than 10,000 watts in high income societies with the inclusion of vast amounts of extra-metabolic energy. Moses and Brown (2003) made the analogy that the amount of energy use per capita consumed by modern humans is equivalent to the metabolic rate of a hypothetical 30-ton ape (see Figure 4.40). The footprint of our colossal appetites certainly has huge effects on ecosystems globally and raises questions for how it can be *sustained* through time (*sensu* Costanza and Patten 1995).

Scaling Energy Use and Densities

The scaling of metabolic rate with the animal size just described has consequences for many aspects of biological rates and times including life history, population dynamics, and ecology where +-¼ and +-3/4 allometries are observed (Brown et al. 2004, 2018; Burger et al. 2019; Peters 1986; Sibly et al. 2012). One ecological consequence is the inverse scaling of densities (D) of wild mammal populations with size (M), where D scales proportional to $M^{-3/4}$ (Eqn. 3; Damuth 1981). Larger organisms such as rhinos and elephants with higher absolute metabolic rates occur at lower densities than small organisms like mice and shrews with small body sizes and lower total metabolic demands. The maximum number of individuals per unit area (D_{max}), therefore, scales as the inverse of per capita energy demands,

$$D_{max} \propto M^{-3/4} \tag{3}$$

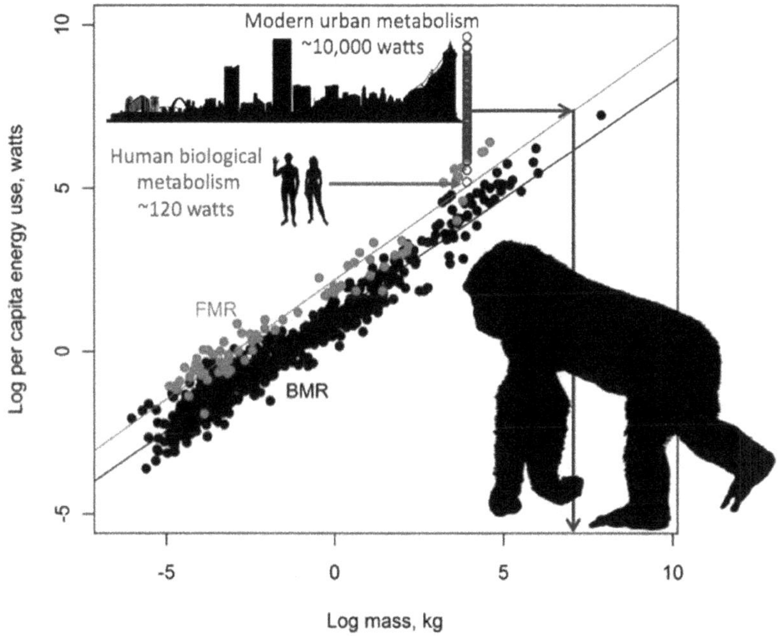

FIGURE 4.40 Energy use as a function of body mass in mammals and human societies. Black circles are basal or resting metabolic rate (BMR) for mammals; gray circles are "field" or active metabolic rate (FMR) for mammals. The open circles are per capita energy use among industrial human populations with little variation in body size. Humans have a metabolic rate of 100 watts at resting (e.g. basal) and 120 watts active. The horizontal arrow represents median values for modern industrial societies. The arrows illustrate the level of per capita energy use and corresponding hypothetical mass (~30- ton ape). (Adapted from Burger et al. 2012.)

The result is a compensation between size and density. However, densities are not actually limited by body mass directly, but instead by the energetic requirements to support an individual, E_i, of a given body size. Rearranging the allometric relationships in Equations 2 and 3, energetic scaling theory predicts density scales proportional to E_i to the −1 power.

$$D_{max} \propto E_i^{-1} \tag{4}$$

This new representation of the scaling of energy use provides a biophysical macro-indicator to understand the emergence of modern cities and energy compared to biodiversity and ecosystem baselines. We previously evaluated this prediction in (Burger et al. 2017) using a combination of open source data for wild mammal populations from the PANTHERIA database (Jones et al. 2009), hunter-gatherer societies (Binford 2001), archeological data (Zimmermann et al. 2009), modern global city data (demographia.com), and energy use data from the World Bank (Worldbank.org). Here we revisit that analysis focusing on Latin American cities to understand the wide variation across countries globally, and how Latin American countries compare. Due to data limitations, we assumed constant per capita resource use across cities within countries. This assumption is supported by studies showing linear scaling of resource use with city population size for a few countries including the United States (Bettencourt et al. 2007; Fragkias et al. 2013). Per capita energy use generally varies greater between countries than within; however, this assumption needs to be evaluated with better data from cities in Latin American and globally.

Comparing density and per capita energy use revealed human hunter-gatherers occurred at population densities comparable to or lower than other mammals (Burger et al. 2017). The prediction of the scaling of D_{max} with E_i is supported for land mammals, where densities scale with individual energy use (E_i) with a slope of −1.08, statistically indistinguishable from the theoretically predicted −1 slope based on the 95% confidence intervals (−0.88, −1.27). H-G densities of <1 Ind/km^2 and 120 watts of energy use are slightly lower than predicted for a mammal of our size but within the scatter of points. In contrast, early agricultural societies pushed these ecological constraints through the incorporation of extra-metabolic energy (Burger et al. 2017). These societies (e.g. ancient Greeks; square symbols), with the inclusion of "beasts of burden" in addition to their own metabolism, increased their per capita energy demands to roughly 600 watts (Moses and Brown 2003) and lived on the order of ~100 Ind/km^2–two orders of magnitude higher densities than hunter-gatherers. In modern urban societies per capita energy use ranges from one to two orders of magnitude greater than those for H-Gs while obtaining unprecedented densities (see Figure 4.41). The trend clearly illustrates the relation between extra-metabolic energy and the extraordinary densities of modern cities (Bettencourt et al. 2010). This rise in urban hyper-densities, which has occurred in just 200 years since the industrial revolution, undoubtedly has major consequences for cultural evolution, economics, and public health including psychology, disease transmission, food security, waste management, transportation, and sustainability (Burger et al. 2012; Hall and Balogh 2019; Hammond et al. 2015; Nekola et al. 2013). For example, growth rates of COVID-19 cases increase as a function of population size and density of cities in the United States (Stier et al. 2020).

Energy Use Is Directly Related to Per Capita Income

In modern human societies, higher per capita energy use corresponds to greater economic activity (i.e. per capita GDP) and with that greater access to education, less infant mortality, morbidity, and mortality rates (Brown et al. 2011, Snyder-Mackler et al. 2020). Brown et al. (2014) described how these highly correlated socio-economic indexes are also strongly associated with increased energy use and the United Nations' Human Development Index (HDI). So, it is reasonable to associate the variability in energy use between countries to differences in socio-economic development (Brown et al. 2011, 2014). Across global cities there is a wide range in energy use per capita, from ~400 watts in Dhaka, Bangladesh to more than 10,000 watts in highly developed countries such as the United States, Canada, UK, and Saudi Arabia. It is interesting to note that all Latin American cities fall within an intermediate range of energy use compared to the global distribution. Peru's cities (with

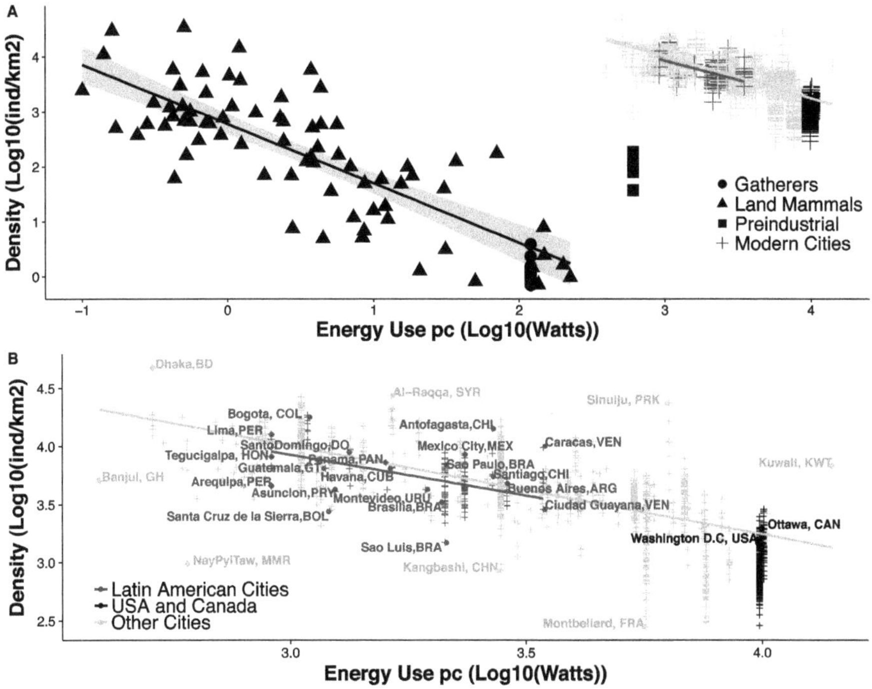

FIGURE 4.41 Scaling density as a function of per capita energy use (watts) comparing (A) land mammals, hunter-gatherers, preindustrial societies, and modern cities. (B) A zoomed-in plot of cities shows Latin America compared to worldwide. See the text for data sources and details.

over 500,000 inhabitants) stand as the cities that consume less energy per capita, but their ~1,000 watts per capita is still ~600 watts higher than the average inhabitant of Dhaka, Bangladesh. Venezuela's cities appear as the cities with the greatest energetic consumption, with over 3,400 watts per capita. Still, this is one order of magnitude less than the ~14,150 watts per capita demand that stands for cities in Kuwait.

The variation in HDI, per capita GDP, and energy use in Latin American countries largely seems to follow the same latitudinal gradient globally (Bonds et al. 2012). This convergent pattern emerges in similar climates in different continents, and in the Northern and Southern hemispheres, and may be driven by ecological drivers of productivity and disease burdens (Bonds et al. 2012; Burger and Fristoe 2018; Tallavaara et al. 2018). Policies related to these constrains mainly relate to greater investment in public health and non-dependence on natural resources for their economy (Diamond 1997). Indeed, countries in higher latitudes of the temperate southern cone, comprising Chile, Argentina and Uruguay, are currently categorized as "Very high human development index," and their 2018 per capita GDP in US dollars are the greatest in the region, with Uruguay ($17,278), Chile ($15,923), and Argentina ($11,683) greatly surpassing the region's average ($7,531) (WorldBank 2019). Moreover, these southern temperate climate countries achieve greater economic development compared to other large economies in the region such as Mexico ($9,673) and Brazil ($9001) that lie at tropical-temperate zones. The Central American countries of Guatemala, El Salvador, Honduras, and Nicaragua have "Medium Development," with nominal per capita incomes of $4,549, $4,058, $2,506, and $2,020, respectively; and only Haiti in the Americas has a "Low Development" based on the HDI, with a per capita GDP of only $868. The poorest countries in South America are Bolivia ($3,548) and Paraguay ($5,821), which one tropical (Bolivia, but with high topography) are both landlocked. In contrast, tropical countries like Cuba ($8,821) and Colombia ($6,667) with extensive coastlines have relatively high per capita incomes despite being tropical and Cuba lacking a free market economy. Also of interest is tropical Panamá which shows a relatively high per capita income ($15,575) and high service economy including trade services via the Panama Canal. Further interesting is the elevational distribution of population and cities throughout the tropics largely along coasts, rivers, or high elevation mountains (Cohen and Small 1998).

Peru, the country with the smallest per capita energy use in Latin America (909.83 watts), shows relatively high levels of urbanization, medium per capita GDP ($6,941), and "High Development" based on HDI. Hydroelectricity accounts for over 25% of Peru's total energy use (BP 2019), a significant contribution of what is considered as renewable energy that might be masking the countries' actual position in the plot and accounting for the greater developmental statistics. However, hydroelectricity in Peru (and other Andean countries) may suffer from uncertainties due to climate change and glacial melt. Also, hydroelectric externalities can also impact the natural system, with potential devastation on local community development if transgressing the ecosystem's resilience and resistance (Downing et al. 2012; Folke et al. 2002; Holling 2001). Even more, if externalities affect global change drivers (e.g. land transformation, freshwater appropriation and degradation, biodiversity loss), the effects could transgress the local to national or bigger spatial scales (Barnosky et al. 2012; Sage 2020). Other Latin American countries with over 25% renewable source contribution (also hydropower) are Colombia, Ecuador, Brazil, and Venezuela (BP 2019).

Certainly neither per capita income nor HDI captures all dimensions of human well-being and other metrics such as the Genuine Progress Indicator (GPI; Kubiszewski et al. 2013) have been proposed as a more comprehensive approach to understanding the optimal size of the human system relative to the finite biosphere (Daly 2005). The GPI incorporates environmental and social factors in addition to GDP. Its 26 indicators seek to determine the net effect of a country's economic production and consumption on its inhabitants' overall health and well-being, by weighting their social costs and environmental impact. Yet, data for GPI and other finer metrics are more challenging to obtain for countries, cities, and across time. Venezuela is a case in point. No Latin American country has per capita energy consumption as high as Venezuela at over 3,400 watts per capita, a country whose economy is based on oil production for export. Venezuela is similar to other energy producing countries including those in the Middle East (e.g. Kuwait) and has higher per capita energy use for their level of economic development (e.g. GDP; Brown et al. 2011, 2014; Moses and Brown 2003). These trends are associated with unequal distributions of wealth and/or the great energetic demand that such economic activities require for production, which are later exported abroad increasing GDP, instead of local economies (ECLAC 2016,

Ocampo 2017). According to the World Bank Development Indicators the last report of Venezuela's per capita GDP was $16,055 in 2014 – higher than any country in Latin America. But it's HDI is medium and the Venezuelan economy has been in economic and political turmoil including hyperinflation and environmental degradation in recent years. So, GDP and HDI clearly do not reflect well-being in Venezuela.

Additional cases in point are Latin American countries that, despite having "High Development" status, have a history of economic crisis (e.g. Argentina) and current signs of large-scale social revolts (e.g. Chile), indicators of great inequalities that are not necessarily accounted for through the HDI index (UNDP 2019). Indeed, the Inequality Adjusted HDI shows Chile to be the second lowest in HDI when measures of inequality (i.e. GINI coefficient) are incorporated. And Chile is not unique. High inequality regions across the globe (e.g. France, Syria, Iran) and in Latin America (e.g. Chile, Ecuador, Venezuela, Colombia, Argentina) are sensitive to changes in energy and transportation prices resulting in recent social revolts (*sensu* Ahmed 2017, 110; Tainter 2019). So, deviations from the macro-trend in the latitudinal gradient in economic productivity offer key insights into understanding how tropical countries have managed to reach higher development status, whether the diversity in their economic activities and/or investment in public health (Table 4.1).

Energy Flux of Modern Cities Far Exceeds Ecosystems and Previous Societies

In addition insight from energy use scaling comes from multiplying densities and individual energy use ($E_i \cdot D_{max}$), providing us with an estimate of population-level energy flux, E_p. Theory predicts E_p to be "invariant" or constant over species varying in per capita energy requirements with a slope $\beta = 0$.

$$E_p \propto E_i^0 \tag{5}$$

Plotting the data this way (Figure 4.42) allows us to compare the E_p across mammals, hunter-gatherers, preindustrial, and modern cities in reference to Net Primary Productivity (NPP) – the rate solar energy is stored in green plant biomass on Earth forming the essential energetic bases of all consumers in food webs in all ecosystems (Figure 4.42) including humans.

As expected from fundamental limits to ecosystem energetics, the energy flux of mammal populations and hunter-gatherers is ~2 orders of magnitude below the reference NPP. With each transfer of energy, most is lost as heat. Preindustrial societies fluxed greater E_p than other mammals and hunter-gatherers, but did not exceed the ecological "green ceiling" of NPP. This increase is likely due to technological and cultural advancements in plant and animal domestication and food storage during good times (Hammond et al. 2015). In contrast, modern cities flux two orders of magnitude greater energy than background ecosystem levels in NPP through the incorporation of vast quantities of extra-metabolic energy use largely in the form of fossil fuels (Burger et al. 2017). This clearly imposes a major challenge for sustainability. Modern cities must currently subsidize the concentrated local demands of hyper-dense citizens by

TABLE 4.1
A Comparison of Highest and Median Density and Energy Use between Latin American Cities and the Rest of the World.

	Highest Population Density in city	Highest Energy Use Country	Median Energy Use [Range]	Median Population density in city [Range]	Median Populations Energy flux [Range]
	Individuals/km²	Watts	Watts	Individuals/km²	watts/km²
Latin America	**Ibagué, Colombia** (18,414 ind/km2)	**Venezuela** (3,438 watts per capita; 140 watts from food)	**Ecuador** 1,337 [909.83–3,438] watts per capita (Lowest: Peru; Highest: Venezuela)	**Buenos Aires, Argentina** 4,832 [1,490–18,414] (Lowest: Sao Luis, Brazil; Highest: Ibagué, Colombia)	**Chiclayo, Peru** 10,450,804 [3,160,568–38,084,871] (Lowest: Sao Luis, Brazil; Highest: Antofagasta, Chile)
World	**Dhaka, Bangladesh** (47,379 ind/km2)	**Kuwait** (14,154 watts per capita; 168 watts from food)	**Iran** 4,030 [391–14,155] watts per capita (Lowest: Gambia; Highest: Kuwait)	**Naha, Japan** 3,911 [278.81–47,379] (Lowest: Montbeliard, France; Highest: Dhaka, Bangladesh)	**Las Palmas, Spain** 12,716,652 [579,030–149,754,304] (Lowest: NayPyiTaw, Myanmar; Highest: Sinuiju, North Korea)

Energy use includes the energy demanded from food consumption plus extra-metabolic energy use in the form of primary energy sources. Note that the demographia.org dataset does not include all cities <50,000 population for all countries.

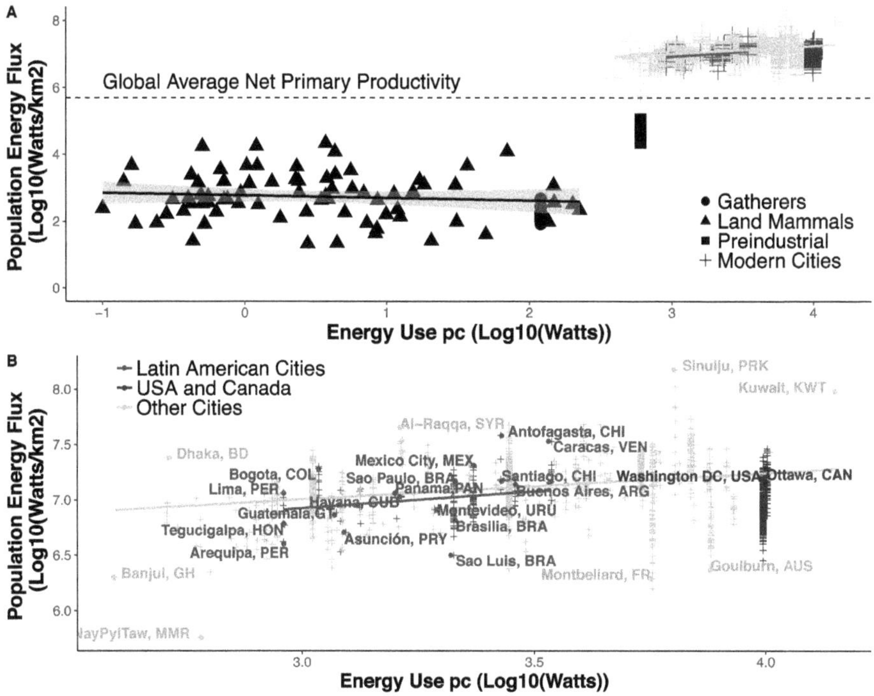

FIGURE 4.42 Population energy flux (E_p) as a function of individual energy use (E_i) for (A) human populations and other land mammals with the global average terrestrial NPP as a reference (503,558.48 watts/km^2; Haberl et al. 2007). (B) Zoomed-in comparison of global cities to Latin American cities and the United States and Canada.

resources far from outside the urban boundaries (Burger et al. 2012, 2017, 2019a, Meyfroidt et al. 2010; Rees and Wackernagel 2008). This requires vast infrastructure networks in the form of roads, shipping, air transportation (Hammond et al. 2015), and the communication and regulatory networks to coordinate market activity, all of which increase the material, energetic, and informational needs per capita, leading to a greater "footprint" (Reese and Wackernagel 2008, Global Footprint Network) and threatening the biophysical capacity of the biosphere.

The global trend – where higher per capita energy use countries have higher E_p – is exhibited similarly in Latin American countries. This includes Caracas in Venezuela – a high energy producing country – as well as different cities within the region's most developed countries such as Antofagasta, Chile; Buenos Aires, Argentina; and Mexico City. In contrast, cities in lower energy use countries such as Lima, Peru or Bogota, Colombia flux lower or similar energy as other more developed countries while acquiring greater population densities. However, none of these cities could support themselves from energy and food from local NPP and currently rely on unsustainable global trade networks dependent on fossil fuels (Burger et al. 2012; Schramski et al. 2019).

There are key insights that can be gained from the data: Antofagasta, Chile, stands as the city with the greatest population energy flux (~38,000,000 watts/km^2), significantly higher than the country's capital (Santiago de Chile, ~15,000,000 watts/km^2). Antofagasta is a port city based on mining and shipping activities, and is an example of how economic activities might be masking the energy flux trends seen for Latin American countries, as these raw materials are exported to global trade markets. It's high density and energy flux could also be due to the constraints on human settlement in the largely inhospitable Atacama Desert. However, even this key economic Latin American city fluxes considerably less energy per area than Sinuiju, North Korea (150,000,000 watts/km^2), the city with the highest population energy flux in the data. Sinuiju is key for economic activities in North Korea, as the country's main port to China and important for key wood and recycling industries. Finer data and analysis of how energy is allocated by different cities are needed to understand whether the energetic costs actually benefit the urban dwellers, or is it lost to global economic trade activities.

Urbanization Trends Across and Within Countries

We can further evaluate the role of energy use in urbanization across decades. Urbanization is a hallmark of the Anthropocene and corresponds with the Great Acceleration – the synchronized surge in a range of measurable human activity including global resource extraction, CO_2 emission, and GDP growth (Schramski et al. 2020; Steffen et al. 2015). The world's population crossed the urban transition around 2010, where >50% of the world's population lives in cities. However, Europe, Oceania, the Caribbean, United States, Canada, and many Latin American countries have been mostly urban for decades prior to 2010 (Figure 4.43). Indeed, the Latin American countries of Uruguay, Argentina, Chile, Brazil, and Venezuela are more urban than the United States and Canada. On the other hand, the Central American countries of Nicaragua, Honduras, and Guatemala have only recently made the 50% urban transition. The rest of Latin American countries transitioned decades before and are generally much more urban than global variation among countries. For example, India with 1.34 billion people is still mostly rural (only ~35% of its population is urban), while China (with 1.43 billion people) and Singapore (with only 5.61 million) are highly urbanized (60% and 100%, respectively). In Africa, most countries remain predominantly rural (Rwanda 82.8% rural, 12,301,970 people) and few such as South Africa (57.7 million people, 66% urban) are urbanized. Egypt with its nearly 100 million people mostly residing along the Nile River and Delta is only 42.7% urban.

Along with urbanization comes increased interdependence of urban populations on rural resources impacting the environment (Burger et al. 2012, 2017, 2019b; Irwin et al. 2009; Rees and Wackernagel 2008; Schramski et al. 2019). Such global trends impose huge pressures on how to subsidize an increasing trend of non-locally energetic sustainable urban centers from other places that seem to lack the technological and/or workforce capacities. Finally, the massive shift of people from rural lands to cities threatens not only the production capacity but also the resilience of trade

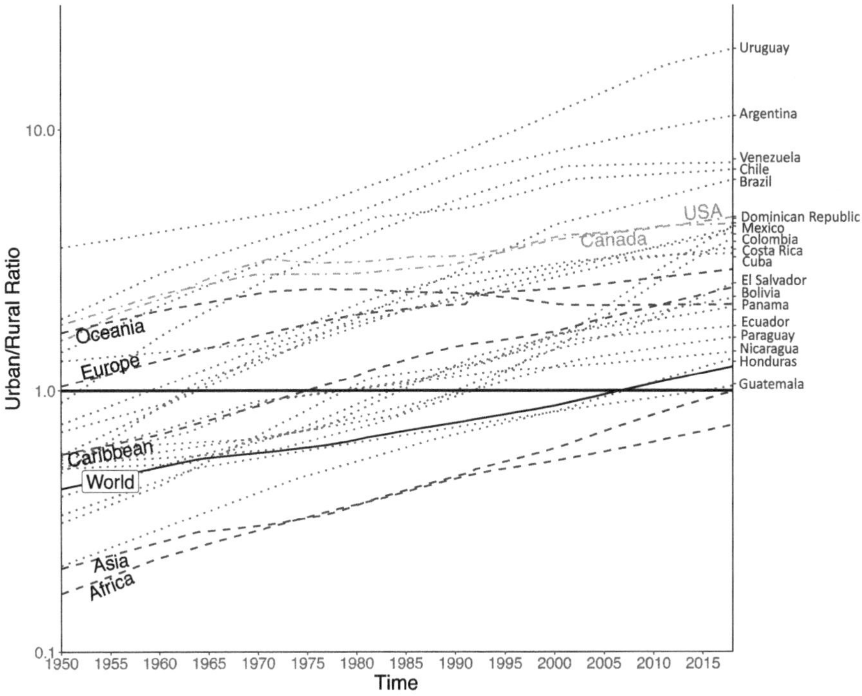

FIGURE 4.43 The urban transition globally, for select regions and Latin American countries. Data from http://data.worldbank.org. Since ~2015 all Latin American countries are predominantly urban. See Burger et al. (2019a) and World Bank for projections to 2050.

networks, as lower income countries are supporting the needs of more high income urban countries (Meyfroidt et al. 2011; Schramski et al. 2019, 2020).

Burger et al. (2019a) analyzed World Bank data among countries and within countries over time showing that per capita GDP, energy use, and CO_2 emissions increased by over two orders of magnitude from low-income, resource-based rural countries to high-income, urbanized countries with service economies. In general, urbanization in Latin American countries is similar (Figures 4.44 and 4.45). However, for the level of urbanization, Latin American countries have lower per capita GDP than the world average, translating to lesser energy use and CO_2 production. This income gap could be reflecting differences in economic activities that are developed by Latin American countries compared to the rest of the world, which has historically been affected by their colonization history and the particular raw material and agricultural development that the region has experienced over the past century (ECLAC 2016; Gudynas 2013; Ocampo 2017). Indeed, for the level of urbanization, Latin American countries have greater employment in resource sectors compared to the rest of the world (Figure 4.45) and much of the national GDP is from resource extraction (Hidalgo et al. 2007; ECLAC 2016), despite being mostly urban societies. Moreover, the World Bank database defines mining and quarrying activities as "industrial employment," when

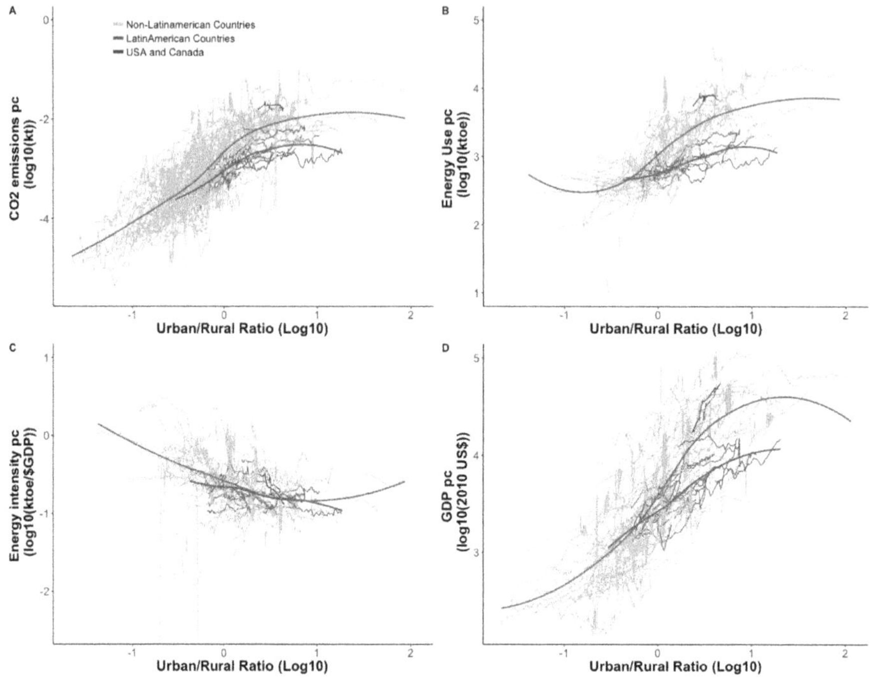

FIGURE 4.44 Global trends in per capita values with urbanization across and within countries over time looking at A) CO2 emissions, B) energy use, C) energy intensity, and D) gross domestic product (GDP). Data from Worldbank.org with the following countries and political entities removed as outliers: Macao; Hong Kong; and Kuwait. Fitted lines represent LOESS spline fits. Modified from Burger et al. (2019a) to highlight Latin American countries, Canada and United States.

there are non-renewable resource extraction activities that are sold to other countries as raw material and not processed within the country. Indeed, Latin American countries export vast quantities of raw materials to other countries abroad (ECLAC 2016; Hidalgo et al. 2007), affecting their potential per capita income, as manufactured goods are more lucrative (Hidalgo et al. 2007).

Challenges for Sustaining Urban Resource Flows

As shown, extra-metabolic energy use has enabled modern societies to temporarily escape ecological constraints in hyper-dense cities of today. However, the footprint of modern cities – the area and resources required to support an urban population and economy – clearly extends well beyond the local urban boundaries (Burger et al. 2012; Rees and Wackernagel 2008). When considering the sustainability of a particular city, it is important to understand the resource flows – both metabolic and extra-metabolic energy – required to sustain a hyper-dense population. This is akin to ecological footprint (Rees and Wackernagel 2008; Meyfroidt et al. 2011; Burger et al.

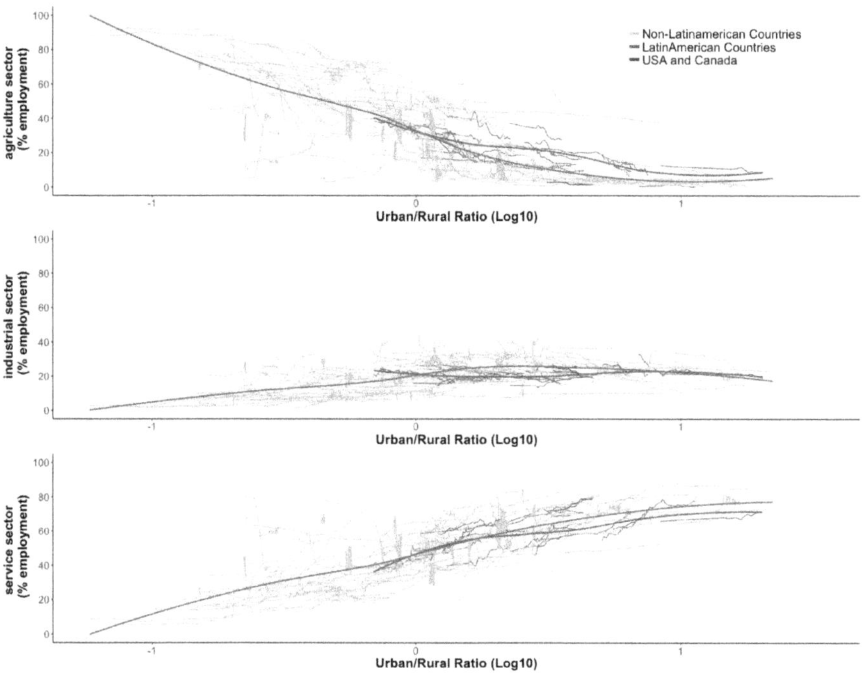

FIGURE 4.45 The combined employment (men and women based on Worldbank.org bigender classification) for countries over time. Resource employment includes the total number of individuals employed in resource-extraction occupations such as agriculture, hunting, forestry, and fishing. Industrial employment includes mining and quarrying, oil and gas production, manufacturing, construction, and public utilities such as water, electricity, and gas. Today, most of Latin American economies are dominated by the service sector including wholesale and retail trade, hotels and restaurants, transportation, storage, communications, finance, insurance, real estate, business services, health care, education, recreation, and community, social, and personal services. Guatemala and Honduras are the two Latin American countries that most recently crossed the 50% urban transition in just the last decade. Fitted lines represent LOESS spline fit to data. Highlighted are Latin American countries, Canada and United States.

2012), which highlights the importance of importing resources far from the urban boundary through fossil fuel-powered global trade networks to sustain localized urban populations (Hammond et al. 2015). The question is how can we sustain hyper-dense cities – and their connections through global trade networks – in a post-fossil fuel world?

One area of increasing interest in Latin America and globally is urban agriculture. But the capacity for urban farms to produce the food demands of modern cities is very low ranging from <1% of caloric demand to 20% under very optimistic scenarios (Day and Hall 2016; Grewal and Grewal 2012; Schramski et al. 2019, 2020). Moreover, the increasing presence of wild and domesticated animals in dense urban areas poses health risks as global pandemics are known to come from spillover zoonotics in dense urban areas (Bradley and Altizer 2007; Di Marco et al. 2020; Kapan et al. 2006; Wu et al. 2017). Thus, urban food production as a solution to less energy-intensive food security is questionable (Schramski et al. 2020). Other solutions deal with the transition toward renewable energies, but both biophysical (Hall and Klitgaard 2018; Schramski et al. 2015) and innovation rates (Burger 2018; Strumsky et al. 2010, Weinberger et al. 2017) put limits to these alternatives. Moreover, it has amply been shown that energy efficiency does not necessarily lead to more sustainable societal outcomes, as societies with greater energy availability, usually achieved through technological development, tend to increase their per capita demand instead of diminishing it, a trend known as "Jevons Paradox" and/or "Rebound effect" (Alcott 2005; Gralla et al. 2017). The future of cities is uncertain. Pandemics, social instability, and environmental impact may result in an exodus from cities and a reversal to resource-employment in local scale farming, hunting, fishing, and collecting (e.g. Day et al. 2016). However, it is important to note that the inevitable long-term de-urbanization may not take the same path down as the trajectory up, a concept in complex systems known as hysteresis (Burger et al. 2019b).

Why Do Latin American Countries Have Lower Economic Development for Urbanization?

Our results show that Latin American countries, for their level or urbanization, have relatively low per capita GDP and energy use, and because these two metrics are so tightly related (Brown et al. 2011, 2014), we would expect that such differences translate into differences in quality of living for the region. Only Argentina, Uruguay, and Chile are currently considered as "Very developed" status, according to the HDI, and these countries have higher urbanization rates than the United States and Canada (Figure 4.43). Interestingly, these are the countries of temperate Latin America, which supports the long observation in economic geography that the temperate zones typically have higher rates of per capita income (Bonds et al. 2012), but also the lack of a proper index for properly quantifying well-being (e.g. GPI). One hypothesis to explain this "Latitudinal gradient in income" is that the ecology of infectious disease drives poverty traps and impacts human distribution, density, and economies (Bonds et al. 2012; Burger and Fristoe 2018; Tallavaara et al. 2018), but also that we are still lacking of a more international-based index that accounts for environmental impacts of economic trade activities. Certainly, studying the Americas holistically from a human macroecological approach, accounting for their economic activities, provides an excellent opportunity to study the environmental and geographic relations to economic development and options to mitigate through policy and technology.

Economic development in Latin American has historically been based on resources as commodities (ECLAC 2016; Ocampo 2017), that is, the extraction of great amounts of raw materials, such as minerals, oil, wood, and foodstuffs, for exportation to international commercial markets. Thus, the region is characterized by urbanization, yet a great dependence on vast natural resource exploitation for exportation for economic growth (i.e. “extractivist economies,” *sensu* Gudynas 2013) and not from industrialization, service, and technology jobs that we see with urbanization in the highest development countries (Burger et al. 2019; Pérez 2010). Indeed, the current social and economic difficulties seen in the region suggest that the benefits of energy use of their countries are not necessarily resulting in greater quality of life for all of their citizens. Countries’ dependence on extractive activities is known to threat local ecosystems, local communities, and economic growth (Gudynas 2013; Hidalgo et al. 2007; Max-Neef 1991), therefore, citizens’ well-being. A more detailed analysis between energetic use, lifestyles, inequality, policy, and economic trade activities for this region is undoubtedly needed.

Finally, knowing that Latin American economies depend on extractive economies, CO_2 emissions for the region are likely severely underestimated. The Worldbank.org database only reflects emissions that refer to “those stemming from the burning of fossil fuels and the manufacturing of cement. They include carbon dioxide produced during consumption of solid, liquid, and gas fuels and gas flaring.” Therefore, it does not include CO_2 emissions as a result of deforestation and land use change. Rapid deforestation, especially in tropical forests, is releasing large quantities of CO_2 that are not accounted for in these estimates. The rate of deforestation in Latin American forests is increasing and represents 25% of all forest loss globally, with fastest rates in Amazonas of Brazil, but increasingly in forests of Colombia, Venezuela, Ecuador, and Central America (Curtis et al. 2018). Moreover, it was demonstrated that 87% of all forest loss in the region was mainly due to commodity-driven deforestation with activities that pertain to agriculture, cattle, mining, and energy infrastructure (Curtis et al. 2018).

Latin American economies mainly depend on natural resource trade; yet there is a great underestimation in the land clearing effects on CO_2 emissions in the region. It is likely Latin American countries have greater CO_2 productions from deforestation activities for their level of urbanization and energy consumption that is not reflected in our analyses or the Worldbank.org data. We suggest future data compilations to include a complete accounting of CO_2 emissions. These data are essential for delivering the science required for sustainable solutions.

Conclusions

Latin America is following the same urbanization trajectories as global trends, albeit with lower levels of per capita GDP, CO_2, and energy use. The region’s trajectory, following common global trends in energy use, poses major challenges for sustainability at local to global scales. Our results clearly show that modern rates of urbanization and the hyper-dense cities now the norm throughout the Americas are largely dependent on fossil fuels and other resource extraction. They are clearly unsustainable. Alternatives to economic growth paradigms and urbanization should be necessary to uncover a sustainable vision within biophysical limits. A rapid cultural transition to a steady-state human system that acknowledges other development paradigms and/or indexes of quality of living is of the essence.

Our human macroecological approach illustrates the utility of using large-scale data and energetic principles to understand the emergence and future sustainability of cities. Fundamental biophysical and ecological constraints are useful macro-indicators of the ecological sustainability of human systems (Burger et al. 2012, 2014, 2017; Hall and Klitgaard 2018; Schramski et al. 2015). We suggest these indicators of strong sustainability rather than economic indicators of human capital (Ayres et al. 2001), because there is no substitute for the current energy basis that powers the economy. A rapid cultural evolution (Burger 2018; Nekola et al. 2013) is needed for society, not based on a fallacy of infinite growth in a finite system (Burger 2020; Malthus 1798), but on improving human well-being (correctly weighting economic, social, and environmental factors) within global biophysical constraints and considering future generations.

A productive way forward is to acknowledge these biophysical constraints and enact policy incentives to transition society to levels within biocapacities of regions, while accounting for impacts made by technology and trades. Latin American is rich in natural resources. A recent global analysis of global priority regions for biodiversity, water, and carbon highlighted the importance of parts of Latin America including Mesoamerica, the Caribbean, Valdivian forests of southern Chile, and tropical mountain regions including the eastern slope of the Andes (Jung et al. 2020). Yet, preserving these resources in perpetuity requires a rapid cultural evolution and paradigm shift to policies prioritizing steady-state healthy populations (especially in the tropics), economies within regional biocapacities, and lifestyles and technology that couple humans to the biophysical limits of Earth.

ACKNOWLEDGMENTS

The authors thank the symposium organizers of the conference and José R. Soto, Travis Knowles, Luis Pacheco, Christopher Scott, John Day, Sebastian Abades, Manuel Muñoz, and Horacio Samaniego for graciously providing discussions and comments on an earlier draft of the manuscript. The authors were supported by a Postdoctoral Fellowship Agencia Nacional de Investigación y Desarrollo de Chile ANID 3190609 and the Bridging Biodiversity and Conservation Science program at the University of Arizona.

REFERENCES

Ahmed, Nafeez. 2017. *Failing States, Collapsing Systems: Biophysical Triggers of Political Violence*. New York: Springer.

Alcott, Blake. 2005. "Jevons' paradox." *Ecological Economics* 54 (1): 9–21.

Ayres, Robert, Jeroen van den Berrgh, and John Gowdy. 2001. "Strong versus weak sustainability: economics, natural sciences, and consilience." *Environmental Ethics* 23 (2): 155–168.

Barnosky, A., Hadly, E., Bascompte, J. et al. 2012. Approaching a state shift in Earth's biosphere. *Nature* 486, 52–58. https://doi.org/10.1038/nature11018

Bettencourt, Luís M.A., Lobo, José, Helbing, Dirk, Kühnert, Christian, and West, Geoffrey B. 2007. "Growth, innovation, scaling, and the pace of life in cities." *Proceedings of the National Academy of Sciences* 104 (17): 7301–7306.

Bettencourt, Luís M. A., José Lobo, Deborah Strumsky, and Geoffrey B. West. 2010. "Urban scaling and its deviations: revealing the structure of wealth, innovation and crime across cities." *PloS One* 5 (11): e13541.

Binford, Lewis R. 2001. *Constructing Frames of Reference: An Analytical Method for Archaeological Theory Building Using Ethnographic and Environmental Data Sets.* Oakland, CA: University of California Press.

British Petroleum BP. 2019. "BP statistical review of world energy." *BP Statistical Review.* London, UK. www.bp.com/en/global/corporate/energy-economics/statistical-review-of-world-energy.html.

Bonds, Matthew H., Andrew P. Dobson, and Donald C. Keenan. 2012. Disease ecology, biodiversity, and the latitudinal gradient in income. *PLoS Biology* 10 (12): e1001456.

Bradley, Catherine A., and Sonia Altizer. 2007. Urbanization and the ecology of wildlife diseases. *Trends in Ecology & Evolution* 22(2): 95-102. DOI:10.1016/j.tree.2006.11.001.

Brown, James H., Joseph R. Burger, William R. Burnside, Michael Chang, Ana D. Davidson, Trevor S. Fristoe, Marcus J. Hamilton, et al. 2014. Macroecology meets macroeconomics: resource scarcity and global sustainability. *Ecological Engineering* 65 (April): 24–32.

Brown, James H., William R. Burnside, Ana D. Davidson, John P. DeLong, William C. Dunn, Marcus J. Hamilton, Norman Mercado-Silva, et al. 2011. Energetic limits to economic growth. *Bioscience* 61 (1): 19–26.

Brown, James H., James F. Gillooly, Andrew P. Allen, Van M. Savage, and Geoffrey B. West. 2004. Toward a metabolic theory of ecology. *Ecology* 85 (7): 1771–1789.

Brown, James H., Charles A. S. Hall, and Richard M. Sibly. 2018. Equal fitness paradigm explained by a trade-off between generation time and energy production rate. *Nature Ecology & Evolution* 2 (2): 262.

Burger, J.R. 2020. "Malthus on population." In *Encyclopedia of Evolutionary Psychological Science*, edited by T.K. Shackelford, and V.A. Weekes-Shackelford. p. 1267. Switzerland: Springer-Nature.

Burger, Joseph R., Allen, Craig D., Brown, James H., Burnside, William R., Davidson, Ana D., Fristoe, Trevor S., Hamilton, Marcus J., et al. 2012. "The macroecology of sustainability." *PLoS Biology* 10 (6): e1001345.

Burger, Joseph R., Brown, James H., Day, John W., Flanagan, Tatiana P., and Roy, Eric D. 2019(a). "The central role of energy in the urban transition: global challenges for sustainability." *BioPhysical Economics and Resource Quality* 4 (1): 5.

Burger, Joseph R., Anderson, Robert P., Balk, Meghan A., and Fristoe, Trevor S. 2019(b). "A constraint-based model of dynamic island biogeography: environmental history and species traits predict hysteresis in populations and communities." *Frontiers of Biogeography* 11 (3): e44383. http://dx.doi.org/10.21425/F5FBG44383.

Burger, Joseph R., and Fristoe, Trevor S. 2018. "Hunter-gatherer populations inform modern ecology." *Proceedings of the National Academy of Sciences* 115 (6): 1137–1139. https://doi.org/10.1073/pnas.1721726115.

Burger, J.R. 2018. "Modeling humanity's predicament." *Nature Sustainability* 1: 15–16. https://doi.org/10.1038/s41893-017-0010-z.

Burger, J.R., Hou, C., and Brown, J.H. 2019. "Toward a metabolic theory of life history." *Proceedings of the National Academy of Sciences* 116 (52): 26653–26661. 10.1073/pnas.1907702116.

Burger, Joseph R., Weinberger, Vanessa P., and Marquet, Pablo A. 2017. "Extra-metabolic energy use and the rise in human hyper-density." *Scientific Reports* 7. doi:10.1038/srep00056.

Burnside, William R., James H. Brown, Oskar Burger, Marcus J. Hamilton, Melanie Moses, and Luis Bettencourt. 2012. Human macroecology: linking pattern and process in big-picture human ecology. *Biological Reviews of the Cambridge Philosophical Society* 87 (1): 194–208.

Cohen, Joel E., and Small, Christopher. 1998. "Hypsographic demography: the distribution of human population by altitude." *Proceedings of the National Academy of Sciences* 95 (24): 14009–14014. doi:10.1073/pnas.95.24.14009.

Costanza, Robert, and Patten, Bernard C. 1995. "Defining and predicting sustainability." *Ecological Economics* 15 (3): 193–196.

Crutzen, Paul J. 2002. "Geology of mankind." *Nature* 415: 23. doi:10.1038/415023a.

Curtis, Philip G., Slay, Christy M., Harris, Nancy L., Tyukavina, Alexandra, and Hansen, Matthew C. 2018. "Classifying drivers of global forest loss." *Science* 361 (6407): 1108–1111.

Daly, Herman E. 2005. "Economics in a full world." *Scientific American* 293 (3): 100–107.

Damuth, John. 1981. "Population density and body size in mammals." *Nature* 290 (5808): 699–700.

Day, John W., and Hall, Charles. 2016. *America's Most Sustainable Cities and Regions.* New York: Springer. https://doi.org/10.1007/978-1-4939-3243-6.

Day, John W., Hall, Charles A., Yáñez-Arancibia, Alejandro, Pimentel, David, Ibáñez Martí, Carles, and Mitsch, William J. 2009. "Ecology in times of scarcity." *Bioscience* 59 (4): 321–31.

Di Marco, Moreno, Baker, Michelle L., Daszak, Peter, De Barro, Paul, Eskew, Evan A., Godde, Cecile M., Harwood, Tom D., et al. 2020. "Opinion: sustainable development must account for pandemic risk." *Proceedings of the National Academy of Sciences* 117 (8): 3888–3892.

Downing, Andrea S., van Nes, Egbert H., Mooij, Wolf M., and Scheffer, Marten. 2012. "The resilience and resistance of an ecosystem to a collapse of diversity." *PloS One* 7 (9): e46135.

ECLAC (Economic Commission for Latin American and the Caribbean). 2016. Economic Survey of Latin America and the Caribbean 2016: The 2030 Agenda for Sustainable Development and the challenges of financing for development. www.cepal.org/en/publications/40327-economic-survey-latin-america-and-caribbean-2016-2030-agenda-sustainable

Ellis, Erle, Maslin, Mark, Boivin, Nicole, and Bauer, Andrew. 2016. "Involve social scientists in defining the Anthropocene." *Nature* 540 (7632): 192.

Folke, Carl, Carpenter, Steve, Elmqvist, Thomas, Gunderson, Lance, Holling, Crawford S., and Walker, Brian. 2002. "Resilience and sustainable development: building adaptive capacity in a world of transformations." *AMBIO: A Journal of the Human Environment* 31 (5): 437–440.

Fragkias, Michail, Lobo, José, Strumsky, Deborah, and Seto, Karen C. 2013. "Does size matter? Scaling of CO_2 emissions and US urban areas." *PloS One* 8 (6): e64727.

Goldberg, Amy, Mychajliw, Alexis M., and Hadly, Elizabeth A. 2016. "Post-invasion demography of prehistoric humans in South America." *Nature* 532 (7598): 232–235.

Gralla, Fabienne, Abson, David J., Møller, Anders P., Lang, Daniel J., and von Wehrden, Henrik. 2017. "Energy transitions and national development indicators: a global review of nuclear energy production." *Renewable and Sustainable Energy Reviews* 70: 1251–1265.

Grewal, Sharanbir S., and Grewal, Parwinder S. 2012. "Can cities become self-reliant in food?" *Cities* 29 (1): 1–11.

Gudynas, Eduardo. 2013. "Debates on development and its alternatives in Latin America: A brief heterodox guide." *Beyond Development* 15. www.rosalux.org.ec/pdfs/BeyondDevelopment.pdf#page=16.

Hall, Charles A.S., and Klitgaard, Kent. 2018. *Energy and the Wealth of Nations: An Introduction to Biophysical Economics*. Switzerland: Springer.

Hall, Myrna H. P., and Balogh, Stephen B. 2019. *Understanding Urban Ecology: An Interdisciplinary Systems Approach*. Switzerland: Springer.

Hamilton, Marcus J., Burger, Oskar, and Walker, Robert S. 2012. "Human ecology." In *Metabolic Ecology*, edited by Richard M. Sibly, James H. Brown, and Astrid Kodric-Brown. pp. 248–257. Hoboken, NJ: John Wiley & Sons, Ltd. https://doi.org/10.1002/9781119968535.ch20.

Hammond, Sean T., Brown, James H., Burger, Joseph R., Flanagan, Tatiana P., Fristoe, Trevor S., Mercado-Silva, Norman, Nekola, Jeffrey C., and Okie, Jordan G. 2015. "Food spoilage, storage, and transport: implications for a sustainable future." *Bioscience* 65 (8): 758–768.

Hidalgo, C.A., B. Klinger, A-L Barabási, and R. Hausmann. 2007. The product space conditions the development of nations. *Science* 317 (5837): 482–487.

Holling, Crawford S. 2001. "Understanding the complexity of economic, ecological, and social systems." *Ecosystems* 4:390–405.

Irwin, Elena G., Jayaprakash, Ciriyam, and Munroe, Darla K. 2009. "Towards a comprehensive framework for modeling urban spatial dynamics." *Landscape Ecology* 24: 1223–1236. https://doi.org/10.1007/s10980-009-9353-9.

Diamond, Jared. 1997. *"Guns, Germs, and Steel: The Fates of Human Societies."* New York, NY: WW Norton & Company 14.

Jones, Kate E., Jon Bielby, Marcel Cardillo, Susanne A. Fritz, Justin O'Dell, C. David L. Orme, Kamran Safi, et al. 2009. "PanTHERIA: a species-level database of life history, ecology, and geography of extant and recently extinct mammals: ecological archives E090-184." *Ecology* 90 (9): 2648–2648.

Jung, Martin, Arnell, Andy, de Lamo, Xavier, García-Rangel, Shaenandhoa, Lewis, Matthew, Mark, Jennifer, et al. 2020. "Areas of global importance for terrestrial biodiversity, carbon, and water." *bioRxiv preprint*. https://doi.org/10.1101/2020.04.16.021444.

Kapan, Durrell D., Bennett, Shannon N., Ellis, Brett N., Fox, Jefferson, Lewis, Nancy D., Spencer, James H., Saksena, Sumeet, and Wilcox, Bruce A. 2006. "Avian influenza (H5N1) and the evolutionary and social ecology of infectious disease emergence." *EcoHealth* 3 (3): 187–194.

Kleiber, Max. 1932. "Body size and metabolism." *Hilgardia* 6 (11): 315–53.

Kubiszewski, Ida, Robert Costanza, Carol Franco, Philip Lawn, John Talberth, Tim Jackson, and Camille Aylmer. 2013. "Beyond GDP: measuring and achieving global genuine progress." *Ecological Economics* 93: 57–68.

Malthus, T.R. 1798. An essay on the principle of population, as it affects the future improvement of society with remarks on the speculations of Mr. Godwin, M. Condorcet, and other writers.

Max-Neef, Manfred A. 1991. *Human Scale Development: Conception, Application and Further Reflections.* USA: Apex Press.

Meyfroidt, Patrick, Rudel, Thomas K., Lambin, Eric F. 2010. "Forest transitions, trade, and the global displacement of land use." Proceedings of the National Academy of Sciences. Dec 2010, 107 (49) 20917-20922; DOI: 10.1073/pnas.1014773107

Meyfroidt, Patrick, and Lambin, Eric F. 2011. "Global Forest Transition: Prospects for an End to Deforestation." *Annual Review of Environment and Resources* 36(1). DOI:10.1146/annurev-environ-090710-143732

Moore, J.D. 2014. *A Prehistory of South America: Ancient Cultural Diversity on the Least Known Continent.* Louisville, CO: University Press of Colorado.

Moses, Melanie E., and Brown, James H. 2003. "Allometry of human fertility and energy use." *Ecology Letters* 6 (4): 295–300.

Nekola, Jeffrey C., Allen, Craig D., Brown, James H., Burger, Joseph R., Davidson, Ana D., Fristoe, Trevor S., Hamilton, Marcus J., Hammond, Sean T., Kodric-Brown, Astrid, and Mercado-Silva, Norman. 2013. "The Malthusian–Darwinian dynamic and the trajectory of civilization." *Trends in Ecology & Evolution* 28 (3): 127–130.

Ocampo, José Antonio. 2017. "Commodity-led development in Latin America." *Revue Internationale de Politique de Développement* 9 (9): 51–76.

Odum, Howard T. 1973. "Energy, ecology, and economics." *Ambio* 2 (6): 220–227.

Pacheco, L.F., Altrichter, M., Beck, H., Buchori, D., and Owusu, E.H. 2018. "Economic growth as a major cause of environmental crisis: comment to Ripple et al." *BioScience* 68 (4): 238. https://doi.org/10.1093/biosci/biy006.

Pacheco, L.F., Altrichter, M., Beck, H., Buchori, D., and Owusu, E.H. 2016. "Conservation as the new paradigm for development." In Aguirre, A.A., Sukumar, R. (Eds.). *Tropical Conservation: Perspectives on Local and Global Priorities*, 390–402. New York, NY: Oxford University Press.

Pérez, Carlota. 2010. "Technological dynamism and social inclusion in Latin America: A resource-based production development strategy." *CEPAL Review*. https://repositorio.cepal.org/handle/11362/11494.

Peters, Robert H. 1986. *The Ecological Implications of Body Size*. Vol. 2. Cambridge: Cambridge University Press. doi:10.1017/CBO9780511608551

Rees, William, and Wackernagel, Mathis. 2008. "Urban ecological footprints: why cities cannot be sustainable – and why they are a key to sustainability." In *Urban Ecology: An International Perspective on the Interaction Between Humans and Nature*, edited by John M. Marzluff, Eric Shulenberger, Wilfried Endlicher, Marina Alberti, Gordon Bradley, Clare Ryan, Ute Simon, and Craig ZumBrunnen, pp. 537–555. Boston, MA: Springer US.

Rothhammer, F., and Dillehay, T.D. 2009. "The late Pleistocene colonization of South America: an interdisciplinary perspective." *Annals of Human Genetics* 73: 540–549.

Sage, Rowan F. 2020. "Global change biology: a primer." *Global Change Biology* 26 (1): 3–30.

Schramski, John R., Woodson, C. Brock, and Brown, James H. 2020. "Energy use and the sustainability of intensifying food production." *Nature Sustainability* 3 (4): 257–259.

Schramski, John R., Gattie, David K., and Brown, James H. 2015. "Human domination of the biosphere: rapid discharge of the earth-space battery foretells the future of humankind." *Proceedings of the National Academy of Sciences of the United States of America* 112 (31): 9511–9517.

Schramski, John R., Woodson, C. Brock, Steck, Garrett, Munn, Dylan, and Brown, James H. 2019. "Declining country-level food self-sufficiency suggests future food insecurities." *BioPhysical Economics and Resource Quality* 4 (3): 12.

Sibly, Richard M., James H. Brown, and Astrid Kodric-Brown (Eds.). 2012. *Metabolic Ecology*. Hoboken, NJ: John Wiley & Sons.

Snyder-Mackler, Noah, Burger, Joseph Robert, Gaydosh, Lauren, Belsky, Daniel W., Noppert, Grace A., Campos, Fernando A., Bartolomucci, Alessandro, et al. 2020. "Social determinants of health and survival in humans and other animals." *Science* 368 (6493): eaax9553. https://doi.org/10.1126/science.aax9553.

Steffen, Will, Broadgate, Wendy, Deutsch, Lisa, Gaffney, Owen, and Ludwig, Cornelia. 2015. "The trajectory of the Anthropocene: the great acceleration." *The Anthropocene Review* 2 (1): 81–98.

Steffen, Will, Persson, Asa, Deutsch, Lisa, Zalasiewicz, Jan, Williams, Mark, Richardson, Katherine, Crumley, Carole, et al. 2011. "The Anthropocene: from global change to planetary stewardship." *Ambio* 40 (7): 739–761.

Stier, Andrew, Berman, Marc, and Bettencourt, Luis. 2020. "COVID-19 attack rate increases with city size." Preprint: www.medrxiv.org/content/10.1101/2020.03.22.20041004v2

Strumsky, Deborah, Lobo, José, and Tainter, Joseph A. 2010. "Complexity and the productivity of innovation." *Systems Research and Behavioral Science* 27 (5): 496–509.
Tainter, J. A. 2019. "Scale and Metabolism in Ancient Cities." In *Understanding Urban Ecology: An Interdisciplinary Systems Approach*, edited by Myrna H. P. Hall and Stephen B. Balogh, pp. 85–99. Springer Nature Switzerland, Cham.
Tallavaara, Miikka, Eronen, Jussi T., and Luoto, Miska. 2018. "Productivity, biodiversity, and pathogens influence the global hunter-gatherer population density." *Proceedings of the National Academy of Sciences* 115 (6): 1232–1237.
Ulanowicz, R.E. 1980. "An hypothesis on the development of natural communities." *Journal of Theoretical Biology* 85 (2): 223–245.
UNDP (United Nations Development Programme). 2019. "Human Development Report 2019: Beyond income, beyond averages, beyond today: Inequalities in human development in the 21st century." New York, NY: AGS. http://hdr.undp.org/sites/default/files/hdr2019.pdf
Weinberger, V. P., Quiñinao, C., and Marquet, P. A. 2017. Innovation and the growth of human population. *Philosophical Transactions of the Royal Society B – Biological Sciences*. https://doi.org/10.1098/rstb.2016.0415
West, Geoffrey. 2017. *Scale: The Universal Laws of Life, Growth, and Death in Organisms, Cities, and Companies*. New York, NY: Penguin Press.
The World Bank. 2019. "GDP per capita (current US$): Chile, Argentina, Uruguay." https://data.worldbank.org/indicator/NY.GDP.PCAP.CD?end=2020&name_desc=false&start=1960&view=chart
Wu, Tong, Perrings, Charles, Kinzig, Ann, Collins, James P., Minteer, Ben A., and Daszak, Peter. 2017. "Economic growth, urbanization, globalization, and the risks of emerging infectious diseases in China: a review." *Ambio* 46 (1): 18–29.
Zimmermann, Andreas, Hilpert, Johanna, and Wendt, Karl Peter. 2009. "Estimations of population density for selected periods between the Neolithic and AD 1800." *Human Biology* 81 (2–3): 357–380.

FRAMEWORK OF URBAN ECOLOGICAL SUSTAINABILITY INDICATORS IN DRYLANDS

Parastoo Parivar
Assistant Professor, School of Natural Resources & Desert Studies, Yazd University

David Quanrud
Research Scientist, School of Natural Resources and the Environment, The University of Arizona

Hassan Vafai
Research Professor, Civil & Architectural Engineering & Mechanics, The University of Arizona

INTRODUCTION

In 2008 for the first time in history, more than half of the world's population of 3.3 billion people lived in urban areas. Urban growth will continue to increase through the twenty-first century. By 2030, the world's population is expected to reach more than five billion and by 2050 reach 9.8 billion, and nearly 70% of this projected population, 6.7 billion people, is expected to live in urban areas (United Nations 2018). On the other hand, roughly 2.5 billion people (30% of the world's population) live in dry areas covering more than 40% of the world's land surface (CGIAR 2014). Scarce natural resources, water scarcity, frequent droughts, high climatic variability, extreme climatic conditions manifested by high temperatures and heat, various forms of land degradation, including desertification, and loss of biodiversity are characteristics of arid regions including desert cities (Chalfoun 2003). Many of the world's cities in arid areas utilize water supplies that originate from outside their own watersheds and rely on technology to import water from adjacent basins. As a result, urban development, food security, and citizens' satisfaction of a city's environment can all be heavily dependent on the importation of water from other watersheds. In addition, climate change has complicated the process of urban development and urban sustainability. For example, droughts caused by climate change have resulted in additional limits on water and food sources in arid areas.

In the absence of appropriate planning, urban development in arid regions can lead to negative consequences. Baker and colleagues showed that rapid expansion of urbanization in the Phoenix, Arizona area caused environmental consequences and had a negative impact on the regional life support system. Specifically, salinity of water and soil has increased and sustainability in general has decreased (Baker, Brazel, and Westerhoff 2004). In contrast, other studies conducted in the Sonoran Desert have shown the potential for sustainable development in arid regions. In 2008, Estelaji found that despite water and climate limitations in arid areas, there are potentials such as desert tourism, wind, and solar energy that can be the focus of sustainable development (Estelaji 2008). In an earlier 2003 study, Chalfoun focused on the design and architecture of urban sustainability in arid and desert regions (Chalfoun 2003). Scientists at the University of Arizona, along with a team of architects, designed strategies for the well-being of citizens and for energy conservation inside and outside of buildings.

Urban development, especially in arid lands, should be based on ecological capacity and on recognizing the limits of basic environmental services available within watersheds. Urban expansion based on borrowed environmental services beyond the ecological capacity of a watershed cannot ensure sustainability. Consequently, considering the constraints on natural resources for growth and sustainability of cities in arid areas, planners should assess sustainability and orientate development through a set of inclusive indicators. The main question in this study is which indicators best determine the sustainability of cities? In this paper, we present a framework for assessing the sustainability of cities in arid areas.

Framework of Urban Sustainable Indicators

Urban systems are understood as open living systems in which natural and social processes constitute mutual and integral subsystems (Ahren 2011; Pickett, Cadenasso, and Grove 2004). Scholars, within the realm of resilience, have created the term of social-ecological system, which can also be applied to cities. The term

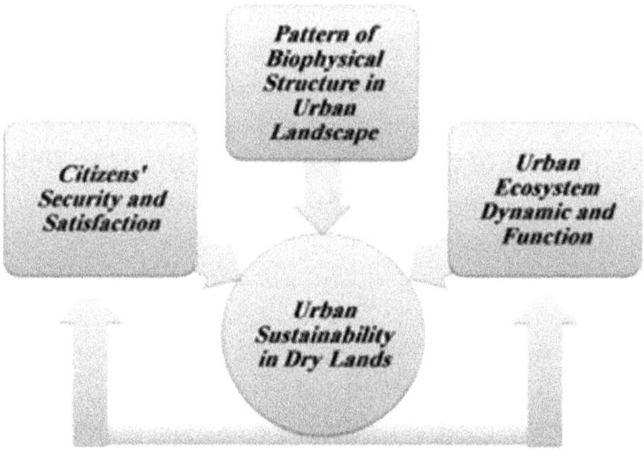

FIGURE 4.46 Framework of urban sustainability in drylands.

TABLE 4.2
Criteria for Assessing Urban Sustainability, Axis One

Criteria	Feature of Criteria with urban landscape
Diversity of urban structural pattern	*Spatial diversity in habitat patches that are sources of supply of ecosystem services in urban landscape*
Connectivity of urban structural pattern	*The presence of ecological networks (greenways, green belt)*
Modularity of urban structural pattern	*Spatial and temporal configuration of the structure that are sources of supply of ecosystem services in urban landscape*

social-ecological system elaborates on the integration between human society and nature, emphasizing the concept of humans-in-nature with mutual feedbacks and interdependence (Alberti and Marzluff 2004; Berkes, Colding, and Folke 2003).

In a socio-ecological system there is a close relationship between structure and performance (Alberti et al. 2003; Grimm et al. 2000; Pickett et al. 2001). In cities and urban regions, with regard to the decisions made related to land use, we experience things such as structure changing, fragmentation, isolation of natural habitats (Marzluff 2001), hydrological system failure (Arnold and Gibbons 1996), and changes of energy flow and food cycles (Grimm et al. 2000). These factors all can contribute to a reduction in resilience on different spatial scales and can make urban systems increasingly vulnerable. Based on this approach, we can improve the ecological performances and sustainability of urban habitats by reforming the urban structure (Alberti and Marzluff 2004). A conceptual model of new urban ecology developed by Alberti and colleagues (2003) shows the interactions between driving forces, structures, functions, and impacts in urban environments. Using this cognition, we can develop suitable strategies and find the location of measures in promoting urban sustainability. On the other hand, sustainable urban land development in drylands should also consider economics, community, and the environment, and achieving a balance among these three components that responds to the present and future needs of the people (Herber 1992). In this paper, based on the new definitions of cities and urban sustainability mentioned earlier, the framework for assessing urban sustainability, especially in arid regions, is expanded (see Figure 4.46).

This framework consists of three main axes of urban sustainability: (1) *Pattern of Biophysical Structure in Urban Landscape* – indicators in this dimension are related to city structure and are defined based on the infrastructure of the city's vital network design; (2) *Urban Ecosystem Dynamic and Function* – this category includes environmental services or slow variables. Regulatory ecosystem services are one of the slow-moving variables that are slowly organized after the intervention in landscape (Isbell et al. 2011; Schouten et al. 2012); and (3) *Citizens' Security and Satisfaction* – this category relates to the positive feelings of citizens about the urban landscape, which is known as biophilia or human and nature relationship. Citizens living in dry cities are confronted with a dry urban landscape, limited vegetation, and fear of lack of adequate water.

Pattern of Biophysical Structure in Urban Landscape (Axis One)

The criteria for assessing pattern of biophysical structure in urban landscapes and their application in urban sustainability are shown in Table 4.2.

Considering these three criteria, appropriate landscape metrics for assessing urban sustainability based on pattern of biophysical structure are listed in Table 4.3. In the following, the criteria for measuring urban sustainability are explained based on the Pattern of Biophysical Structure.

Diversity of Urban Structural Pattern

Diversity plays a crucial role in sustainability. When high diversity exists in a system, different parts can replace or compensate for each other at times of disorder (Walker and Salt 2006). When there is high uncertainty in the system, diversity promotes decentralization of disaster responses, creation of buffers, and development of multiple strategies through learning. Diversity is also effective for reorganizing the processes following disturbances or events that cause changes in social ecological systems by creating a framework for innovation or adaptive capacity (Berkes, Colding, and Folke 2003). Low diversity in a system leads to vulnerability or failure in one of the special applications of the system. That is why the processes contributing to the decrease in system diversity must be recognized (Gunderson et al. 2010). Biodiversity and social, physical, and economic diversity are important strategies to support urban resilience and increase system capacity against disorders (Alberti and Marzluff 2004). Diversity is related to spatial capacity and creates multiple performance spaces such as shelter and supporting area. Oliver et al. (2010) showed that spatial heterogeneity and diversity in landscapes create resources for species that increase their resilience at the time of a crisis such as climate crisis.

Connectivity of Urban Structural Pattern

Ecological connectivity aims to protect natural corridors in order to maintain potential for transfer of biodiversity among habitants in environments that have been

TABLE 4.3
Landscape Metrics for Assessing Urban Sustainability, Axis One

Pattern of biophysical structure criteria	Measurable variables or data	Metric	Definition of metric
Diversity or spatial heterogeneity	*The amount of green and open patches coverage* *Number of green and open patches* *Richness of green and open patches*	*Patch Density (PD)*	*Patch density is the most useful metric as the basis for computing other more interpretable metrics and is determined as the total number of patches in the city divided by the total area of the landscape and is expressed as Number per 100 hectares. Through this metric we can compare different landscapes.*
		Percentage of Landscape (PLAND)	*This metric measures landscape composition; specifically, how much of the landscape comprises a particular patch type as percentage.*
Connectivity	*Mean distance between each class of patches*	*Euclidean Nearest-Neighbor Distance (ENN)*	*This metric shows the mean nearest neighbor distance between one type patch. This metric computes the shortest straight-line distance between the focal patch and its nearest neighbor of the same class.* *The unit of this metric is meter.*
		Radius of Gyration (GYRATE)	*This metric, for each patch, is calculated as the average distance between each cell in the patch to the center of the patch.* *The unit of this metric is meter.*
Modularity	*Amounts of heterogeneity or homogeneity of urban green and open patches*	*Contagion index (CONTAG)*	*This metric shows the composition and texture of land cover. Contagion metric measures both patch type interspersion and patch dispersion at the landscape level. Based on the result of this metric recognized the extent to which patch types are aggregated or clumped.* *The unit of this metric is percentage*

TABLE 4.4
Criteria for Assessing Urban Sustainability, Axis Two

Criteria	Feature of Criteria with urban landscape
Urban air quality	*One of the characteristics of the drylands is the occurrence of dust storms which in recent years, due to the impacts of climate change, have had many environmental impacts, especially in cities*
Urban water resources	*Most cities in arid lands depend on water resources outside the watershed. Inter-watershed water transfer is a way to meet the water needs of these cities, which causes instability in the watersheds of the source and destination (sink)*
Urban soil	*Urban growth and construction are among the main reasons for the soil destruction and shifting of the soil surfaces to impermeable layers*
Urban climate	*Drought is one of weather phenomena and one of the occurrences that cause a lot of damage each year. It also increases the dependence of cities in arid lands on water resources outside their watershed*

dramatically altered (Gunderson et al. 2010). In urban environments, the connectivity of built elements is relatively high, but the connectivity of natural elements is relatively low. The low connectivity of natural systems in urban landscapes leads to fragmentation, separation, and isolation of these elements. In addition, it greatly influences specific ecological processes that need connectivity such as movement of species. Revival or regeneration of structural connectivity of natural elements at different spatial scales in urban environments can improve resilience capacity through repeating the orbiting that results in keeping performance connectivity. Structural connectivity of ecological service producer elements is the origin of urban resilience (Ahren 2011).

Modularity of Urban Structural Pattern

Modularity is defined as the division of a system into parts with respect to space, time, or organizational structure and refers to the degree to which a system's components may be separated and recombined (Schouten et al. 2012). The parts of a modular system perform specific actions and operate independently (Berkes 2007). In other words, modularity is the lack of connectivity in spatial-temporal system distribution (Levin 2008). There is no optimal level for modularity but a system dependent on completely related parts can be compromised by a contagion (such as a disease or adverse administrative action) affecting any specific part of the whole system (Earn, Levin, and Rohani 2000; Gunderson et al. 2010). This criterion is true for both social-economic and ecological systems. In a resilient social-ecological system, the different parts are related to each other in an irregular way but are also independent from each other so that disorders do not compromise the integrity of the whole system.

Urban Ecosystem Dynamic and Function (Axis Two)

Urban ecosystem dynamic and function is about diversity of types of ecosystem services or slow variables that are beneficial for urban landscapes. Biophysical variables such as air quality, water resources, soil, and climate are considered as the basic factors of urban ecosystem services. Urban ecosystem dynamic and function criteria and their application in urban sustainability are shown in Table 4.4.

In the following, urban sustainability metrics based on the urban ecosystem dynamic and function are explained.

- **Urban Air Quality**

Air pollution is a serious problem in many urban environments. In general, air pollution can cause serious effects on people's health and may also contribute to economic, social, and cultural damage. The most common air quality measurement indices is the Air Quality Index (AQI), which is based on the five air pollution measurements CO, NO_2, SO_2, O_3, and PM2.5 (fine-particulate matter with a diameter of 2.5 microns or less). In recent years, air pollution from dust storms has increased in many urban areas of the world. Dust storms can be very vast and can travel long distances prior to affecting metropolitan areas.

- **Urban Water Resources**

Reliable water supply is the most important issue and challenge facing many countries, especially in arid and desert regions. By the year 2050, more than 66 countries, containing about two thirds of the world's population, will face water shortage (CGIAR 2014). Water demands for cities in arid zones are often supplied from resources located outside of a city's watershed. Moreover, reliance on groundwater has increased in many areas and aquifers are considered strategic resources in arid regions. Due to reduced rainfall and shortage of surface water resources in many arid and semiarid areas in recent years, groundwater has become an increasingly important and primary source of water supply for various uses, especially agriculture. This increased reliance on aquifers has strained groundwater resources, leading to groundwater overdraft and unsustainable actions such as drilling of ever deeper wells (CGIAR 2014; Perrone and Jasechko 2019).

- **Urban Soil**

Soil is a living environment for many types of organisms and is also essential for human existence. The presence of suitable soils along with sufficient water supply has been the basis for the formation and maintenance of human civilizations (Aghelikohneh and Sadeghi 2005). The most important land use considered as a major cause of soil degradation is urbanization and related activities such as building roads, industries, and farming. The increase in construction results in an increase of occupied land and decrease of pervious surfaces in urban regions. Such changes in urban environmental structure have environmental impacts such as decreasing

TABLE 4.5
Metrics for Assessing Urban Sustainability, Axis Two

Urban ecosystem dynamic and Function	Measurable variables or data	Metric	Definition of metric
Urban air quality	*Air quality*	*Concentration of natural pollutants*	*AQI* *The ratio of the days of the dust phenomenon occurrence in the year*
Urban water resources	*Water resources quantity*	*The volume of inter-basin water transfer*	*Changing in volume of inter-basin water transfer*
		Groundwater	*Changing in volume of groundwater*
Urban/Soil	*Pollution and soil destruction*	*Impervious surfaces*	*Change in ratio of impervious surfaces*
Urban Climate	*Drought*	*Drought severity*	*SPI*
	Climate regulation	*UV index*	*UV radiation indicator*

TABLE 4.6
Criteria for Assessing Urban Sustainability, Axis Three

Criteria	Feature of Criteria with urban landscape
Health	*The existence of a connected and integrated environmental network for citizens communication with nature*
Urban governance and biophilic institute	*Investing to facilitate citizens' communication with nature*
Biophilic activities	*The existence of infrastructures facilitating the communication of citizens with nature*

groundwater recharge, increasing storm flow rate, increasing flooding frequency, creating heat island effect, and changing hydrological cycle (Marzluff 2001). Table 4.5 shows the appropriate landscape metrics for assessing urban sustainability based on urban ecosystem dynamic and function criteria.

- **Urban Climate**

The Standardized Precipitation Index (SPI) is widely used to characterize meteorological drought on a range of timescales and was first used by McKee et al. (1993). This index can be calculated for any geographic location based on long-term recorded rainfall data. According to the SPI, the average standardized precipitation for each region and for each arbitrary period is assigned a value of zero and a standard deviation value of one. Positive SPI values indicate above average rainfall and negative values indicate below average rainfall. According to this index, a drought period begins when the SPI is continuously negative and reaches a value of –1 or less and a drought period ends when the SPI again becomes positive (Akbari et al. 2016).

The ultraviolet (UV) radiation index is an international standard for measuring the intensity of UV radiation emitted by the sun, which in excess is harmful to human health. The value of this index varies by time of day and time of year. The UV index is an appropriate tool for raising public awareness about the dangers of over-exposure to ultraviolet rays and adopting protective methods (Bouzarjomehri and Tsapaki 2012).

Satisfaction and Security of Citizens (Axis Three)

The satisfaction and security of citizen's criteria is related to Biophilia theory. According to this theory (Kellert and Calabrese 2015), two points are important in urban planning and design: (1) citizen interaction with nature should be frequent and sustainable, and (2) citizen health is based on a relationship with nature. Biophilic cities are those where citizens' inherent tendency to life and life-like processes concentrate and provide useful communication between people and nature (Derr and Lance 2012). The inclusion of water, plants, and animals in the construction of the urban environment is important for the communication of citizens with nature (Rakhshandehroo et al. 2015). Criteria for citizens' security and satisfaction index and their application in urban sustainability are shown in Table 4.6.

In the following, urban sustainability metrics based on citizen's security and satisfaction criteria are explained.

Health

Daily urban life, in a harsh and busy environment, is very stressful. The human brain is negatively affected by noise and air pollution and studies have shown this may lead to headache, depression, and anger. There is growing evidence that access to nature in urban areas has a positive impact on human health and well-being, and that it also

TABLE 4.7
Metrics for Assessing Urban Sustainability, Axis Three

Security and satisfaction of the citizen	Measurable variables or data	Metrics definition
Health	*Urban citizens access to nature*	*Percent of citizens that have at least 30 minutes access to nature*
Urban governance and biophilic institute	*Directive of urban design and planning*	*Local budgets allocated to nature conservation, recreation, education, and related activities by municipalities*
		Number of NGOs (local biophilic supporting organizations)
Biophilic activities	*Recreation*	*Percentage of population active in clubs or nature or outdoor organizations*

increases the quality of life and behavioral well-being of residents. People exposed to natural environments have significantly reduced stress levels compared to those exposed to urban environments (Beatley and Peter 2013).

Organizations and Biophilic Governance

Biophilic cities should be identified not only with the presence or absence of nature, green space and green infrastructure, but also through other forms of investment that facilitate biophilic life. A biophilic city has a strong network of public and private institutions whose purpose is to invest in the effective communication of citizens with nature, as well as education for the reconstruction and conservation of nature (Derr and Lance 2012).

Biophilic Activity

Citizens use urban green spaces based on the needs, preferences, time, and physical capabilities for different purposes and recreation (Rakhshandehroo et al. 2015; Sadeghian and Vardanyan 2013). These recreations are both active and inactive. Active recreation usually involves sports facilities, while inactive recreation is more an environment for relaxation (Derr and Lance 2012). In sustainable cities, the necessary facilities for both groups of activities must be available. Table 4.7 shows the appropriate measures for each urban sustainability metric based on citizen's security and satisfaction criteria.

Conclusion

The most important limitation on development in urbanized arid regions is availability of water resources. In order to meet the water needs of cities in these areas, inter-basin water transfers are often necessary and rely on technological solutions. These borrowed water resources can provide a false sense of security for residents and planners, leading to greater development of services and industries in the city and suburbs. Droughts in recent years have potentially worsened due to climate change and have caused water shortages in many urbanized areas. Inter-basin water transfers have increased to satisfy water demands in dry cities, increasing the potential for instability and conflict among regions competing for the same water resources. Hence, wars and tensions have occurred in several parts of the world, especially in the Middle East, because of water. The major concern of many citizens in these areas is the lack of water, fierce urban landscape, and lack of connection with nature. It is therefore necessary to assess the sustainability of cities in these areas. In this paper, we developed a three-axis indicator framework to assess the sustainability of cities in arid areas, considering the definition of the city as a social ecological system.

- First axis: pattern of the city's biophysical structure. This axis is related to the development of a vital network throughout the city. Based on this group of criteria, planners could determine whether ecosystem services expanded effectively throughout the city and can be effective in the whole city.

- Second axis: performance and dynamics of the urban ecosystem. This axis is related to the life support system or the base ecosystem services in the city and assesses whether the city is sustainable in terms of its life support system.
- Third axis: citizens' security and satisfaction. This axis relates to the reflection of the previous axes on citizens. Based on the indicators of this axis, a city can be evaluated in terms of expected performances of citizens. A sustainable city is where its citizens feel safe and happy with their present and future. A sustainable city should also inspire citizens to feel a relationship with nature.

Using the metrics described in this proposed framework along with a multi-criteria ranking method that combines expert opinion with data surveying, planners can assess the level of sustainability in their cities. If the assessment suggests a city's sustainability ranking is lacking, we propose that planners should adopt an approach to urban growth and development that increases their sustainability, based on the three indicator axes in this study. General strategies for achieving urban sustainability in arid regions include the following:

- Planning to modify the composition and configuration of open and green patches as a sustainable urban refuge network;
- Planning to revitalize the vital functions of the city (urban ecosystem services);
- Increasing the resiliency of the city from risks like droughts and dust;
- Investing in creating recreational infrastructure for citizens;
- Developing institutions and NGOs to promote citizen's environmental knowledge and engage them in urban sustainable plans;
- Developing sustainable urban design strategies to create a platform for the citizens' biophilic activities.

In conclusion, urban development should be based on ecological capability and basic environmental services available within watersheds. Reliance on external water resources and environmental services leading to urban expansion beyond the ecological capacity cannot ensure sustainability.

REFERENCES

Aghelikohneh, S.L., and Sadeghi, H. 2005. "Estimation of economic effects of soil erosion in Iran." *The Economic Research* 5: 8–100.

Ahren, J. 2011. "From fail to safe: sustainability and resilience in the new urban world." *Landscape and Urban Planning* 100(4): 341–343.

Akbari, M., Ownegh, M., Asgari, H.R., Sadoddin, A., and Khosravi, H. 2016. "Drought monitoring based on the SPI and RDI indices under climate change scenarios (case study: semi-arid areas of West Golestan Province)." *Ecopersia* 4 (4): 1585–1602.

Alberti, M., and Marzluff, J. 2004. "Ecological resilience in urban ecosystems linking urban patterns to human and ecological functions." *Urban Ecosystems* 7 (3): 241–265.

Alberti, M., Marzluff, J.M., Shulenberger, E., Gordon, B., Ryan, C., and Zumbrunnen, C. 2003. "Integrating humans into ecology: opportunities and challenges for studying urban ecosystems." *BioScience* 53 (12): 1169–1179.

Arnold Jr., C.L., and Gibbons, C.J. 1996. "Impervious surface coverage: the emergence of a key environmental indicator." *Journal of the American Planning Association* 62 (2): 243–258.

Baker, L.A., Brazel, A.T., and Westerhoff, P. 2004. "Environmental consequences of rapid urbanization in warm, arid lands: case study of Phoenix, Arizona (USA)." In *The Sustainable City III, Urban Regeneration*, edited by C. Marchettini, A. Brebbia, E. Tiezzi, and L. C. Wadh, 155–164. Boston: WIT Press.

Beatley, T., and Peter, N. 2013. "Biophilic cities are sustainable, resilient cities." *Sustainability* 5: 3328–3345.

Berkes, F. 2007. "Understanding uncertainty and reducing vulnerability: lessons from resilience thinking." *Natural Hazards* 41 (2): 283–295.

Berkes, F., Colding, J., and Folke, C. 2003. *Navigating Social-Ecological Systems: Building Resilience for Complexity and Change.* Cambridge: Cambridge University Press.

Bouzarjomehri, F., and Tsapaki, V. 2012. "Measurement of solar ultraviolet radiation in Yazd, Iran." *Iranian Journal of Radiation Research* 10: 187–191.

CGIAR Research Program on Dryland Systems. 2014. *Annual Report 2014: Pathways to Lasting Impact for Rural Dryland Communities in the Developing World.* Amman, Jordan: CGIAR.

Chalfoun, N.V. 2003. "Sustainable urban design in arid regions; integrating energy and comfort." *Collaborative Symposium: Urban Design in Arid Regions.* Tucson, AZ: The University of Arizona and The Pontificia Universidad Catolica De Chile.

Derr, V., and Lance, K. 2012. "Biophilic boulder: children's environments that foster connections to nature." *Children, Youth and Environments* 22: 112–143.

Earn, D.J., Levin, S.A., and Rohani, P. 2000. "Coherence and conservation." *Science* 290: 1360–1364.

Estelaji, A. 2008. "Approaches to sustainable development of arid regions (A case study in arid regions of Iran)." *Desert* 13: 53–58.

Grimm, N.B., Grove, J.G., Pickett, S.T., and Redman, C.L. 2000. "Integrated approaches to long-term studies of urban ecological systems." *BioScience* 50 (7): 571–584.

Gunderson, L., Kinzig, A., Quinlan, A., and Walker, B. 2010. *Assessing Resilience in Social-Ecological Systems: A Workbook for Practitioners. Revised Version 2.0.* Resilience Alliance. [Online] URL: www.resalliance.org/files/ResilienceAssessmentV2_2.pdf

Herber, G. 1992. *New Direction for Sustainable Urban Living*. London: Gaia Books.

Isbell, F., Calcagno, V., Hector, A., Connolly, J., and Harpole, W.S. 2011. "High plant diversity is needed to maintain ecosystem services." *Nature* 477: 199–202.

Kellert, S.R., and Calabrese, E.F. 2015. *The Practice of Biophilic Design.* Paris: www.biophilic-design.com.

Levin, S. 2008. "Resilience, robustness, and marine cosystem-based management." *BioScience* 58: 27–32.

Marzluff, J.M. 2001. "Worldwide urbanization and its effects on birds." In *Avian Ecology and Conservation in an Urbanizing World*, edited by J.M. Marzluff, R. Bowman, and R. Donnelly, pp. 19–47. New York: Academic Publishers.

McKee, T.B., Doesken, N.J., and Kliest, J. 1993. "The relationship of drought frequency and duration to time scales." In *Proceedings of the 8th Conference of Applied Climatology, 17–22 January, Anaheim, CA*, American Meteorological Society, Boston, MA, pp. 179–184.

Oliver, T., Roy, D.B., Hill, J.K., Brereton, T., and Thomas, C.D. 2010. "Heterogeneous landscapes promote population stability." *Ecology Letters* 13 (4): 473–484.
Perrone, D. and Jasechko, S. 2019. "Deeper well drilling an unsustainable stopgap to groundwater depletion." *Nature Sustainability* 2: 773–782. www.nature.com/articles/s41893-019-0325-z.
Pickett, S.T., Cadenasso, M.L., and Grove, J.M. 2004. "Resilient cities: meaning, model and metaphor for integrating the ecological, socio-economic and planning realms." *Landscape and Urban Planning* 69: 369–384.
Pickett, S., Cadenasso, M.L., Grove, J.M., Nilon, C.H., Pouyat, R.V., Zipperer, W.C., and Costanza, R. 2001. "Urban ecological systems: linking terrestrial ecological, physical, and socioeconomic components of metropolitan areas." *Annual Review of Ecology and Systematics* 32: 127–157.
Rakhshandehroo, M., Mohdyusof, M.J., Tahirholder, O.M., and Yunos, M.Y. 2015. "The social benefits of urban open green spaces: a literature review." *Management Research and Practice* 7 (4): 60–71.
Sadeghian, M.M., and Vardanyan, Z. 2013. "The benefits of urban parks, a review of urban research." *Journal of Novel Applied Sciences* 2 (8): 231–237.
Schouten, M., Van der Heide, C.M., Heijman, W.J., and Opdam, P.F. 2012. "A resilience-based policy evaluation framework: application to European rural development policies." *Ecological Economics* 81 (C): 165–175.
United Nations. 2018. *World Urbanization Prospects*. https://population.un.org/wup/.
Walker, B.H., and Salt, D. 2006. *Resilience Thinking: Sustaining Ecosystems and People in a Changing World*. Washington, DC: Island Press.

A STRATEGIC FRAMEWORK FOR CREATING RESILIENT, ECOLOGICAL AND SUSTAINABLE INFRASTRUCTURE SYSTEMS – AN OVERVIEW

Ying Zhao
International Institute for Urban Systems Engineering, Southeast University

Mohammad Noori
Department of Mechanical Engineering, California Polytechnic State University

Wael Altabey
International Institute for Urban Systems Engineering, Southeast University

Zhishen Wu
International Institute for Urban Systems Engineering, Southeast University; Department of Urban and Civil Engineering, Ibaraki University

INTRODUCTION

Rapid economic development and extensive globalization have expanded the scale of infrastructure systems, which needs a closer connectivity of subsystems to deal with natural and human triggered disasters unrestricted to geographic vicinity. These factors lead to increasing risks, sensitivity, and vulnerability of infrastructure systems. A resilient infrastructure system can be viewed as a sophisticated time-varying, multi-dimensional, and interconnected system, and it is a system that is capable of surviving and recovering from the likelihood of damage due to unexpected disruptions. These changes or disruptions may improve, maintain, or lower system performance within a short period of time or a long term. In most cases, disasters or extreme events generally increase system risk, decrease system robustness, and have a negative impact on population and their support system (Lounis and McAllister 2016). Increased emphasis with combined strategies was put on vulnerability and protection of infrastructure systems over the past few decades. These infrastructure systems mainly include buildings, bridges, highways, electric power grids, dams, and traffic networks that provide the backbone of the network of society, sustaining the lifeline system. To meet the global challenges of infrastructure provision and the efficient allocation of limited resources, it is increasingly imperative to incorporate resilience, ecology, and sustainability as key attributes, and explore and propose a strategy for infrastructure development conceived from a whole life cycle perspective (Lounis and McAllister 2016).

Background

An infrastructure system is defined as "a network of human-made systems and processes that function cooperatively and synergistically to produce and distribute a continuous flow of essential goods and services." Another definition is "a network of independent, mostly privately-owned, manmade systems and processes that function collaboratively and synergistically to produce and distribute a continuous flow of essential goods and services." It is a cyber-physical-social system that have not only interconnected physical components but also include processes as an integral constituting part (Kwasinski 2014). Natural support, economic metabolism, and social regulation and complementary relationships, and the relationships among these three subsystems in time, space, quantity, and structure, constitute the coupled social–economic–natural complex ecosystem (Wang et al. 2018).

Smart infrastructure have become important carriers in the context of smart cities including five main components, i.e. smart architecture, intelligent transportation system, smart energy, intelligent manufacturing, and shared economy. Smart architecture refers to the effective interaction between architectures, energy to make the architecture function efficiently. Intelligent transportation systems include vehicle-to-vehicle (V2V), vehicle-to-infrastructure (V2I), and infrastructure-to-infrastructure (I2I), and are the application of sensing, analysis, control, and communication technologies in order to improve safety, mobility, and efficiency of ground transportation systems. Smart energy mainly focuses on the use of renewable resources such as solar, wind, hydroelectric, and geothermal energy to achieve a high-quality and comfortable lifestyle. Intelligent manufacturing is to optimize manufacturing resource allocation, improve business value and safety by using combined intelligence of people, processes, and machines. Sharing economy provides an online platform to sustain a sharing access to superior goods and services, and sustaining a reliable balance of demand and supply. With the limited space, energy and resources, risk management, and optimization of environmental performance are directly related to resilient and sustainable infrastructure systems and key resources (Diab 2017).

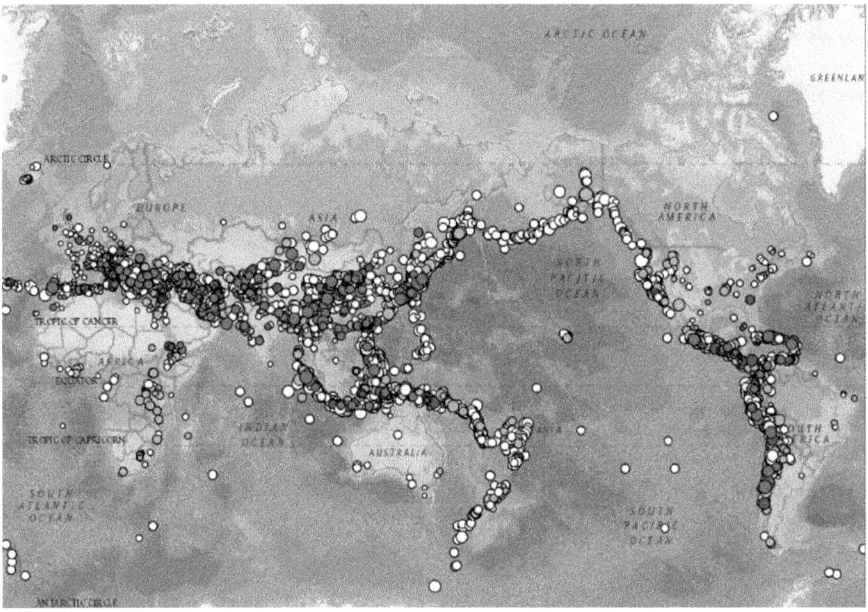

FIGURE 4.47 Global earthquake distribution map. (NOAA 2018.)

Infrastructure networks are large scale and complex systems characterized by unpredictable, unexpected, and uncertain, internal and external disturbances, interdependent and interactive components or subsystems, static and dynamic variable states with different resilient cycle period. Over 50% of the global population consumes 75% of natural resources of the Earth, where large infrastructure systems support the complex lifeline ecosystem (Chopra et al. 2016). It is imperative to face and cope with the threat of both local and global natural disasters and man-made disruptions, and create resilient and sustainable infrastructure networks and systems (Pandit et al. 2014).

Influence of Disturbances on Infrastructure Systems

The world has witnessed a number of natural and man-made disasters with unexpected forms, magnitudes, or locations. Preventative measures have been employed to mitigate and keep from these disturbances (Matthews 2015). Some of the most iconic examples of these disasters include the 2001 World Trade Center attack in New York City, 2017 tropical cyclone at Puerto Rico, 2009 bushfires in Victoria, 2009 Swine flu in Canada, 1996 USAF CT-43 crash in Croatia, 2003 heat wave in Europe, 2004 Indian Ocean earthquake in Sumatra, 2011 earthquake in Christchurch, 2011 meteorite impact at Tablas Island and Romblon, 2011 Fukushima Daiichi nuclear disaster initiated primarily by the tsunami following the Tōhoku earthquake, 2008 great earthquake at Wenchuan, China, to cite a few examples. These disasters

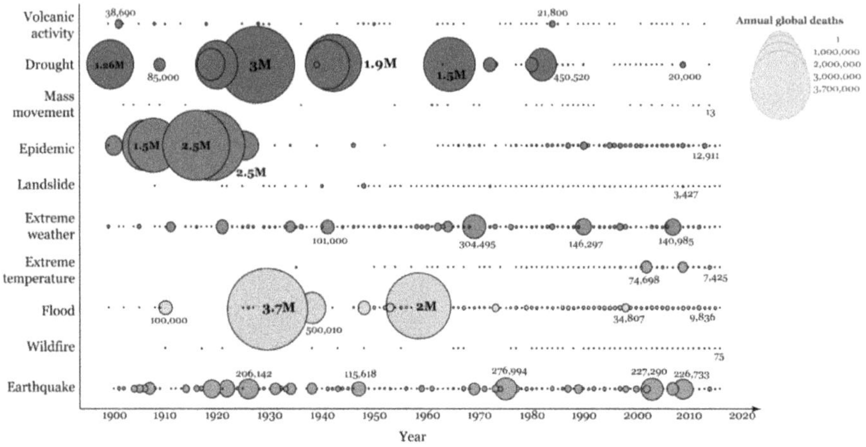

FIGURE 4.48 Global deaths from natural disasters. (Data sourced from Roser and Ritchie 2018.)

have caused severe damage and cascading failures to infrastructure systems, lifeline support systems and ecological, economic, and social chain, sending ripple effects throughout regional, national, and worldwide scales. The United Nations Office for Disaster Risk Reduction reported that the 2011 natural disasters, including the earthquake and tsunami that struck Japan, resulted in $366 billion in direct damages and 29,782 fatalities worldwide. Disasters are exacting a huge toll with hundreds of thousands of lives and US$1.5 trillion lost in the last decade alone. Economic losses from disasters are now averaging US$250 billion to 300 billion each year (UNISDR 2015).

Figure 4.47 shows the distribution of historically significant earthquakes worldwide, where the darker color represents the higher magnitude of earthquake (NOAA, National Centers for Environmental Information 2018). Figure 4.48 shows the relationship between the number of deaths and global natural disasters, where the bigger circle represents more death number (Roser and Ritchie 2018), where the flood disaster in about 1930 reached the maximum death number.

These disturbances and changes make infrastructure systems become increasingly sensitive and vulnerable, and have created challenges to regional stability, national security, and global sustainability. Specifically, severe disruptions may lead to unforeseen influences on interconnected infrastructure systems, which will result in cascading failures. Therefore, how to endow systems with capabilities to withstand adverse events, respond to risks, and facilitate effective recovery of disrupted systems become prompt issues in regard to sustainable resilience design.

Significance of Investing in Infrastructure Systems

Infrastructure systems constitute the backbone of society and economy, facilitating public's health and welfare and national prosperity. Massive important data from these infrastructure systems are analyzed (Mostafavi et al. 2012) by innovative

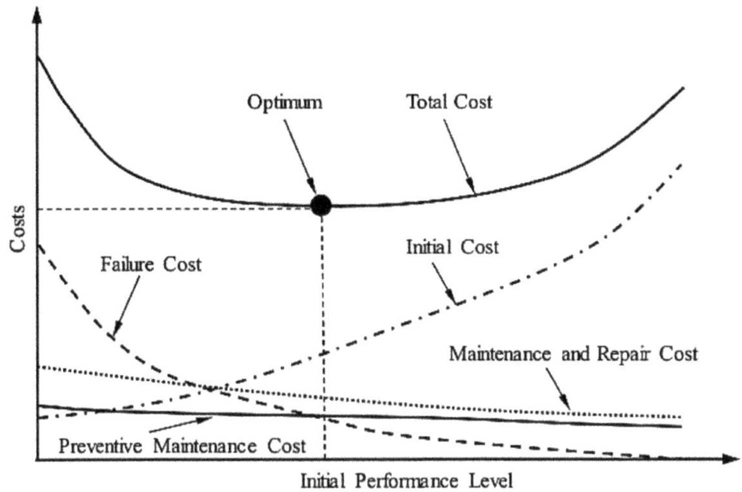

FIGURE 4.49 Balance between initial performance level and cost. (Qualitative illustration adapted from Frangopol and Saydam 2011.)

policy makers from government for investment, financing, and risk decision-making, multidomain experts and scholars from universities for research and development of new knowledge, tools and curricula, and engineers from industries for operation, maintenance, protection, and improvement.

"Infrastructure systems must be designed to protect the natural environment and to withstand both natural and man-made hazards, using sustainable practices, to ensure that future generations can use and enjoy what we build today, as we have benefited from past generations." The 2013 and 2017 ASCE Infrastructure Report Card's cumulative GPA of D^+ reflects the significant backlog of emerging needs facing the nation's infrastructures (ASCE Report Card 2013; ASCE Report Card 2017). Many factors that affect the creation, diffusion, and innovations of complex infrastructures comprise ecological periods, social networks, economic conditions, public attitudes, political priorities, and business and market dynamics.

Therefore, a comprehensive framework should be proposed regarding resilience design and enhancement for time-varying networked infrastructure systems with vulnerability and uncertainty.

Life-cycle cost and performance level of infrastructure systems constitute two conflicting criteria that should be balanced, as qualitatively illustrated in Figure 4.49 (Frangopol and Saydam 2011). The expected total cost of a general structure during the whole lifecycle can be expressed as follows:

$$C_{ET} = C_T + C_{PM} + C_{INS} + C_{REP} + C_F \quad (6)$$

where C_{PM} is the initial cost, C_{INS} is the expected cost of routine maintenance cost, C_{REP} is the expected cost of inspections, C_F is the expected cost of repair, and $t_0 < t < t_d$ is the expected failure cost. By minimizing the total cost while keeping structural safety at a desired level, the proper allocation of resources can be achieved at an optimum (Frangopol et al. 1997).

The Global Infrastructure Outlook forecasts infrastructure investment needs across 56 countries, seven sectors, and five regions up to 2040 (GIO 2018). According to a report by the World Bank Group (Heathcote, 2017), over half of global infrastructure investment needs is in Asia. The United States is forecast to have the largest infrastructure investment gap – the difference between investment needs and current trends in investment – of $3.8 trillion, double the next largest of China at $1.9 trillion, followed by Brazil at $1.2 trillion and Russia at $0.7 trillion. Achieving adaptive, robust infrastructures and sustaining their reliability, robustness, and efficiency are critical long-term issues that require strategic investments in research and development (Amin 2002).

Resilience, Ecology, and Sustainability

Understanding the relationship, and the impact, of sustainability and ecology in resilience study is important. The following sections elaborate this relationship.

TABLE 4.8
Resilience Definitions

Year	Author	Definition
1973	Holling	Ecological systems resilience is a measure of persistence of systems and of their ability to absorb change and disturbance and still maintain the same relationships between populations or state variables.
1997	Horne and Orr	Resilience is the ability of a system to withstand stresses of "environmental loading"... [it is] a fundamental quality found in individuals, groups, organizations, and systems as a whole.
1998	Haimes et al.	Resilience is the ability of system to return to its optimal condition in a short period time. Consider resilience as one of four strategies for hardening a system, together with security, redundancy and robustness.
1999	Mileti	Local resiliency with regard to disasters indicates that a locale is able to withstand an extreme natural event without suffering devastating losses, damage, diminished productivity, or quality of life and without a large amount of assistance from outside the community.
1999	Comfort	Resilience is the capacity to adapt existing resources and skills to new situations and operating conditions.
2003	Bruneau	Resilience is defined in terms of three stages: the ability of a system to reduce the probability of an adverse event, to absorb the shock if the adverse event occurs, and to quickly reestablish normal operating conditions.

TABLE 4.8 (Continued)
Resilience Definitions

Year	Author	Definition
2005	Allenby and Fink	Resilience is defined as the capability of a system to maintain its functions and structure in the face of internal and external change and to degrade gracefully when it must.
2005	Rose and Liao	Regional economic resilience is the inherent ability and adaptive response that enables firms and regions to avoid maximum potential losses.
2006	Kinzig et al.	Resilience is defined as the capacity of a system to absorb disturbance and reorganize while undergoing change so as to still retain essentially the same function, structure, identity, and feedbacks.
2006	Perrings	Resilience is the ability of the system to withstand either market or environmental shocks without losing the capacity to allocate resources efficiently.
2007	Holmgren	Resilience is the ability of the system to return to a stable condition after a disruption.
2007	Tierney and Bruneau	Resilience is both the inherent strength and ability to be flexible and adaptable after environmental shocks and disruptive events.
2007	Masse et al.	Resilience is the ability of systems, infrastructures, government, business and citizenry to resist, absorb, recover from, or adapt to an adverse occurrence that may cause harm, destruction, or loss of national significance.
2009	Haimes	Resilience is defined as the ability of the system to withstand a major disruption within acceptable degradation parameters and to recover within an acceptable and time and composite costs and risk.
2010	Vugrin et al.	Given the occurrence of a particular disruptive event (or set of events), the resilience of a system to that event (or events) is the ability to efficiently reduce both the magnitude and duration from targeted system performance levels.
2017	Woods	Resilience concerns the ability to recognize and adapt to handle unanticipated perturbations that call into question the model of competence, and demand a shift of processes, strategies, and coordination.
2017	Hollnagel	Resilience is defined as intrinsic ability of an organization (system) to maintain or regain a dynamically stable state, which allows it to continue operations after a major mishap and/or in the presence of continuous stress.

Disturbance, Disruption, Uncertainty, and Risk

Disturbances and disruptions to a system refer to the incorporation of inherent degree of uncertainty and unreliability, reallocation of natural and social resources, transition of sustainable management approaches, variation of supply chain of ecosystem services (López et al. 2013). A multitude of equivalent or similar terms relative to a stable system include disturbances, disruptions, perturbations, hazards, disasters, threats, shocks, and anomalies. Disturbances from nature include earthquake, fire, landslide, disease, flood, drought, slope failure, tsunami, storms, snow, ice, derechos, hurricanes, tornadoes, and natural aging. Disturbances from human intervention are load cycles from vehicles, blast, epidemic, accident, land expansion, petroleum production, building of irrigation canals, intensive fishing, or mining operations and space exploration. A disturbance that occurs as a relatively discrete event in time is referred to as a "pulse" disturbance, while more gradual or cumulative pressure on a system can be regarded as a "press" disturbance (Resilience Alliance 2007, 2010). The main characteristics of disturbance are frequency, duration, severity, and predictability (Nourzad and Pradhan 2014). Disturbances to a system are potential, unpredictable, and uncontrollable, can be either deterministic or probabilistic, and they generally increase risks of vulnerability and instability with a certain probability.

Resilience of Infrastructure Systems

The term resilience is originated from the Latin word "*resiliere*," which means to "bounce back." Resilience concept covers a wide range of viewpoints (Woods 2015) across professional disciplines, i.e. engineering, social, ecological, economic perspectives, and its definitions can be summarized as shown in Table 4.8.

Ecology of Infrastructure Systems

Ecology is the relation and pattern between environment and organisms, sustaining life support system as an organized whole (Pandit et al. 2011; Tao et al. 2013). Ecological systems or networks are composed of interlinked, adaptive, and complex components or subsystems that exchange material, energy, and information among themselves from and to the environment, exhibiting characteristic scaling properties, ultimately achieving the sustainability of the ecological chain. Infrastructure ecology can be viewed interlinked and interdependent components and systems constituting a "Material-Water-Energy-Land Use-Transportation-Socioeconomic Nexus" (Chang 2017; Li and Wang 2014).

Most ecological systems are dynamic and vary over time, following four phase patterns: growth, conservation (maintenance), release (collapse), and renewal (reorganization). The adaptive cycles describe how a human-nature system or social-ecological system is originally established, develops and stabilizes, undergoes changes due to disturbances, and then reorganizes itself via reallocation of resources to begin a new sequence again, forming a panarchy, characterized by cross-scale, interdisciplinary, and dynamic nature (Holling 2001). In an adaptive cycle, a system can be disrupted by disturbances and either regenerates to a similar state or be transformed to some new state. What is happening at multiple scales in a dynamic way requires understanding on how the focal system responds to innovation from smaller nested scales and to constraints imposed from larger-scale systems. The property of varying lens of different adaptive cycles suggests that an effective long-term operation, management, and policy-making for infrastructure and lifeline systems must be highly flexible and adaptive. Different frameworks are further compared to analyze the social-economic system by establishing the corresponding criteria (Binder et al. 2013).

Sustainability of Infrastructure Systems

The word "sustainability" is derived from the Latin *sustinēre*, meaning to "hold up, endure and support." It is defined by the United Nations Environment Programme

(UNEP): the use of the environment and resources to meet the needs of the present without compromising the ability of future generations to meet their own needs (Adams 2006; Boz and El-adaway 2014). The term sustainability can also be described as enduring, resilient, reliable, functional, inclusive, affordable, supportable, permittable, adaptable, implementable, scalable, or pragmatic (Binney 2010). Sustainability requires a long-term strategy to meet a stable and compatible human-nature environment, infrastructure, and lifeline systems by using basic material and energy in order to achieve a good life style, freedom and choice, social networks and regional security, economic and culture variation. Sustainability assessment over the life cycle should be considered as a method for integrating decisions throughout the project development cycle (Binney 2014; Gharehbaghi and Raso 2013; Gopalakrishnan and Peeta 2010; Reiner et al. 2014). "Flow benchmark" is used to demonstrate technological integration and introduce sustainability indicators between macro-level system dynamics modeling, micro-level agent-based simulation, and multi-objective optimization (El-Adaway 2013). Four levels of sustainability are given to show the hierarchy including the categorization of key assessment of indicators and criteria (Shen et al. 2010), and three dimensions (economy, ecology, and society). Timeline of the development of sustainability and related important events are also presented (Bocchini et al. 2013; Wang et al. 2016). Managing resources to foster resilience to respond to and shape change in ways that both sustain and develop the same fundamental function, structure, identity, and feedbacks is crucial to the future of humanity and the life support system (Chapin et al. 2009).

Qualitative and Quantitative Approach for Resilience Assessment

Potential performance of infrastructure systems can be measured both in a qualitative and quantitative way due to the deterioration of physical networks, breakage of geographic networks, failure of cyber networks, and disconnection logical networks by considering it in a component level, system level, or system-of-systems level.

Features and Properties of Resilience

System structure, characteristics, and features are taken into consideration by using three system features (absorptive capacity, adaptive capacity, and restorative capacity) and four properties (robustness, redundancy, resourcefulness, and rapidity) to formulate how inherent properties of a system can determine system resilience (Vugrin et al. 2010).

Absorptive capability refers to an endogenous ability of the system to reduce the negative impacts caused by disruptive events and minimize consequences. The absorptive capacity is an endogenous feature of the system. Adaptive capability refers to an endogenous ability of the system to adapt to disruptive events through its self-organization capabilities in order to minimize consequences. It reflects actions that result from ingenuity or extra effort over time, often in response to a crisis situation. It also reflects a dynamic ability of the system to change endogenously throughout the recovery period. Restorative capability refers to an exogenous ability of the system to be repaired by external actions throughout the recovery period. These repairs usually restore the system to near its original, pre-event state, but can also restore the system to a completely new state or regime that anticipates future system requirements (Francis and Bekera 2014).

Robustness refers to the strength or the ability of elements, systems, and other units of analysis to withstand a given level of stress or demand without suffering degradation or loss of function. Redundancy refers to the extent to which elements, systems, or other units of analysis exist that are substitutable, i.e. capable of satisfying functional requirements in the event of disruption, degradation, or loss of functionality. Resourcefulness refers to the capacity to identify problems, establish priorities, and mobilize resources when conditions exist that threaten to disrupt some element, system, or other unit of analysis. It can be further conceptualized as consisting of the ability to apply material (i.e. monetary, physical, technological, and informational) and human resources to meet established priorities and achieve goals. Rapidity refers to the capacity to meet priorities and achieve goals in a timely manner in order to contain losses and avoid future disruption (Chang and Shinozuka 2004).

Quantification of Resilience Capabilities

Resilience engineering is a paradigm for function and safety evaluation, uncertainty and risk management, and assessment of complex lifeline systems (Patriarca et al. 2018; Righi et al. 2015). Unforeseen disturbances can cause ecosystems manifold responses including the transient change of processes and properties of ecosystem, or even lead to complete reorganization of an ecosystem by causing a shift to an alternative stable state. The way resilience is quantified affects the degree to which different trajectories of ecosystem recovery from disturbance are represented as "resilient." It gives an overview of indices for engineering resilience, grouped into different categories according to the component of recovery they address. Normalized impact of disturbance and the normalized recovery rate are defined as bivariate space for quantification of resilience (Ingrisch and Bahn 2018).

Resilience is a dynamic multi-faceted process and its assessment should cover all the evolutionary phases and essential system features (absorptive, adaptive, and restorative capability). Figure 4.50 shows a general illustration of system resilience. The x-axis is time and y-axis represents the measurement of performance (MOP) of a system. Examples of MOP include critical facilities, connectivity of networks, the

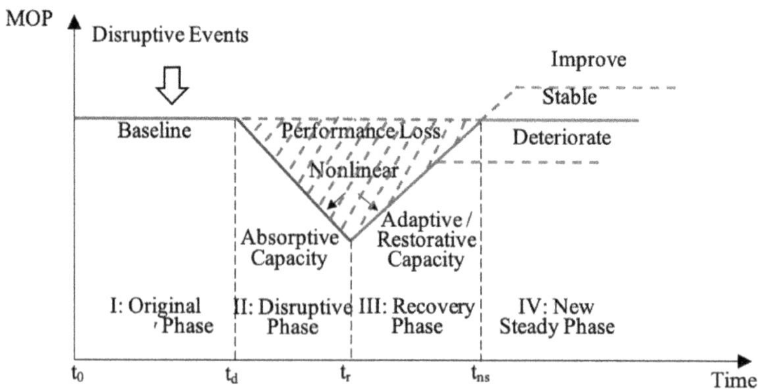

FIGURE 4.50 System resilience transitions and phases.

number of customers served or the level of functionality, social and economic activities. The first phase is the original steady phase ($t_d < t < t_r$), in which the system performance assumes its target value. The second phase is the disruptive phase (t_d), in which system performance begins to drop (in most cases) due to the disruptive event(s) at time t_r until it reaches the lowest level at time R. During this phase, the system absorptive capability can be assessed by Robustness ($RAPI_{DP}$) combined with two complementary measures Rapidity (PL_{DP}) and Performance Loss ($TAPL$) to identify the maximum impact caused by disruptive events. The system performance loss can be interpreted and quantified as the region bounded by the graph of the MOP before and after the occurrence of negative effects caused by disruptive events, which can also be referred to as the system impact area. Time Averaged Performance Loss ($t_r < t < t_{ns}$) is introduced to encompass the time of appearance of negative effects due to disruptive events up to full system recovery and provides a time-dependent indication of both adaptive and restorative capabilities as responses to the disruptive events. The third phase is the recovery phase ($RAPI_{RP}$), in which the system performance starts increasing until the new steady level. Rapidity (PL_{RP}) and Performance Loss (RA) are developed to assess system adaptive and restorative capability. The newly attained steady level may equal to, higher (resilience), or lower than (degradation, collapse) the previous steady state. A resilient system possesses the ability to recover its normal performance state level from the disruptive state, while non-resilient system may gradually decline towards a low performance level with a certain of magnitude due to an unexpected event. Recovery Ability (RA) is quantitatively developed. Loss function, recovery function model, and fragility function were established in comparison to a mechanical analogy (Cimellaro et al. 2010). A general resilience metric (GR) is developed by incorporating time effect and three capabilities, as shown below:

$$GR = f\left(R, RAPI_{DP}, RAPI_{RP}, TAPL, RA\right) = R \times \left(\frac{RAPI_{RP}}{RAPI_{DP}}\right) \times \left(TAPL\right)^{-1} \times RA$$

$$= R \times \left(\frac{\dfrac{\left| \sum_{i=1}^{K_{RP}} \dfrac{MOP(t_i) - MOP(t_i - \Delta t)}{\Delta t} \right|}{K_{RP}}}{\dfrac{\left| \sum_{i=1}^{K_{DP}} \dfrac{MOP(t_i) - MOP(t_i - \Delta t)}{\Delta t} \right|}{K_{DP}}} \right)$$

$$\times \left(\frac{\int_{t_d}^{t_{ns}} \left(MOP(t_0) - MOP(t)\right)}{t_{ns} - t_d} \right)^{-1} \times \left| \frac{MOP(t_{ns}) - MOP(t_r)}{MOP(t_0) - MOP(t_r)} \right| \quad (7)$$

where $TAPL$ represents the time average performance loss, K_{DP} and K_{RP} represent the number of detected ramps in the disruptive phase and the recovery phase, respectively, $MOP(t_0)$ represents the performance level at the original steady phase. *GR* can be used to compare resilience of various systems under the same or different disruptive events. Higher GR value indicates that the system is more resilient to certain disruptive events (Nan et al. 2014). A proposed model of resilience measure according to its definition and the system performance curve is presented in a more detailed way (Ayyub 2014, 2015; Sánchez-Peral 2017). Rule-based model is built to describe system resilience (Reed et al. 2011). A weighted indicator based on the priority of risks is given to evaluate resilience at a country level (Kammouh et al. 2017). By estimating, quantifying, and planning for interdependent infrastructures, a risk-based economic input-output model is employed for static and dynamic measures of economic resilience (Pant et al. 2014). General approaches to measure resilience are converted from qualitative definitions to qualitative ones (Wang et al. 2017).

By considering dynamics of infrastructure, time-dependent metrics for resilience quantification, and the benefits of resilience improving strategies, a multilayer hybrid modeling approach is developed by considering infrastructure systems as a coupling of interacting subsystems characterized by layers, i.e. system under control, operational control system, and human operator level system. It provides an effective way in capturing the effects of interdependencies and in quantifying the coupling strength between interdependent infrastructures (Nan and Sansavini 2017). The classification of resilience approaches is summarized according to different application domains. A quantitative approach is classified as conceptual and semi-qualitative methods, while a qualitative approach is classified as generic (deterministic and probabilistic) and structural-based modeling (optimization, simulation, and fuzzy) methods (Hosseini et al. 2016). Resilience metrics are concretely discussed based on resilience curve, pre- and post-disruption performances, and reliability and restoration. The engineering design implication is further clarified by discussing how a system responds in the aftermath of a disruption which will largely determine the resiliency level of the system; thus, one of the essential and challenging tasks in predictive resilience analysis is being able to analyze system disruption responses at an early system design stage (Yodo and Wang 2016; Uday and Marais 2015). By establishing a Hybrid Hidden Markov Model a resilience metric for dynamic infrastructure systems based on the interaction of resilience capacities, disruption scenario, and dispatch strategies is developed (Zhao et al. 2017).

Critical Infrastructure Systems and Interdependency

Interdependency of critical infrastructure, pertaining to resilience, is an important and rapidly emerging area that is reviewed in the following.

Performance Metrics of Infrastructure Systems

The U.S. Patriot Act defines critical infrastructures as "systems and assets, whether physical or virtual, so vital to the U.S. that the incapacity or destruction of such systems, capitals and assets would have a debilitating impact on security, national economic security, national public health or safety, or any combination of those matters." Critical Infrastructure Security and Resilience advances a national policy

TABLE 4.9
Critical Infrastructure and Key Resources

CIKR	Sector-specific Agency	Performance Metric
Chemical	Department of Homeland Security	Chemical property, health hazard, storage risk
Commercial Facilities	Department of Homeland Security	Commercial value of goods, commercial competitiveness
Communications	Department of Homeland Security	Signal stability, communication mode
Critical Manufacturing	Department of Homeland Security	Manufacturing raw materials, fabrication process, productivity
Dams	Department of Homeland Security	Storage capacity, discharge quantity, impact on surroundings
Defense Industrial Base	Department of Defense	Equipment complexity, confidentiality
Emergency Services	Department of Homeland Security	Emergency case, supplies, methods and risk
Energy	Department of Energy	Energy production, efficiency, waste and sustainability
Financial Services	Department of Treasury	Currency circulation, investment and financing management, stock market
Food and Agriculture	Department of Agriculture and Department of Health and Human Services	Food safety, agricultural production, consumption and demand
Government Facilities	Department of Homeland Security and General Services Administration	Facility safety, functionality, serviceability, reliability
Healthcare and Public Health	Department of Health and Human Services	Sources of infection, illness severity and drug safety
Information Technology	Department of Homeland Security	Innovativeness of information technology, information security
Nuclear Reactors, Materials and Waste	Department of Homeland Security	Nuclear security, emergency protection
Transportation Systems	Department of Homeland Security and Department of Transportation	Intelligent monitoring system and intelligent networking system
Water and Wastewater Systems	Environmental Protection Agency	Water circulation efficiency and waste water treatment

to strengthen and maintain secure, functioning, and resilient critical infrastructure (Directive PP 2013). The 16 critical infrastructure (Habibian and Minaie 2017) and key resources (CIKRs) identified by sector-specific agencies and their performance metrics are shown in Table 4.9, and the corresponding performance measures and tracking parameters are presented as well (Bruneau et al. 2003; Minsker et al. 2015;

Vugrin et al. 2010). Topology-based performance metrics, flow-based functional performance metrics, metrics for power-distribution networks, metrics for water distribution networks, and metrics for transportation networks are given to measure the performance level of infrastructure networks. Individual component assessment of the performance of structural systems has limited evaluation value, and is gradually being replaced by measures of the reliability of entire interested systems (Ghosn et al. 2016). Comprehensive and revealing performance measures are critical for the assessment of long-term investment plans (Mejia-Giraldo et al. 2012).

Interdependency of Networked Infrastructure Systems

The security and welfare of a region mainly rely on a continuous flow of essential goods (such as energy, food, water, oil, electricity, and gas) and services (such as banking, health care, communication, and public administration) provided by a set of systems called critical infrastructures. The same type and different types of critical infrastructures are viewed as system components or subsystems. The input and output flow dynamics, and the mutual dependency relationships of each subsystem of the whole are interdependency problem. Intersections and interdependencies between infrastructural components are recognized as a unique feature in that these attributes impact the behaviors of system-of-systems for sparse or dense infrastructure systems subjected to emergencies and disruptions (Reed et al. 2009). The positive impact is that these unforeseen interdependencies may provide the tolerance and redundancy to attacks and failures if well managed. However, the negative effect is that they might also be a source of threat, creating new unknown risks and vulnerability such as cascading failures. The major differences between single system and multiple systems are the interconnection characteristics between systems. Interdependencies of systems manifest in multiple ways:

1) The failure or disruption of partial function in one system may probably propagate to other associated systems in a cascading manner.
2) An event exerting on one subsystem may result in adverse or unfavorable influences on several subsystems simultaneously, i.e. "conjoined symbiosis effect."
3) The effects on one subsystem may build up over time, and then they transfer to other subsystems depending on the consecutiveness strength of mutual subsystems, correlation degree, and component importance (Fang et al. 2016).

Identifying, understanding, and analyzing dependency and interdependency of infrastructure systems are challenging. These interdependencies and resultant infrastructure topologies can create interactions, feedforward and feedback mechanisms (interaction between subsystems and *m*th orders of effects) that often lead to unintended behaviors and consequences during disruptions (Rinaldi et al. 2001). These linkages showing interdependency include physical, geographic, cyber, and logical. Physical interdependency is the physical linkage between the output of one subsystem and input of another. Geographic interdependency describes the spatial proximity between elements of infrastructure systems. Cyber interdependency refers to the interrelated state depending on the information transmitted through information

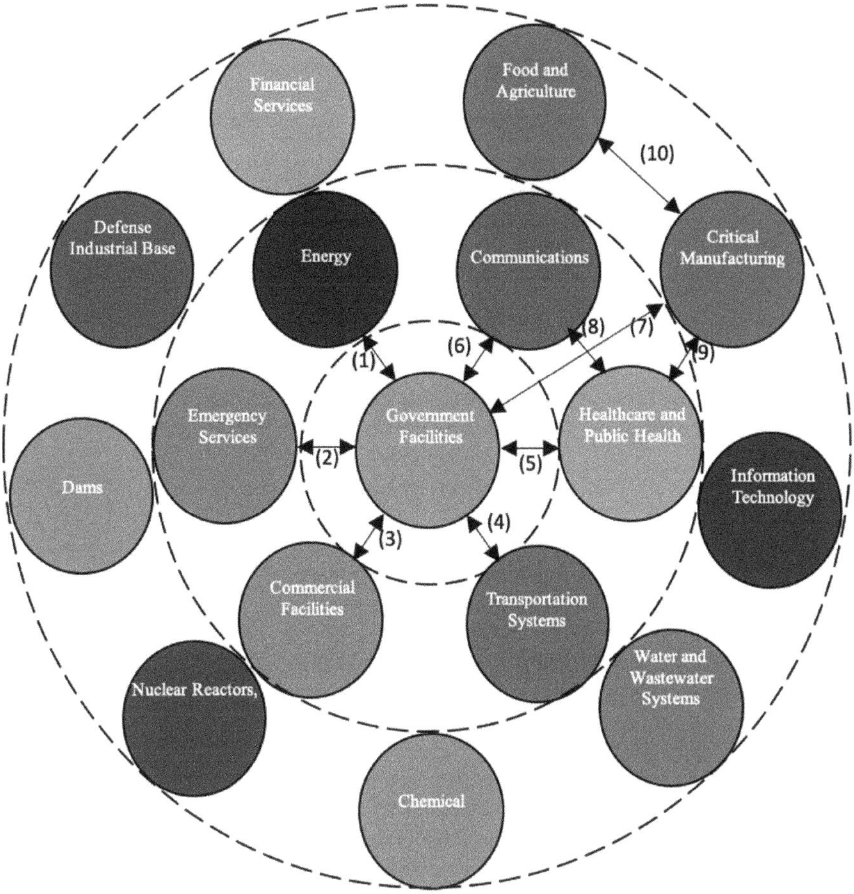

FIGURE 4.51 Interdependencies between critical infrastructure systems.

infrastructure systems. Logical interdependency establishes a logical mechanism that is neither physical, cyber, nor geographic connections. One property in common of these interdependencies of infrastructures is that they are all complex collections of interactively unidirectional dependent or mutually dependent components changing as a result of adaptively learning processes.

Figure 4.51 illustrates the key interdependent relationships among several infrastructure subsystems when an unexpected event (e.g. earthquake) occurs. For 16 critical infrastructure systems, there are theoretically A_{16}^{2} strong and weak interdependencies in total. Generally, they are spatially distributed and have multistep impact effects. Ten important interdependencies of all are listed as follows. The "links" with a diverse and complex "network" provide various ways to transmit capitals between "nodes," which determine the vulnerability and robustness of the network (Anderies et al. 2004). Three typical topological structures can be constructed, namely the random network model, the core/periphery model, and the center-periphery model. Network connection is used to illustrate the relationship between "nodes" and "links"

TABLE 4.10
Interdependency among Infrastructure Networks

Year	Author	Problem	Approach
2001/2004	Haimes and Jiang	Evaluate the intraconnectedness within and interconnectedness among infrastructures	Input-output risk model
2002	Amin	Maintaining reliability, security, and robustness of critical national and international infrastructures	Agent-based approach, system dynamics, distributed control
2005	Zhang, Peeta and Friesz	Identify and understand dynamic multi-layer infrastructure networks for the automobile, urban freight and data networks	Network flow equilibrium model, game theory
2005	Rose and Liao	Business interruption stemming from damaged infrastructure, water service disruptions	Input-output analysis, computable general equilibrium
2006	Jeong et al.	Minimizes the consequences of water shortage for water infrastructures	Network optimization model
2006	Conrad et al.	Coupled infrastructure consequence models and cascading of impacts	Network flow
2007	Patterson and Apostolakis	Rank geographic regions influencing multiple infrastructures	Importance measures, performance index, geographic information system
2007	Lee et al.	Restoration of services for infrastructures	Network flows mathematical representation
2007	McDaniels et al.	Characterize infrastructure failure interdependencies	Empirical approach
2007	Dueñas-Osorio et al.	Interdependencies of electric power and water distribution systems and risk management	Network, topology and graph theory
2007	Friesz et al.	Modeling large-scale and complex infrastructure systems as computable games	Dynamic game theory
2007	Svendsen and Wolthusen	Interdependencies and cascading failure modes	Topology, group theory, flow
2008	Nagurney and Qiang	Component impact among transportation networks, Internet and electric power supply chain networks	Network efficiency/ performance

TABLE 4.10 (Continued)
Interdependency among Infrastructure Networks

Year	Author	Problem	Approach
2008	Sanford Bernhardt and McNeil	Improvement of infrastructure decision-making	Agent-based modeling
2009/2011(a & b)	Ouyang et al.	Structural and functional vulnerability of interdependent infrastructures	Topology analysis
2010	Johansson and Hassel	Electrified railway network modeling for functional and geographic interdependency analysis	Graph theory, vulnerability analysis
2011	Zhang and Peeta	Modeling of various infrastructure systems	A multilayer infrastructure network concept, the computable general equilibrium theory, its spatial extension, and market mechanisms
2011	Gao et al.	Performance response for n interdependent networks under random hazards	Percolation approach
2011	Zio and Sansavini	Modeling of interdependencies among network systems and of their effects on failure propagation	Flow-based approach
2012	Henry et. al	Quantification of system resilience	Flow approach
2012	Trucco et. al	Vulnerability and interoperability of Critical Infrastructures	Dynamic functional modeling
2012	Di Giorgio and Liberati	Reliability study, adverse events propagation study and a failure prediction analysis of interdependency	Dynamic Bayesian network
2013	Zio and Ferrario	Seismic risk for a nuclear power plant embedded in the connected power and water distribution, and transportation networks (system-of-systems)	Muir web and Monte Carlo
2013	Ng and Cai	Joint behavior of interdependent systems	Network flow, reliability and vulnerability theory

(continued)

TABLE 4.10 (Continued)
Interdependency among Infrastructure Networks

Year	Author	Problem	Approach
2014	Huang, Liu and Chuang	Interdependency and the feedback effects between different types of critical infrastructures	A hybrid model which combining decision-making trial and evaluation laboratory and the analytic network process
2015	Baroud et al.	Quantification of interdependent impacts of network resilience	Inherent cost metrics
2015/2017	Nan and Sansavini	Modeling systems of systems and characterizing their failure behaviors	A multilayer hybrid modeling framework including systems under control, operational control system and human-organizational social system.
2016	Golara and Esmaeily	Resilience quantification and characterization of infrastructures	Delivery importance index and flow
2016	Fang et al.	Component important measures of network resilience	Optimal repair time, resilience reduction worth
2017	Huang et al.	Design protection mechanisms for large-scale interdependent critical infrastructures against cyber and physical attacks	Dynamic game model
2017	Heracleous et al.	Cascading effects of interdependencies using hybrid modeling	Open hybrid automata
2017	Thacker et. al	Multi-scale disruption analysis for system-of-systems	Network and flow
2018	Zhang et al.	Quantification of resilience for system design under uncertainty	Flow, optimization
2018	Ramirez-Marquez et al.	Quantification of resilience of community structures	Network and flow
2018	Galbusera and Giannopoulos	Resilience assessment of economic systems and disaster impact analysis	Demand- and supply-driven input-output models

in the spatial dimension (Peng et al. 2017). Recent advances in the research on interdependency of infrastructure networks can be shown in Table 4.10.

1) Government facilities provide sustainable energy supply for inhabitants and consumers, while energy service companies generate oil, gas, wind, water, and nuclear energy when government invest a lot on facilities.
2) Government facility provides the basic coverage to build emergency service centers, while emergency centers ensure the life and property safety of patients.
3) Government provides viable facilities and platforms for the commercialization of innovative products by allocating social resources such as education, culture, and research. These commercialized new tools and technologies can provide government with policies for better decision-making.
4) Government improves transportation systems with matching facilities such as bridge monitoring stations, traffic light control systems, and subway network systems. Big data monitored from these transportation systems provide government with better operation and maintenance.
5) Government invests on hospital facilities for different types of patients, and these health care centers can also increase productivity by improving national health index.
6) Government establishes reliable, secure, and large-scale communication facilities. Stable signal transmission can sustain a good communication network between people.
7) Government improves national competitiveness by developing, or supporting the development of, cutting-edge technologies for critical manufacturing, while these manufactured products provide government, universities, and industries with better life qualities.
8) Communication operators provide specific services for health care service centers, and the hospitals need reliable and high-quality communication networks in the event of emergencies.
9) Healthy and educated people can make better products in the critical manufacturing industry, while these high-quality medical apparatus and instruments in the manufacturing companies are developed to better serve patients at the health care centers.
10) Highly automated production lines can improve crop productivity, while people with high-quality diets and good educational background can generally create better developed and manufactured products.

According to the advances on interdependency of infrastructure systems and the associated problems, eight key points regarding a holistic and comprehensive modeling framework (Cimellaro 2016; Eusgeld and Kröger 2008; Ouyang 2014; Zhang and Peeta 2011) can be summarized as follows. Resilient, self-healing, self-diagnosing infrastructures regarding threat identification, vulnerability and risk analysis, loss analysis, and the feedback loop of physical, cyber, and human interactive systems are further clarified (Peerenboom and Fisher 2008; Domich 2008). The theories behind include but are not limited to complex nonlinear dynamics, percolation, chaos, self-similarity, fractals, self-organization, criticality, self-organized criticality, highly optimized tolerance, graph theory, group theory, statistical learning, network topology, system reliability and robustness, micro-, meso- and macro-economic, control domain, social embeddedness and adaptation (Cimellaro et al. 2016; Glass et al. 1987).

1) Uncertainty, probability, and risk quantification of internal and external disturbances
2) Disturbance propagation mechanisms for both intra-dependences and inter-dependences
3) Interdependency, vulnerability, and complexity of networks
4) Scale effects including component level, system level, and system-of-systems level
5) Six dimensions of resilience including technical, organizational, economic, social, environmental, and ecological dimensions
6) Three capacities of resilience including absorptive capacity, adaptive capacity, and restorative capacity, and four properties of resilience including robustness, redundancy, resourcefulness, and rapidity
7) Period cycle of resilience involving static and dynamic, temporal and spatial characteristics
8) Four types of interdependency of networks including physical, geographic, cyber, and logical types.

Organizations Devoted to Resilience of Infrastructure Systems

Comprehensive infrastructure modeling and evaluation systems are presented to show a framework regarding "greenness," "resilience," and "sustainability" of infrastructure systems. These systems mainly include *BEST-in-Highways, I-LAST, STARS, Envision, INVEST, Greenroads, GreenLITES, AGIC IS, CEEQUAL, ASPIRE, GreenPave, MARS, GIS, MADM, CIMS, ResilUS, and USRC Ratings* (Amini et al. 2018; Dudenhoeffer et al. 2006; Mayes and Reis 2015; Mehany 2016; Miles 2011; Miles and Chang 2007, 2011; Mills and Attoh-Okine 2014; Papajohn et al. 2016; Reis and Mayes 2015; Shields 2016). Table 4.11 presents main academic and industrial research organizations, institutions, and communities worldwide devoted to designing, improving, and facilitating resilience and sustainability of infrastructure systems.

Promise of an AI-Based Platform for Health Monitoring of Infrastructure Systems to Improve Resilience

SHM has become a critical field of research and technology development in recent decades. The broad concept of structural health monitoring (Aktan et al. 2000; Christodoulou 2000) can be defined as "by monitoring the health state of materials, structures and systems, effective models and data interpretation algorithms are developed to track and evaluate the symptoms of operational incidents, anomalies, and deterioration that may affect operation, durability, serviceability, safety and reliability of infrastructure systems." An overview of performance limit states associated with return periods is presented with diverse performance criteria. It indicates that

TABLE 4.11
Resilience Research Organization

Organization	Mission / Goal
Resilience Alliance (RA)	The Resilience Alliance is a research organization that focuses on resilience in social-ecological systems as a basis for sustainability.
Resilience Engineering Association (REA)	Resilience Engineering Association looks for ways to enhance the ability at all levels of organizations to create processes that are robust yet flexible, to monitor and revise risk models, and to use resources proactively in the face of disruptions or ongoing production and economic pressures.
U.S. Resiliency Council (USRC)	The US Resiliency Council's mission is to establish and implement meaningful rating systems that describe the performance of buildings during earthquakes and other natural hazard events, to educate the general public to understand these risks, and to thereby improve societal resilience.
100 Resilient Cities (100RC)	100 Resilient Cities is dedicated to helping cities around the world become more resilient to the physical, social, and economic challenges that are a growing part of the twenty-first century.
ASCE Infrastructure Report Card	The American Society of Civil Engineers' Report Card for America's Infrastructure depicts the condition and performance of American infrastructure based on the physical condition and needed investments for improvement.
International Council on Systems Engineering (INCOSE)	INCOSE has a dedicated working group for Resilient Systems that shapes research on the use of systems engineering practices to achieve resilience
International Institute for Sustainable Development (IISD)	The IISD is an independent think tank championing sustainable solutions to promote human development and environmental sustainability.
International Council for Local Environmental Initiatives (ICLEI)	ICLEI is an international organization of local governments and national and regional local government organizations that have made a commitment to sustainable development.
United Nations International Strategy for Disaster Reduction (UNISDR)	UNISDR defines itself through its multi-stakeholder coordination approach based on the relationships it has developed with national and local governments, intergovernmental organizations, and civil society, including the private sector, and by its mode of operating through a network of global partners.
National Oceanic and Atmospheric Administration (NOAA)	NOAA is an agency that enriches life through science by using cutting-edge research and high-tech instrumentation to provide citizens, planners, emergency managers, and other decision-makers with reliable information of changing environment.

(continued)

TABLE 4.11 (Continued)
Resilience Research Organization

Organization	Mission / Goal
Multidisciplinary and National Center for Earthquake Engineering Research (MCEER)	MCEER is a national multidisciplinary and multi-hazard earthquake engineering research center, helping communities by developing new knowledge, tools and technologies to increase disaster resiliency in the face of earthquakes and other extreme events.
Centre of Excellence for National Security (CENS)	CENS aspires to be an international research and thought leader in the multidisciplinary study of the concept of resilience and its policy-relevant application to promote security within and beyond Singapore.
National Infrastructure Simulation and Analysis Center (NISAC)	NISAC provides strategic, multidisciplinary analyses of interdependencies and the consequences of infrastructure disruptions across all 16 critical infrastructure sectors at national, regional, and local levels.
Department of Homeland Security (DHS)	DHS is dedicated to preventing terrorism and enhancing security, managing our borders, administering immigration laws, securing cyberspace and ensuring disaster resilience.
The Infrastructure Security Partnership (TISP)	TISP is to lead public and private sector collaboration that advances the practice and policy of infrastructure security and resiliency.
Critical Infrastructure Resilience Institute (CIRI)	To conduct research and education to enhance the resiliency of the Nation's critical infrastructures and the businesses and public entities that own and operate those assets and systems.
Brook Byers Institute for Sustainable Systems (BBISS)	BBISS promotes comprehensive and innovative systems-based approaches to address the challenges and opportunities inherent in achieving a sustainable and prosperous future.
Institute for Sustainable Infrastructure (ISI)	ISI provides a comprehensive framework of 60 sustainability criteria that address the full range of environmental, social, and economic impacts to sustainability in project design, construction, and operation.
Resilient Infrastructure Networks Lab (RINL)	The goal of RINL is to design network control algorithms and economic incentive schemes that help infrastructure users and operators make better decisions in the face of uncertainties, both random and adversarial.

resilience limit state of infrastructure systems needs a longer and comprehensive design period compared to the general utility limit state, serviceability limit state, and safety limit state (Aktan et al. 2016). The SHM system should be able to provide, on demand, reliable information pertaining to the safety and integrity of a structure (Daum 2013). AI has been developed and advanced thanks to the interaction of various disciplines, namely computer science, information theory, computational intelligence (Parmee 2004), cybernetics, linguistic, and neurophysiology. It is defined as the ability of a machine to mimic intelligent human behavior, seeking to use

TABLE 4.12
Leading Scholars in the Domain of AI and SHM

Name	Research Focus	Affiliation
Ajit Mal	Mechanics of solids, composite materials, structural health monitoring, and biomechanics.	Mechanical and Aerospace Engineering, UCLA
Ali Mosleh	Reliability engineering, resilient systems design, prognostics and health monitoring, hybrid systems simulation	Materials Science and Engineering, UCLA
Anne Kiremidjian	Wireless sensor networks, structural health monitoring, system reliability	Civil and Environmental Engineering, Stanford
Andrew Ng	Machine learning, deep learning, computer vision, artificial intelligence	Computer Science Department, Stanford
Baher Abdulhai	Intelligent transportation systems, advanced traffic management and information systems, modeling and simulation for dynamic transportation networks	Department of Civil and Mineral Engineering, University of Toronto
B.F. Spencer	Structural health monitoring, structural control, cyberinfrastructure applications	Department of Civil and Environmental Engineering, UIUC
Branko Glisic	Advanced sensing technologies, smart structures, fiber optic sensors, materials & structures, structural health monitoring	Department of Civil and Environmental Engineering, Princeton
Burcin Becerik	User-centered sustainable built environments, situational awareness for built environments, building/infrastructure information modeling and visualization	Department of Civil and Environmental Engineering, USC
Charles R Farrar	Solid mechanics, structural dynamics	Structural Engineering Department, UCSD
Chatzi Eleni	Structural health monitoring, damage detection and nonlinear dynamics	Department of Civil, Environmental and Geomatic Engineering, ETH Zürich
Christopher S. Lynch	Mechanics of materials, characterization of ferroelectric materials	Mechanical of Aerospace Engineering, UCLA
Daniele Zonta	Smart structures, structural health monitoring, structural reliability, Bayesian decision-making, infrastructure management, fiber optic sensors	Department of Civil, Environmental and Mechanical Engineering, University of Trento

(continued)

TABLE 4.12 (Continued)
Leading Scholars in the Domain of AI and SHM

Name	Research Focus	Affiliation
Daniel J. Inman	Structural health monitoring, smart materials and structures	Department of Aerospace Engineering, University of Michigan
Elchin Khalilov	Earthquakes forecast, seismic mitigation	Global Network for the Forecasting of Earthquakes
Ertugrul Taciroglu	Topology optimization of smart material systems, soil-structure interaction in deep and shallow foundation systems, wave propagation	Department of Civil and Environmental Engineering, UCLA
Feifei Li	Machine vision, human vision	Computer Science Department, Stanford
Feniosky Pena-Mora	Information technology support, conflict resolution, sustainable construction, visualization, augmented reality, and processes integration	Department of Civil Engineering and Engineering Mechanics, Columbia
Fuh-Gwo Yuan	Structural health monitoring, damage tolerance of composite structures, smart materials and structures	Department of Mechanical and Aerospace Engineering, NC State
Fukuo Chang	Structural health monitoring, smart structures, composite and multifunctional materials	Department of Aeronautics and Astronautics, Stanford
Gang Wu	Structural concept design, prestressed structural systems, prefabricated construction, composite structures	Department of Civil Engineering, Southeast University
Gregory Buckner	Electromechanical systems, manufacturing automation, intelligent control, mechatronics	Department of Aeronautics and Astronautics, NC State
Hae Young Noh	Structural health monitoring, risk analysis, smart structures, energy management	Civil and Environmental Engineering, Carnegie Mellon
Hoon Sohn	Structural health monitoring, sensing technology, statistical pattern recognition	Department of Civil and Environmental Engineering, KAIST
Huajiang Ouyang	Structural dynamics and control, structural identification	Department of Mechanical, Materials and Aerospace Engineering, University of Liverpool
Ian Smith	Biomimetic structures, advanced computer-aided engineering applications, stochastic optimization and search, multi-criteria analysis, system uncertainties	Ecole Polytechnique Fédérale de Lausanne

TABLE 4.12 (Continued)
Leading Scholars in the Domain of AI and SHM

Name	Research Focus	Affiliation
Jerome P. Lynch	Intelligent sensing, structural health monitoring, smart materials and structures	Department of Civil and Environmental Engineering, University of Michigan
Jian Zhang	Earthquake engineering, structural dynamics and mechanics	Department of Civil and Environmental Engineering, UCLA
Keith Worden	Advanced signal processing, machine learning, structural dynamics	Department of Mechanical Engineering, Sheffield
Kenichi Soga	Infrastructure sensing, geotechnical structures, energy and engineering sustainability	Department of Civil Engineering, Berkeley
Lizhi Sun	Micro/nano-mechanics, multiscale materials modeling, nanocomposites	Civil and Environmental Engineering, UC Irvine
Lori Graham-Brady	Stochastic finite elements, probabilistic mechanics, structural reliability	Department of Civil Engineering, Johns Hopkins
Lucio Soibelman	Information technology, artificial intelligence, data mining, knowledge discovery, image reasoning, text mining, machine learning	Department of Civil and Environmental Engineering
Mahendra Singh	Vibrations and control, earthquake engineering, probabilistic mechanics	Department of Biomedical Engineering and Mechanics
Maria Q. Feng	Sensors, smart structures, structural health monitoring, system control	Department of Civil Engineering and Engineering Mechanics
Massoud Amin	System identification and control, transportation, optimization and scheduling theory, learning control for game theory	Department of Science and Engineering, University of Minnesota
Matteo Pozzi	Bayesian methods, risk assessment, structural health monitoring, resilient structures	Department of Civil and Environment Engineering, Carnegie Mellon
Michael C. Constantinou	Structural and earthquake engineering, bridge engineering	Department of Civil, Structural and Environmental Engineering, Buffalo
Mircea Dan Grigoriu	Random vibration, stochastic calculus, structural reliability and safety	Department of Civil and Environmental Engineering, Cornell
Mohammad Noori	Structural health monitoring with an emphasis on AI applications	Department of Mechanical Engineering, Cal Poly

(continued)

TABLE 4.12 (Continued)
Leading Scholars in the Domain of AI and SHM

Name	Research Focus	Affiliation
Mohammed Elshafie	Underground construction, foundations and soil-structure Interaction	Department of Engineering, Cambridge
Philip Guildford	Energy efficiency and renewable energy	Department of Engineering, Cambridge
Pol Spanos	Digital signal processing, dynamical systems, Performance/Safety Assessment for Statics and Dynamics Problems	Department of Mechanical Engineering, Rice
Roger Georges Ghanem	Probabilistic modeling and computational stochastic mechanics, quantitative models for the propagation of uncertainty in physical systems	Department of Civil and Environmental Engineering, USC
Sami F Masri	Analysis, control, modeling of nonlinear dynamic systems	Department of Civil and Environmental Engineering, USC
Scott Doebling	Computational physics, national security	Los Alamos National Laboratory
Tsu-Chin Tsao	Modeling and control of dynamic systems with applications in mechanical systems, manufacturing processes, automotive systems, and energy systems	Department of Aerospace and Mechanical Engineering, UCLA
Zhikun Hou	Engineering mechanics, dynamics and vibration, structural health monitoring and control	Department of Mechanical Engineering, WPI
Zhishen Wu	Fiber reinforced polymer, intelligent sensing technology	Department of Urban and Civil Engineering, Ibaraki University

human-inspired algorithms to solve problems (Khan and Yairi 2018; Muspratt 1986; Salehi and Burgueño 2018; Smarsly et al. 2007). Recent development of research in the domain of deep learning and transfer learning focuses on the convolutional layers for feature extraction from pixel level, component level, and structural level, pooling layer for data dimension reduction, and mapping a transfer relationship from the pretrained model in source domain to the target domain (Gao and Mosalam 2018; Lin et al. 2017; Nguyen et al. 2017; Rafiei and Adeli 2017, 2018). Scholars have carried out pioneering work in the domain of AI and SHM as it pertains to resilience. A few examples of the application of these research in infrastructure systems are listed in Table 4.12. We should indicate by no means is this a comprehensive list.

Visualization of AI in a pyramid-like space (Shakeri 2016), which can be further studied for its potential in resilience and SHM-related application, is depicted along the three axes of *learning*, *autonomy*, and *reasoning*, as shown in Figure 4.52.

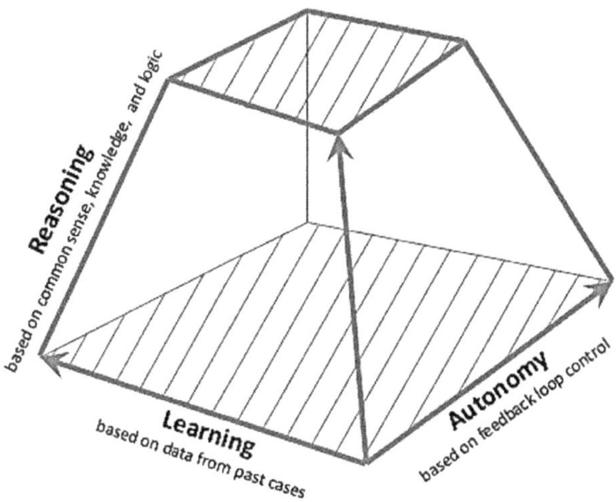

FIGURE 4.52 Pyramid of artificial intelligence. (Adapted from Shakeri 2016.)

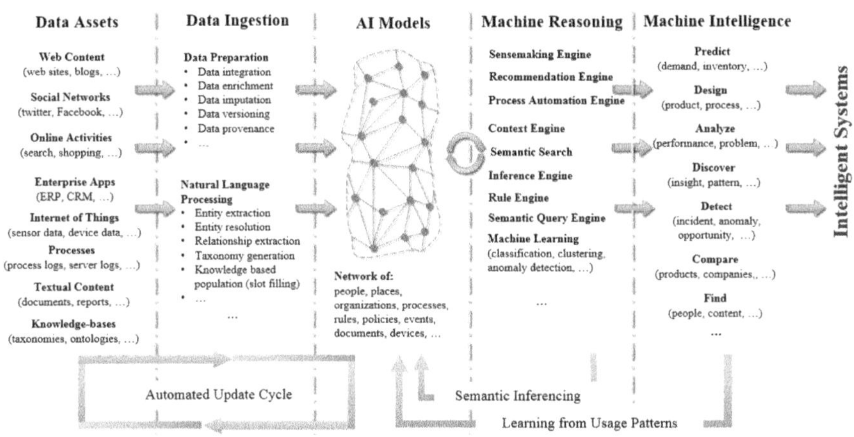

FIGURE 4.53 Architecture of AI engine. (Based on Shakeri 2016.)

The AI engine of a platform developed by Shakeri et al. (2016) based on the above pyramid strategy is shown in Figure 4.53, of which the key components include a self-learning AI engine that extracts signals and recommends answers, a number of curated knowledge bases, integrations with internal and external data sources, a big data analysis engine based on Apache Spark, a cloud computing technology. This strategy is offered as a software-with-a-service (SwaS) that relies on a model or design replica of an AI solution to analyze the entire lifecycle of the new solution from technology feasibility to customer requirements, product features and pricing, business model, cost analysis, etc. Joint efforts of industry and academic are needed to promote a highly innovative and unique AI platform that can assist the researchers, the government, and the asset owners to take the lead on the forefront of AI revolution in order to address the technological and decision-making challenges. As stated in the National Artificial Intelligence Research and Development Strategic Plan, "Artificial intelligence is a transformative technology that holds promise for tremendous societal and economic benefit," (Biegel and Kurose 2016). Capital Ideas describes that "AI is poised to fuel the next wave of innovation, and potentially provide unprecedented opportunities for companies and investors." Thus, potential application of AI in resilience is an important and rapidly emerging area of research.

A Framework for Creating Resilient Infrastructure Systems

Based on the extensive review of various aspects of resilience and emerging research areas, the following framework is suggested.

Dimension of Resilience

Given the occurrence of particular disruptive events, system resilience to the events is the ability to efficiently absorb, adapt, and restore the existing resources to develop and sustain individuals, groups, organizations, and systems and operate coherently and synergistically as a whole performance level. Human behavior is closely related to natural resources, the integrity of which is a dynamic process of coupled human-nature system or social-ecological system that is interacting, updating, and redistributing resources when undergoing unknown changes and disruptions (Liu et al. 2007). Linkages among various elements of the system play an extremely important role since they establish feedback loops or mechanisms by updating and modifying the input and output behaviors of each subsystems. These important linkages not only influence the connectivity of subsystems themselves but also associate one dimension of system with another.

System resilience can be regarded as the interconnected combination of natural components including environment and ecological systems, and human components including technical, organizational, social, and economic systems. The integration of all these multiple domains or components exerts an influence on system resilience, ecology, and sustainability. It is also described as an integrative, multi-system model of core resilience (intra-individual factors), internal resilience (inter-personal factors), and external resilience (social-ecological factors) (Liu et al. 2017).

For natural systems, the environmental dimension of resilience refers to the ability of natural systems to recover from disturbances such as natural disasters and to tolerate or adapt to changing climate. The ecological dimension of resilience is the capacity of an ecosystem to respond to a perturbation or disturbance by resisting damage and recovering quickly. Such perturbations and disturbances can include stochastic events such as fires, flooding, windstorms, insect population explosions, and human activities such as deforestation, fracking of the ground for oil extraction, pesticide sprayed in soil, and the introduction of exotic plant or animal species.

For human systems, the technical dimension of resilience refers to the ability of physical systems to perform to acceptable or desired safety, serviceability, and reliability levels when subjected to hazards. The organizational dimension of resilience refers to the capacity of organizations that coordinate, operate, and manage critical facilities and have the responsibility for carrying out critical disaster-related

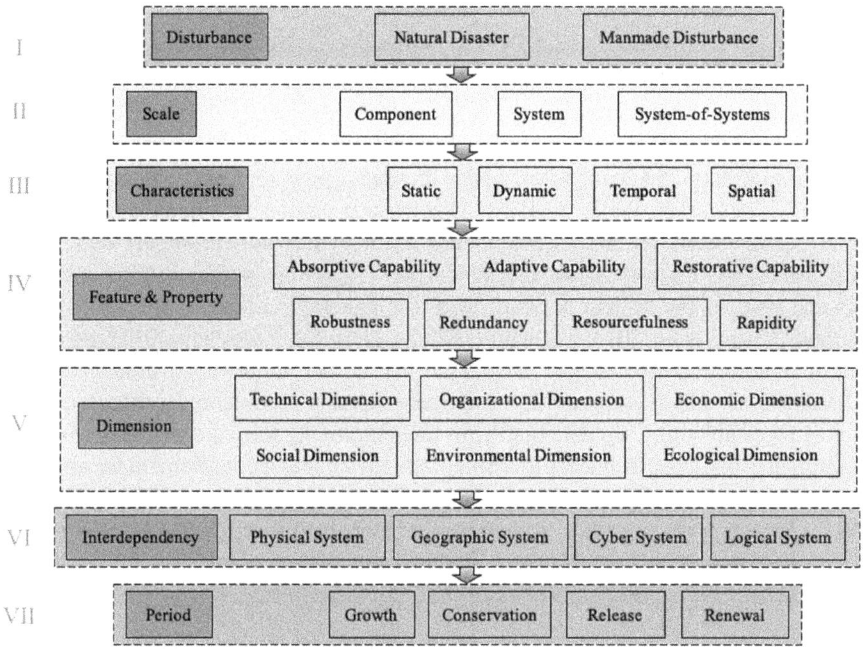

FIGURE 4.54 System resilience assessment framework.

functions to make decisions and take actions that contribute to achieving the properties of resilience to help to achieve greater robustness, redundancy, resourcefulness, and rapidity. Robustness and rapidity are essentially the ends that are accomplished through resiliency-enhancing measures and are the outcomes that more deeply affect decision-makers and stakeholders, while redundancy and resourcefulness are measures employed ready to maintain, improve, or enhance the performance of infrastructure systems. The social dimension of resilience consists of measures specifically designed to lessen the impact extent of individuals or communities to which disaster-stricken communities and governmental jurisdictions suffer negative consequences due to the loss of critical resources and services as a result of disasters. The economic dimension of resilience refers to the capacity to reduce both direct and indirect, static and dynamic economic losses resulting from disturbances. The six dimensions integrate both human and natural factors that influence system resilience, and they should be well designed, optimized, and reorganized to enhance system performance and accommodate likely changes and disturbances.

A Proposed Framework

Figure 4.54 presents a general framework for resilience modeling, design, and assessment of critical infrastructures and key resources. It includes seven phases, i.e. phase I: disturbance, phase II: scale, phase III: characteristics, phase IV: features and property, phase V: dimension, phase VI: interdependency, and phase VII: period. Assume that the targeted model to be analyzed is a network with various types of "edges" and "nodes."

Phase I

In this phase, disturbances resulted from natural disasters or man-made disruptions have an impact on "edges" or "nodes." The type, numbers, distribution, and intensity of disturbances and their stochastics, randomness are the key variables (Shafieezadeh and Burden 2014). The infrastructure disruption model is developed to predict the cascading effects that may occur, given the damage to one or more infrastructure systems (Loggins & Wallace 2015). Disturbance (disaster) impact mechanism is described as a basic cell model with increased direct and indirect impact degrees (Oh et al. 2010, 2012). These disturbances incorporate and may increase higher uncertainties and risks of the system network, and it is generally assessed by a probabilistic method (Chang et al. 2014; Franchin and Cavalieri 2015; Ouyang 2012). Structural health monitoring (Brownjohn and Aktan 2013) and adaptive governance system (Djalante et al. 2011) can provide a long-term reliable prediction for these uncertain factors by establishing big data platforms and employing the state-of-the-art artificial intelligence analytics technologies (Sun et al. 2017). It is a complementary approach for tacking and intelligent management of infrastructure projects (Zhang and Luo 2017). Proactive maintenance decisions can be enabled through the development of prognostics and health management methods that detect, diagnose, and predict the effects of adverse events (Mehrpouyan et al. 2014; Youn et al. 2011).

Phase II

In this phase, the scale of a system to be analyzed is chosen from: component, system, or system-of-systems. For components, safety index, failure probability, and reliability are used as measures for a general structure (Frangopol and Saydam 2011; Ghosn et al. 2016). System-level performance metrics and characteristics such as reliability, redundancy, robustness, resilience, and risk are developed as integral parts of resilient communities and lifeline systems (Ghosn et al. 2016). System-of-systems is used to denote networks that are formed from the integration of independently operating complex systems that interact with one another to provide an overall capability, which cannot be achieved by the individual systems alone. As systems continue to grow in scale and complexity, the "connectivity effect" plays a more important role. As infrastructure system-of-systems such as buildings, powers, energy networked system, transportation and communication networks, grow in interconnectivity and complexity, measuring and improving the resilience of system-of-systems is vital in terms of safety and reliability, and providing uninterrupted services. Resilience of system-of-systems depends on the reliable interconnection of constituent systems (Mostafavi and Abraham 2014; Uday and Marais 2015). Higher level of interdependency between subsystems may give rise to increased risks of failures cascading through system-of-systems depending on the increasingly evolutionary topology of the "network," or can result in the collaborative efficiency of the "network."

Phase III

In this phase, resilience is described as a static or dynamic, temporal and spatial system. Static resilience can be regarded as a way to maintain its desired function of a system, while dynamic resilient network is a recovery process of a system after a disruption (Anderies 2015). The infrastructure network evolves when going through different stages of a period cycle. Spatial scales of the network range from individual parts to the metastructure composed of interdependent infrastructures and the environment. The size, quality, and distribution of open spaces may change following the occurrence and the influence of disturbances (Rus et al. 2018).

Phase IV

In this phase, absorptive, adaptive, and restorative capabilities are closely related to system interdependency. If one subsystem *A* is dependent upon subsystem *B* to operate, the relationship will lower *A*'s absorptive capability in scenarios that negatively affect *B*. If subsystem *A* can reorganize and redistribute resources in an automatic way rather than has dependency upon subsystem *B*, *A* has a strong adaptive capacity. If subsystems *A* and *B* can operate in a collaborative way within a relatively short time when subjected to disturbances, the integral of them will have a strong restorative capability (Bruneau et al. 2003). Robustness and redundancy provide a pre-disaster estimate of the inherent capabilities of the infrastructure systems to absorb perturbations, while resourcefulness and rapidity reflect the post-disaster adaptive activities of the system (Minsker et al. 2015). Responsive, insurance-related tactics are important to the overall performance of resilient infrastructure systems.

Phase V

In this phase, six dimensions affect how a system reacts to disturbances. When an unexpected event occurs (e.g. earthquake), the physical infrastructures (e.g. buildings, electrical transmission stations, transportation networks, underground pipelines) are influenced with a "network effect" due to the change of environmental and ecological system. Business is disrupted and the related emergency services are provided in time, which needs the coordination of both organizational and social communities. Specifications, codes, and standards of the construction and design of infrastructures and related technologies (e.g. design of new materials and devices to absorb energy) need to be improved to resist these unforeseen disruptions, which can be viewed as the technical dimension. Firms, market, demand-supply chain, or policy are influenced due to the disruption of the economic system (Mujumdar 2014). Further, seven dimensions are integrated as a multilayered approach to identify and measure community resilience "PEOPLES": population and demographics, environmental and ecosystem, organized governmental services, physical infrastructures, lifestyle and community competence, economic development, and social-cultural capital (Cimellaro et al. 2016). Economic and business opportunities, public policy, government investment, legal and regulatory concerns, technical and security issues, and social and political concerns are all determinates that have impacts on the degree of coupling (tightness or looseness), the coupling order, and the linearity or complexity of the interactions (Rinaldi et al. 2001; Rose 2007).

Phase VI

In this phase, intervening resources are reorganized to provide continuous interconnections among subsystems. The subsystems can be regarded as "nodes," and the interconnections are viewed as "edges." Important information and control flows are passed between "nodes" to maintain the necessary functions of the whole infrastructure system. An "edge" is a physical or virtual entity that represents a direct level of dependence between two "nodes." Several key determinants are the growing size of nodes, network evolution or optimization, connection diversity, dynamical complexity, node diversity. Physical, cyber, geographic, and logical dependency are the four types of interdependency. When a system is subjected to a disruptive event, cascading consequence transfers from the "node" which is the most venerable to the disruption to other "nodes" (Dudenhoeffer et al. 2006). Interdependency and dependency of "nodes" are largely related to the recovery time, sequence, and trajectory of system "nodes" (Zimmerman et al. 2017).

Phase VII

In this phase, a resilient system undergoes four stages, i.e. growth, conservation, release, and renewal. System state shifts due to risks of disturbances, and it also inherently absorbs these disturbances and reorganizes to maintain critical functions to reach a new desired and stable state (Sasaki et al. 2015). By appropriately designing event-based constraints, functional dependencies, adjustment of capabilities and strategy, a resilient model was developed with an ultimate goal of self-monitoring (Lundberg and Johansson 2015). The critical infrastructures and key resources establish a chain of coupling natural-human system. The coupled human-nature system has a nested hierarchy structural characteristics. Three "hands" play important roles in the coupled human-nature system, i.e. the 'invisible hand' of markets, the 'visible hand' of governments, and the "third hand" human value contribute to the growth, development, and well-being of societies. Human interpersonal networks, intelligence, capacity for innovation, and characteristics as a resilient complex system are only a few examples of how humans are not only an integral component of the traditional infrastructures but also infrastructures in and of themselves (Barnes and Newbold 2005). Life-cycle management of infrastructure systems should be established through probabilistic performance assessment and prediction, integration of information from inspection and structural health monitoring, and multi-criteria optimization for the optimal scheduling of interventions under uncertainties (Frangopol et al. 2014). By building an interaction and feedbacks between interdependent subsystems, i.e. natural support, economic metabolism, social regulation, and complementary relationships, the system can be designed to achieve the integrity of structure, function and dynamic mechanism, as well as system resilience, ecology, and sustainability (Fischer and Amekudzi 2011) in a period cycle.

Conclusions

In summary, this paper is an attempt, with a limited scope, to present a systematic framework and backbone for measuring resilience of infrastructure systems and key resources in both quantitative and qualitative ways. Resilience research of infrastructure systems covers a wide range of multi-disciplinary domains. Well-designed resilient systems can achieve an optimum system performance by distributing limited resources and exchanging energy in a limited time. With the scale of a system increasing, a good interdependency property determines an optimal network evolutionary path and the performance of a typology structure. Establishing consistent criteria for assessing the performance of infrastructure networks or lifeline systems is critical to optimize investment decisions, designing infrastructure structure, and managing the coupled human-nature ecological systems (Aktan et al. 2016; Chang 2017; Ouyang 2014). Therefore, some key suggestions and directions are summarized as follows:

(1) Environmental and ecological systems establish the material basis for infrastructure or lifeline systems. Socioeconomics is the decision drive rather than technical or organizational aspects. Resilience is a multidisciplinary, transdisciplinary and multicultural (government, academia, and industry) research. Infrastructure owners, regulatory agencies, state and local governments and managers from industry are critical stakeholders who can be brought together for a coordinated and productive discussion of infrastructure issues and options for possible solutions.
(2) Infrastructure systems can be regarded as interconnected topology networked structures where material, energy, and resources are transmitted through the input or output physical or cyber flows, to reach an optimized, reliable, and robust status of the network. Inherent complexity, dynamic property, and adaptability are important factors that determine the resilience performance of the network structure.
(3) Data accessibility and reliability, comprehensive integrated modeling of system resilience should be further developed. New technologies and innovations, codes and standards for infrastructures should be transferred to real applications to reduce lifecycle cost, increase operation and maintenance efficiency, and improve risk-based decision-making ability.

Critical infrastructure systems are interdependent domains and systems; thus, small failures in one subsystem may spread to other subsystems and lead to catastrophic events, largely affecting the social and economic level. To understand performance response of interdependent infrastructures due to different disturbances, scholars have proposed numerous modeling and simulation approaches, with an emphasis on identifying effective ex ante and ex post mitigation measures. This paper reviews the conceptual, qualitative, and quantitative metrics and develops a systems thinking strategy (Fiksel 2003) about resilient infrastructure systems. The ultimate goal of the study on the critical infrastructure and key resources is to coordinate the human–nature relationship with a focus on the structural health monitoring of infrastructure systems, reasonable allocation of resources, optimization of the human–nature system performance, and enable the implementation of regional and global resilient, ecological, and sustainable development.

REFERENCES

Adams, W.M. 2006. "The future of sustainability: re-thinking environment and development in the twenty-first century." *Report of the IUCN Renowned Thinkers Meeting* 29 (31).

Aktan, A.E., Catbas, F.N., Grimmelsman, K.A., and Tsikos, C.J. 2000. "Issues in infrastructure health monitoring for management." *Journal of Engineering Mechanics* 126 (7): 711–724.

Aktan, A.E., Moon, F.L., Bartoli, I., and Sjoblom, K. 2016. "Identification of infrastructure systems." *Journal of Infrastructure Systems* 22 (3): 02516002.

Allenby, B., and Fink, J. 2005. "Toward inherently secure and resilient societies." *Science* 309 (5737): 1034–1036.

Amin, M. 2002. "Toward secure and resilient interdependent infrastructures." *Journal of Infrastructure Systems* 8 (3): 67–75.

Amini, M., Zhang, B., and Chang, S. 2018. "Selecting building designs with consideration of sustainability and resiliency." *Journal of Architectural Engineering* 24 (1): 04018001.

Anderies, J.M. 2015. "Understanding the dynamics of sustainable social-ecological systems: human behavior, institutions, and regulatory feedback networks." *Bulletin of Mathematical Biology* 77 (2): 259–280.

Anderies, J.M., Janssen, M.A., and Ostrom, E. 2004. "A framework to analyze the robustness of social-ecological systems from an institutional perspective." *Ecology and Society* 9 (1).

ASCE. 2013. "ASCE 2013 report card for Americas infrastructure."

ASCE. 2017. "ASCE 2017 report card for Americas infrastructure."

Ayyub, B.M. 2015. "Practical resilience metrics for planning, design, and decision making." *ASCE-ASME Journal of Risk and Uncertainty in Engineering Systems, Part A: Civil Engineering* 1 (3): 04015008.

Ayyub, B.M. 2014. "Systems resilience for multihazard environments: Definition, metrics, and valuation for decision making." *Risk Analysis* 34 (2): 340–355.

Barnes, J, and Newbold, K. 2005. "Humans as a critical infrastructure: public-private partnerships essential to resiliency and response." *Proceedings, First IEEE International Workshop on Critical Infrastructure Protection, 75–83.*

Baroud, H., Barker, K., Ramirez-Marquez, J.E., and Rocco, C.M. 2015. "Inherent costs and interdependent impacts of infrastructure network resilience." *Risk Analysis* 35 (4): 642–662.

Biegel, B., and Kurose, J.F. 2016. "The national artificial intelligence research and development strategic plan."

Binder, C.R., Hinkel, J., Bots, P.W., and Pahl-Wostl, C. 2013. "Comparison of frameworks for analyzing social-ecological systems." *Ecology and Society* 18 (4): 26.

Binney, P.D. 2014. "Applying sustainability principles to benefit the overall project delivery cycle of infrastructure systems." *Proceedings, International Conference on Sustainable Infrastructure,* 69–78.

Binney, P.D. 2010. "Transitional approaches to the development of sustainable civil engineering infrastructure." *Proceedings, World Environmental and Water Resources Congress: Challenges of Change,* 504–513.

Bocchini, P., Frangopol, D. M., Ummenhofer, T., and Zinke, T. 2013. "Resilience and sustainability of civil infrastructure: toward a unified approach." *Journal of Infrastructure Systems* 20 (2): 04014004.

Boz, M.A., and El-adaway, I.H. 2014. "Creating a holistic systems framework for sustainability assessment of civil infrastructure projects." *Journal of Construction Engineering and Management* 141 (2): 04014067.

Brownjohn, J., and Aktan, E. 2013. "Improving resilience of infrastructure: the case of bridges." *Proceedings, Structures Congress: Bridging Your Passion with Your Profession,* 1812–1821.

Bruneau, M., Chang, S.E., Eguchi, R.T., Lee, G.C., O'Rourke, T.D., Reinhorn, A.M., Shinozuka, M., Tierney, K., Wallace, W.A., and Von Winterfeldt, D. 2003. "A framework to quantitatively assess and enhance the seismic resilience of communities." *Earthquake Spectra* 19 (4): 733–572.

Chang, S.E., McDaniels, T., Fox, J., Dhariwal, R., and Longstaff, H. 2014. "Toward disaster-resilient cities: characterizing resilience of infrastructure systems with expert judgments." *Risk Analysis* 34 (3): 416–434.

Chang, S.E., and Shinozuka, M. 2004. "Measuring improvements in the disaster resilience of communities." *Earthquake Spectra* 20 (3): 739–755.

Chapin, F.S., Folke, C., and Kofinas, G.P. (eds). 2009. "A framework for understanding change." *Principles of Ecosystem Stewardship*, pp. 3–28. New York: Springer.

Chopra, S.S., Dillon, T., Bilec, M.M., and Khanna, V. 2016. "A network-based framework for assessing infrastructure resilience: a case study of the London metro system." *Journal of The Royal Society Interface* 13 (118): 20160113.

Christodoulou, S. 2000. "Integrated computer-aided monitoring, management and maintenance of civil infrastructure." In: *Computing in Civil and Building Engineering*, edited by Renate Fruchter, Feniosky Peña-Mora, and W. M. Kim Roddis, pp. 580–587. Stanford, CA: American Society of Civil Engineers.

Cimellaro, G.P. 2016. "A comprehensive methodology for the evaluation of infrastructure interdependencies." In *Urban Resilience for Emergency Response and Recovery*, edited by Gian Paolo Cimellaro, pp. 139–223. Cham: Springer.

Cimellaro, G.P, Dueñas-Osorio, L., and Reinhorn, A.M. 2016. "Special issue on resilience-based analysis and design of structures and infrastructure systems." *Journal of Structural Engineering* 142 (8). doi.org/10.1061/(ASCE)ST.1943-541X.0001592

Cimellaro, G.P., Reinhorn, A.M., and Bruneau, M. 2010. "Framework for analytical quantification of disaster resilience." *Engineering Structures* 32 (11): 3639–3649.

Cimellaro, G.P., Renschler, C., Reinhorn, A. M., and Arendt, L. 2016. "PEOPLES: a framework for evaluating resilience." *Journal of Structural Engineering* 142 (10): 04016063.

Comfort, L. 1999. *Shared Risk: Complex Systems in Seismic Response*. Bingley, UK: Emerald Group Publishing Limited.

Conrad, S.H., LeClaire, R.J., O'Reilly, G.P., and Uzunalioglu, H. 2006. "Critical national infrastructure reliability modeling and analysis." *Bell Labs Technical Journal* 11 (3): 57–71.

Daum, W. 2013. "Principles, concepts and assessment of structural health monitoring." In *Handbook of Technical Diagnostics*, edited by H. Czichos, pp. 413–424. Berlin: Springer.

Di Giorgio, A., and Liberati, F. 2012. "A Bayesian network-based approach to the critical infrastructure interdependencies analysis." *IEEE Systems Journal* 6 (3): 510–519.

Diab, Y. 2017. "Sustainable and resilient cities: Is it an infrastructure problem?" *International Conference on Sustainable Infrastructure*, pp. 11–19.

Directive Presidential Policy. 2013. "Critical infrastructure security and resilience." *PPD-21*, Released February 12.

Djalante, R., Holley, C., and Thomalla, F. 2011. "Adaptive governance and managing resilience to natural hazards." *International Journal of Disaster Risk Science* 2 (4): 1–4.

Domich, P.D. 2008. "System and sector interdependencies: An overview of research and development." In *Wiley Handbook of Science and Technology for Homeland Security,* edited by J.G. Voeller, pp. 1–5. New Jersey: John Wiley & Sons, Inc.

Dudenhoeffer, D.D., Permann, M.R., and Manic, M. 2006. "CIMS: a framework for infrastructure interdependency modeling and analysis." *Proceedings of the 38th conference on Winter simulation,* 478–485.

Dueñas-Osorio, L., Craig, J.I., and Goodno, B.J. 2007. "Seismic response of critical interdependent networks." *Earthquake Engineering & Structural Dynamics* 36 (2): 285–306.

Dueñas-Osorio, L., Craig, J.I., Goodno, B.J., and Bostrom, A. 2007. "Interdependent response of networked systems." *Journal of Infrastructure Systems* 13 (3): 185–194.

El-Adaway, I. 2013. "Systems modeling approach for sustainable infrastructure." In *Computing in Civil Engineering*, edited by I. Brilakis, S.H. Lee, and B. Becerik-Gerber, pp. 137–144. Los Angeles, CA: American Society of Civil Engineers.

Eusgeld, I., and Kröger, W. 2008. "Comparative evaluation of modeling and simulation techniques for interdependent critical infrastructures." *9th International Conference on Probabilistic Safety Assessment and Management*, pp. 484–491.

Fang, Y.P., Pedroni, N., and Zio, E. 2016. "Resilience-based component importance measures for critical infrastructure network systems." *IEEE Transactions on Reliability* 65 (2): 502–512.

Fiksel, J. 2003. "Designing resilient, sustainable systems." *Environmental Science & Technology* 37 (23): 5330–5339.

Fischer, J.M., and Amekudzi, A. 2011. "Quality of life, sustainable civil infrastructure, and sustainable development: strategically expanding choice." *Journal of Urban Planning and Development* 137 (1): 39–48.

Franchin, P., and Cavalieri, F. 2015. "Probabilistic assessment of civil infrastructure resilience to earthquakes." *Computer-Aided Civil and Infrastructure Engineering* 30 (7): 583–600.

Francis, R., and Bekera, B. 2014. "A metric and frameworks for resilience analysis of engineered and infrastructure systems." *Reliability Engineering & System Safety* 121: 90–103.

Frangopol, D.M., Lin, K.Y., and Estes, A.C. 1997. "Life-cycle cost design of deteriorating structures." *Journal of Structural Engineering* 123 (10): 1390–1401.

Frangopol, D.M., and Saydam, D. 2011. "Performance indicators for structures and infrastructures." In *Structures Congress 2011,* edited by D. Ames, T. L. Droessler, and M. Hoit, pp. 1215–1226. Reston: American Society of Civil Engineers.

Frangopol, D.M., Soliman, M., and Dong, Y. 2014. "Practical applications of life-cycle considerations in sustainable development of infrastructure." *Sustainable Development of Critical Infrastructure - Proceedings of the 2014 International Conference on Sustainable Development of Critical Infrastructure,* pp. 17–36. Shanghai, China: American Society of Civil Engineers.

Friesz, T.L., Mookherjee, R., and Peeta, S. 2007. "Modeling large scale and complex infrastructure systems as computable games." *Network Science, Nonlinear Science and Infrastructure Systems,* edited by T.L. Friesz. International Series in Operations Research & Management Science, vol 102, pp. 53–75. Boston, MA: Springer.

Galbusera, L., and Giannopoulos, G. 2018. "On input-output economic models in disaster impact assessment." *International Journal of Disaster Risk Reduction* 30 (Part B): 186-198.

Gao, J., Buldyrev, S.V., Havlin, S., and Stanley, H.E. 2011. "Robustness of a network of networks." *Physical Review Letters* 107 (19): 195701.

Gao, Y., and Mosalam, K.M. 2018. "Deep transfer learning for image-based structural damage recognition. *Computer-Aided Civil and Infrastructure Engineering* 33: 748–768.

Gharehbaghi, K., and Raso, V. 2013. "Infrastructure sustainability assessment: A systematic approach." *Developing the Frontier of Sustainable Design, Engineering, and Construction, Proceedings of the 2012 International Conference on Sustainable Design and Construction*, pp. 319–326.

Ghosn, M., Dueñas-Osorio, L., Frangopol, D.M., McAllister, T.P., Bocchini, P., Manuel, L., Ellingwood, B.R., et al. 2016. "Performance indicators for structural systems and infrastructure networks." *Journal of Structural Engineering* 142 (9): F4016003.

Ghosn, M., Frangopol, D. M., McAllister, T. P., Shah, M., Diniz, S. M., Ellingwood, B. R., Manuel, L., et al. (2016). "Reliability-based performance indicators for structural members." *Journal of Structural Engineering* 142 (9): F4016002.

GIO (Global Infrastructure Outlook). 2018. "Forecasting infrastructure investment needs and gaps." https://outlook.gihub.org/.

Glass, R.J., Beyeler, W.E., Conrad, S.H., Brodsky, N.S., Kaplan, P.G., and Brown, T.J. 1987. "Defining research and development directions for modeling and simulation of complex, interdependent adaptive infrastructures." *Physical Review Letters* 59 (4): 381–384.

Golara, A., and Esmaeily, A. 2016. "Quantification and enhancement of the resilience of infrastructure networks." *Journal of Pipeline Systems Engineering and Practice* 8 (1): 04016013.

Gopalakrishnan, K., and Peeta, S. 2010. *"Sustainable and Resilient Critical Infrastructure Systems: Simulation, Modeling, and Intelligent Engineering."* Berlin: Springer Science & Business Media.

Habibian, A., and Minaie, E. 2017. "Exploring the six dimensions of lifeline infrastructure resiliency." In: *Pipelines 2017: Planning and Design*, edited by A. Pridmore and J. Geisbush, pp. 431–441.

Haimes, Y.Y. 2009. "On the definition of resilience in systems." *Risk Analysis: An International Journal* 29 (4): 498–501.

Haimes, Y.Y., Matalas, N.C., Lambert, J.H., Jackson, B.A., and Fellows, J.F. 1998. "Reducing vulnerability of water supply systems to attack." *Journal of Infrastructure Systems* 4 (4): 164–177.

Haimes, Y.Y., and Jiang, P. 2001. "Leontief-based model of risk in complex interconnected infrastructures." *Journal of Infrastructure systems* 7 (1): 1–12.

Heathcote, C. 2017. "Forecasting infrastructure investment needs for 50 countries, 7 sectors through 2040." *World Bank Group, Infrastructure and Public and Private Partnership.*

Henry, D., and Ramirez-Marquez, J.E. 2012. "Generic metrics and quantitative approaches for system resilience as a function of time." *Reliability Engineering & System Safety* 99: 114–122.

Heracleous, C., Kolios, P., Panayiotou, C.G., Ellinas, G., and Polycarpou, M.M. 2017. "Hybrid systems modeling for critical infrastructures interdependency analysis." *Reliability Engineering & System Safety* 165: 89–101.

Holling, C.S. 1973. "Resilience and stability of ecological systems." *Annual Review of Ecology and Systematics* 4 (1): 1–23.

Holling, C.S. 2001. "Understanding the complexity of economic, ecological, and social systems." *Ecosystems* 4 (5): 390–405.

Hollnagel, E. 2017. "Resilience: The challenge of the unstable." In: *Resilience Engineering*: Concepts and Precepts, edited by E. Hollnagel, D.D. Woods, and N. Leveson, pp. 21–30. London: CRC Press.

Holmgren, Å.J. 2007. "A framework for vulnerability assessment of electric power systems." In: *Critical Infrastructure: Advances in Spatial Science*, edited by A.T. Murray and T.H. Grubesic, pp. 31–55. Berlin: Springer.

Home, III., J.F., and Orr, J.E. 1997. "Assessing behaviors that create resilient organizations." *Employment Relations Today* 24 (4): 29–39.

Hosseini, S., Barker, K., and Ramirez-Marquez, J.E. 2016 "A review of definitions and measures of system resilience." *Reliability Engineering & System Safety* 145: 47–61.

Huang, C.N., Liou, J.J., and Chuang, Y.C. 2014. "A method for exploring the interdependencies and importance of critical infrastructures." *Knowledge-Based Systems* 55: 66–74.

Huang, L., Chen, J., and Zhu, Q. 2017. "A large-scale Markov game approach to dynamic protection of interdependent infrastructure networks." In *International Conference on Decision and Game Theory for Security*, pp. 357–376. Cham: Springer.

Ingrisch, J., and Bahn, M. 2018. "Towards a comparable quantification of resilience." *Trends in Ecology & Evolution* 33 (4): 251–259.

Jeong, H.S., Qiao, J., Abraham, D.M., Lawley, M., Richard, J.P., and Yih, Y. 2006. "Minimizing the consequences of intentional attack on water infrastructure." *Computer-Aided Civil and Infrastructure Engineering* 21 (2): 79–92.

Jiang, P., and Haimes, Y.Y. 2004. "Risk management for Leontief-based interdependent systems." *Risk Analysis: An International Journal* 24 (5): 1215–1229.

Johansson, J., and Hassel, H. 2010. "An approach for modelling interdependent infrastructures in the context of vulnerability analysis." *Reliability Engineering & System Safety* 95 (12): 1335–1344.

Kammouh, O., Dervishaj, G., and Cimellaro, G.P. 2017. "Quantitative framework to assess resilience and risk at the country level." *ASCE-ASME Journal of Risk and Uncertainty in Engineering Systems, Part A: Civil Engineering* 4 (1): 04017033.

Khan, S., and Yairi, T. 2018. "A review on the application of deep learning in system health management." *Mechanical Systems and Signal Processing* 107: 241–265.

Kinzig, A.P., Ryan, P., Etienne, M., Allison, H., Elmqvist, T., and Walker, B.H. 2006. "Resilience and regime shifts: assessing cascading effects." *Ecology and Society* 11 (1): 20.

Kwasinski, A. 2014. "Field technical surveys: An essential tool for improving critical infrastructure and lifeline systems resiliency to disasters." *IEEE Global Humanitarian Technology Conference (GHTC)*, pp. 78–85.

Lee, II., E.E., Mitchell, J.E., and Wallace, W.A. 2007. "Restoration of services in interdependent infrastructure systems: a network flows approach." *IEEE Transactions on Systems, Man, and Cybernetics, Part C (Applications and Reviews)* 37 (6): 1303–1317.

Li, F., and Wang, R. 2014. "Urban ecological infrastructure: challenges, construction, and management." *International Conference on Sustainable Infrastructure*, 604–608.

Lin, Y.Z., Nie, Z.H., and Ma, H.W. 2017. Structural damage detection with automatic feature-extraction through deep learning. *Computer-Aided Civil and Infrastructure Engineering* 32 (12): 1025–1046.

Liu, J., Dietz, T., Carpenter, S.R., Alberti, M., Folke, C., Moran, E., Pell, A.N., et al. 2007. "Complexity of coupled human and natural systems." *Science* 317 (5844): 1513–1516.

Liu, J.J., Reed, M., and Girard, T.A. 2017. "Advancing resilience: an integrative, multi-system model of resilience." *Personality and Individual Differences* 111: 111–118.

Loggins, R.A., and Wallace, W.A. 2015. "Rapid assessment of hurricane damage and disruption to interdependent civil infrastructure systems." *Journal of Infrastructure Systems* 21 (4): 04015005.

López, D.R., Brizuela, M.A., Willems, P., Aguiar, M.R., Siffredi, G., and Bran, D. 2013. "Linking ecosystem resistance, resilience, and stability in steppes of North Patagonia." *Ecological Indicators* 24: 1–11.

Lounis, Z., and McAllister, T.P. 2016. "Risk-based decision making for sustainable and resilient infrastructure systems." *Journal of Structural Engineering* 142 (9): F4016005.

Lundberg, J., and Johansson, B.J. 2015. "Systemic resilience model." *Reliability Engineering & System Safety* 141: 22–32.

Masse, T., O'Neil, S., and Rollins, J. 2007. *"The Department of Homeland Security's Risk Assessment Methodology: Evolution, Issues, and Options for Congress."* Library of Congress. Washington D.C.: Congressional Research Service.

Matthews, J.C. 2015. "Disaster resilience of critical water infrastructure systems." *Journal of Structural Engineering* 142 (8): C6015001.

Mayes, R.L., and Reis, E. 2015. "The US Resiliency Council (USRC) and the building rating system." *Second ATC & SEI Conference on Improving the Seismic Performance of Existing Buildings and Other Structures*, 754–764.

McDaniels, T., Chang, S., Peterson, K., Mikawoz, J., and Reed, D. 2007. "Empirical framework for characterizing infrastructure failure interdependencies." *Journal of Infrastructure Systems* 13 (3): 175–184.

Mehany, M.S. 2016. "Resilient, sustainable and green infrastructure concepts–A literature review and proposed preliminary framework for a comprehensive infrastructure evaluation system (CIES)." *Construction Research Congress,* pp. 1609–1618.

Mehrpouyan, H., Haley, B., Dong, A., Tumer, I.Y., and Hoyle, C. 2014. "Resiliency analysis for complex engineered system design." *Artificial Intelligence for Engineering Design, Analysis and Manufacturing* 29 (1): 93–108.

Mejia-Giraldo, D., Villarreal-Marimon, J., Gu, Y., He, Y., Duan, Z., and Wang, L. 2012. "Sustainability and resiliency measures for long-term investment planning in integrated energy and transportation infrastructures." *Journal of Energy Engineering* 138 (2): 87–94.

Miles, S.B. 2011. "The role of critical infrastructure in community resilience to disasters." In *Structures Congress 2011,* edited by D. Ames, T. L. Droessler, and M. Hoit, pp. 1985–1995. Reston: American Society of Civil Engineers.

Miles, S.B., and Chang, S.E. 2007. "A simulation model of urban disaster recovery and resilience: Implementation for the 1994 Northridge earthquake." Technical Report MCEER-07-0014. University of British Columbia.

Miles, S.B., and Chang, S.E. 2011. "ResilUS: A community based disaster resilience model." *Cartography and Geographic Information Science* 38 (1): 36–51.

Mileti, Dennis. 1999. *Disasters by Design: A Reassessment of Natural Hazards in the United States.* Washington, DC: Joseph Henry Press.

Mills, L.O., and Attoh-Okine, N. 2014. "Sustainability of civil infrastructure systems: The past, the present, and the way forward." In *Sustainable Development of Critical Infrastructure,* edited by X. Liu and A. H-S Ang, pp. 79–90. Shanghai, China: American Society of Civil Engineers. doi.org/10.1061/9780784413470.005

Minsker, B., Baldwin, L., Crittenden, J., Kabbes, K., Karamouz, M., Lansey, K., Malinowski, P., et al. 2015. "Progress and recommendations for advancing performance-based sustainable and resilient infrastructure design." *Journal of Water Resources Planning and Management* 141 (12): A4015006.

Mostafavi, A., and Abraham, D.M. 2014. "Resilience-based planning in civil infrastructure using system-of-systems analysis." *Construction Research Congress: Construction in a Global Network - Proceedings of the 2014 Construction Research Congress*, pp.1249–1258. DOI: 10.1061/9780784413517.0128.

Mostafavi, A., Abraham, D., Delaurentis, D., Sinfield, J., and Queiroz, C. 2012. "Innovation policy assessment for civil infrastructure system-of-systems." *Construction Research Congress: Construction Challenges in a Flat World,* pp. 2300–2309.

Mujumdar, V. 2014. "Developing resiliency measures to reduce seismic hazard impact in the central United States." In *Seismic Hazard Design Issues in the Central United States,* edited by James E. Beavers and Nasim Uddin, pp. 143–156. Memphis, TN: American Society of Civil Engineers. doi.org/10.1061/9780784413203.ch10

Muspratt, M.A. 1986. "Artificial intelligence." *Journal of Professional Issues in Engineering* 112 (3): 158–169.

Nagurney, A., and Qiang, Q. 2008. "A network efficiency measure with application to critical infrastructure networks." *Journal of Global Optimization* 40 (1–3): 261–275.

Nan, C., and Sansavini, G. 2017. "A quantitative method for assessing resilience of interdependent infrastructures." *Reliability Engineering & System Safety* 157: 35–53.

Nan, C., and Sansavini, G. 2015. "Multilayer hybrid modeling framework for the performance assessment of interdependent critical infrastructures." *International Journal of Critical Infrastructure Protection* 10: 18–33.

Nan, C., Sansavini, G., and Kröger, W. 2014. "Building an integrated metric for quantifying the resilience of interdependent infrastructure systems." *International Conference on Critical Information Infrastructures Security,* 159–171.

Ng, T.L., and Cai, X. "Relationships between interdependency, reliability, and vulnerability of infrastructure systems: case study of biofuel infrastructure development." *Journal of Infrastructure Systems* 20 (1): 04013008.

Nguyen, T.T., Silander, T., Li, Z., and Leong, T.Y. 2017. "Scalable transfer learning in heterogeneous, dynamic environments." *Artificial Intelligence* 247: 70–94.

NOAA (National Oceanic and Atmospheric Administration). 2018. "*Significant Earthquake Distribution.*" National Centers for Environmental Information. <https://maps.ngdc.noaa.gov/viewers/hazards/>.

Nourzad, S.H., and Pradhan, A. 2014. "Resiliency of intelligent transportation systems to critical disruptions: an eigenvalue-based viewpoint." *Computing in Civil and Building Engineering (2014),* edited by R. Issa Issa and I. Flood, pp. 1731–1738. Orlando, FL: American Society of Civil Engineers.

Oh, E.H., Deshmukh, A., and Hastak, M. 2012. "Criticality assessment of lifeline infrastructure for enhancing disaster response." *Natural Hazards Review* 14 (2): 98–107.

Oh, E.H., Deshmukh, A., and Hastak, M. 2010. "Vulnerability assessment of critical infrastructure, associated industries, and communities during extreme events." In *Construction Research Congress 2010: Innovation for Reshaping Construction Practice,* edited by J. Ruwanpura, Y. Mohamed, and S.H. Lee, pp.449–458. Banff, Canada: American Society of Civil Engineers. DOI:10.1061/41109(373)45

Ouyang, M. 2014. "Review on modeling and simulation of interdependent critical infrastructure systems." *Reliability Engineering & System Safety* 121: 43–60.

Ouyang, M, and Dueñas-Osorio, L. 2011(a). "An approach to design interface topologies across interdependent urban infrastructure systems." *Reliability Engineering & System Safety* 96 (11): 1462–1473.

Ouyang, M., and Dueñas-Osorio, L. 2011(b). "Efficient approach to compute generalized interdependent effects between infrastructure systems." *Journal of Computing in Civil Engineering* 25 (5): 394–406.

Ouyang, M., Dueñas-Osorio, L., and Min, X. 2012. "A three-stage resilience analysis framework for urban infrastructure systems." *Structural Safety* 36: 23–31.

Ouyang, M., Hong, L., Mao, Z.J., Yu, M. H., and Qi, F. 2009. "A methodological approach to analyze vulnerability of interdependent infrastructures." *Simulation Modelling Practice and Theory* 17 (5): 817–828.

Pandit, A., DesRoches, R., Rix, G.J., and Crittenden, J. C. 2014. "A framework to identify the sustainable and resilient zone of urban infrastructure system planning and design." *Creating Infrastructure for a Sustainable World, Proceedings of the International Conference on Sustainable Infrastructure,* pp. 434–441.

Pandit, A., Jeong, H., Crittenden, J.C., and Xu, M. 2011. "An infrastructure ecology approach for urban infrastructure sustainability and resiliency." *IEEE/PES Power Systems Conference and Exposition,* 1–2.

Pandit, A., Minné, E.A., Li, F., Brown, H., Jeong, H., James, J.A., Newell, J.P., et al. 2017. "Infrastructure ecology: an evolving paradigm for sustainable urban development." *Journal of Cleaner Production* 163: S19–S27.

Pant, R., Barker, K., and Zobel, C.W. 2014. "Static and dynamic metrics of economic resilience for interdependent infrastructure and industry sectors." *Reliability Engineering & System Safety* 125: 92–102.

Papajohn, D., Brinker, C., and El Asmar, M. 2016. "MARS: Metaframework for assessing ratings of sustainability for buildings and infrastructure." *Journal of Management in Engineering* 33 (1): 04016026.

Parmee, I.C. 2004. "Computational intelligence and civil engineering-perceived problems and possible solutions." *Towards a Vision for Information Technology in Civil Engineering,* pp. 1–12.

Patterson, S.A., and Apostolakis, G.E. 2007. "Identification of critical locations across multiple infrastructures for terrorist actions." *Reliability Engineering & System Safety* 92 (9): 1183–1203.

Patriarca, R., Bergström, J., Di Gravio, G., and Costantino, F. 2018. "Resilience engineering: current status of the research and future challenges." *Safety Science* 102: 79–100.

Peerenboom, J.P., and Fisher, R.E. 2008. "System and sector interdependencies: an overview." In *Wiley Handbook of Science and Technology for Homeland Security,* edited by J.G. Voeller, pp. 1–20. New Jersey: John Wiley & Sons, Inc. doi.org/10.1002/9780470087923.hhs218

Peng, C., Yuan, M., Gu, C., Peng, Z., and Ming, T. 2017. "A review of the theory and practice of regional resilience." *Sustainable Cities and Society* 29: 86–96.

Perrings, C. 2006. "Resilience and sustainable development." *Environment and Development Economics* 11 (4): 417–427.

Rafiei, M.H., and Adeli, H. 2017. "A novel machine learning-based algorithm to detect damage in high-rise building structures." *The Structural Design of Tall and Special Buildings* 26 (18): 1–11.

Rafiei, M.H., and Adeli, H. 2018. "A novel unsupervised deep learning model for global and local health condition assessment of structures." *Engineering Structures* 156: 598–607.

Ramirez-Marquez, J.E., Rocco, C.M., Barker, K., and Moronta, J. 2018. "Quantifying the resilience of community structures in networks." *Reliability Engineering & System Safety* 169: 466–474.

Reed, D.A., Kapur, K.C., and Christie, R.D. 2009. "Methodology for assessing the resilience of networked infrastructure." *IEEE Systems Journal* 3 (2): 174–180.

Reed, D.A., Zabinsky, Z.B., and Boyle, L.N. 2011. "A framework for optimizing civil infrastructure resiliency." In *Structures Congress 2011,* edited by D. Ames, T. L. Droessler, and M. Hoit, pp. 2104–2112. Reston: American Society of Civil Engineers. doi.org/10.1061/9780784411711.

Reiner, M.B., Fisher, S., and Sperling, J. 2014. "Evaluation of sustainable infrastructure: development context matters." *Proceedings of the 2014 International Conference on Sustainable Infrastructure,* pp. 420–433.

Reis, E., and Mayes, R.L. 2015. "The US Resiliency Council's (USRC) goals, objectives, and founding principles." In *Improving the Seismic Performance of Existing Buildings and Other Structures*, edited by Roberto T. Leon, pp. 765–774. San Francisco, CA: American Society of Civil Engineers.

Resilience Alliance,. 2010. *"Assessing Resilience in Social-Ecological Systems: A Workbook for Practitioners (Revised Version 2.0)." Resilience Alliance.* [online] URL: www.resalliance.org/files/ResilienceAssessmentV2_2.pdf

Resilience Alliance. 2007. *"Assessing Resilience in Social-Ecological Systems: A Workbook For Scientists (Version 1.1)." Resilience Alliance.* [online] URL: www.seachangecop.org/sites/default/files/documents/2007%2006%20RA%20assesing%20Resilience%20in%20SES%20-%20Scientists%20Workbook.pdf

Righi, A.W., Saurin, T.A., and Wachs, P. 2015. "A systematic literature review of resilience engineering: research areas and a research agenda proposal." *Reliability Engineering & System Safety* 141: 142–52.

Rinaldi, S.M., Peerenboom, J.P., and Kelly, T.K. 2001. "Identifying, understanding, and analyzing critical infrastructure interdependencies." *IEEE Control Systems* 21 (6): 11–25.

Rose, A.Z. 2007. "Economic resilience to natural and man-made disasters: multidisciplinary origins and contextual dimensions." *Environmental Hazards* 7 (4): 383–398.
Rose, A.Z., and Liao, S.Y. 2005. "Modeling regional economic resilience to disasters: a computable general equilibrium analysis of water service disruptions." *Journal of Regional Science* 45 (1): 75–112.
Roser, M., and Ritchie, H. 2018. "Natural catastrophes." *OurWorldInData.org.* https://ourworldindata.org/natural-catastrophes.
Rus, K., Kilar, V., and Koren, D. 2018. "Resilience assessment of complex urban systems to natural disasters: a new literature review." *International Journal of Disaster Risk Reduction,* 31: 311–330.
Salehi, H., and Burgueño, R. 2018. "Emerging artificial intelligence methods in structural engineering." *Engineering Structures* 171: 170–189.
Sánchez-Peral, B.C. 2017. "Infrastructure recovery for resilience quantification." *Structures Congress 2017*, edited by J.G. Soules, pp. 264–278. Reston, VA: American Society of Civil Engineers.
Sanford Bernhardt, K.L., and McNeil, S. 2008. "Agent-based modeling: Approach for improving infrastructure management." *Journal of Infrastructure Systems* 14 (3): 253–261.
Sasaki, T., Furukawa, T., Iwasaki, Y., Seto, M., and Mori, A.S. 2015. "Perspectives for ecosystem management based on ecosystem resilience and ecological thresholds against multiple and stochastic disturbances." *Ecological Indicators* 57: 395–408.
Shafieezadeh, A., and Burden, L.I. 2014. "Scenario-based resilience assessment framework for critical infrastructure systems: case study for seismic resilience of seaports." *Reliability Engineering & System Safety* 132: 207–219.
Shakeri, C. 2016. "We are on the 'ground floor' of the Artificial Intelligence Pyramid (and no one knows how to get to the next floor)." *LinkedIn.*
Shen, L., Wu, Y., and Zhang, X. 2010. "Key assessment indicators for the sustainability of infrastructure projects." *Journal of Construction Engineering and Management* 137 (6): 441–451.
Shields, G.M. 2016. "Resiliency planning: Prioritizing the vulnerability of coastal bridges to flooding and scour." *Procedia Engineering* 145: 340–347.
Smarsly, K., Lehner, K., and Hartmann, D. 2007. "Structural health monitoring based on artificial intelligence techniques." *Computing in Civil Engineering,* 111–118. doi.org/10.1061/40937(261)14
Sun, P., Zhang, Y., and Cao, Z. 2017. "Intelligent transportation decision system based on big data mining." *CICTP: Transportation Reform and Change-Equity, Inclusiveness, Sharing, and Innovation.*
Svendsen, N.K., and Wolthusen, S.D. 2007. "Connectivity models of interdependency in mixed-type critical infrastructure networks." *Information Security Technical Report* 12 (1): 44–55.
Tao, P., Li, Y., and Yan, J. 2013. "Research on the evaluation of investment benefits in ecological infrastructure project." *ICCREM: Construction and Operation in the Context of Sustainability,* 1318–1325.
Thacker, S., Pant, R., and Hall, J.W. 2017. "System-of-systems formulation and disruption analysis for multi-scale critical national infrastructures." *Reliability Engineering & System Safety* 167: 30–41.
Tierney, K., and Bruneau, M. 2007. "Conceptualizing and measuring resilience: a key to disaster loss reduction." *TR News.*
Trucco, P., Cagno, E., and De Ambroggi, M. 2012. "Dynamic functional modelling of vulnerability and interoperability of critical infrastructures." *Reliability Engineering & System Safety* 105: 51–63.

Uday, P., and Marais, K. 2015. "Designing resilient systems-of-systems: A survey of metrics, methods, and challenges." *Systems Engineering* 18 (5): 491–510.

UNISDR (Global Assessment Report on Disaster Risk Reduction). 2015. "Making Development Sustainable: The Future of Disaster Risk Management."

Vugrin, E.D., Warren, D.E., Ehlen, M.A., and Camphouse, R.C. 2010. "A framework for assessing the resilience of infrastructure and economic systems." *Sustainable and Resilient Critical Infrastructure Systems,* edited by G. K. Peeta S, pp. 77–116. Berlin: Springer. doi.org/10.1007/978-3-642-11405-2_3

Wang, L., Xue, X., and Zhang, X. 2016. "The sustainability and resilience of infrastructure: critical review and prospects." *ICCREM 2016: BIM Application and Offsite Construction,* pp. 659–667.

Wang, S., Fu, B., Zhao, W., Liu, Y., and Wei, F. 2018. "Structure, function, and dynamic mechanisms of coupled human–natural systems." *Current Opinion in Environmental Sustainability* 33: 87–91.

Wang, Z., Nistor, M.S., and Pickl, S.W. 2017. "Analysis of the definitions of resilience." *IFAC-PapersOnLine* 50 (1): 10649–10657.

Woods, D.D. 2017. "Essential characteristics of resilience." In: Resilience Engineering: Concepts and Precepts, edited by E. Hollnagel, David D. Woods, and Nancy Leveson, pp. 33–46. London: CRC Press.

Woods, D.D. 2015. "Four concepts for resilience and the implications for the future of resilience engineering." *Reliability Engineering & System Safety* 141: 5–9.

Yodo, N., and Wang, P. 2016. "Engineering resilience quantification and system design implications: A literature survey." *Journal of Mechanical Design* 138 (11): 111408.

Youn, B.D., Hu, C., and Wang, P. 2011. "Resilience-driven system design of complex engineered systems." *Journal of Mechanical Design* 133 (10): 101011.

Zhang, P., and Peeta, S. 2011. "A generalized modeling framework to analyze interdependencies among infrastructure systems." *Transportation Research Part B: Methodological* 45 (3): 553–579.

Zhang, P., Peeta, S., and Friesz, T. 2005. "Dynamic game theoretic model of multi-layer infrastructure networks." *Networks and Spatial Economics* 5 (2): 147–178.

Zhang, W., and Luo, X. 2017. "Framework of life-cycle intellectual management platform for infrastructure projects." In *ICCREM: Real Estate and Urbanization*, edited by Y. Wang, Y. Pang, G.Q.P. Shen, and Y. Zhu. 270–287. doi.org/10.1061/9780784481073

Zhao, S., Liu, X., and Zhuo, Y. 2017. "Hybrid hidden Markov models for resilience metrics in a dynamic infrastructure system." *Reliability Engineering & System Safety* 164: 84–97.

Zhang, X., Mahadevan, S., Sankararaman, S., and Goebel, K. 2018. "Resilience-based network design under uncertainty." *Reliability Engineering & System Safety* 169: 364–379.

Zimmerman, R., Zhu, Q., De Leon, F., and Guo, Z. 2017. "Conceptual modeling framework to integrate resilient and interdependent infrastructure in extreme weather." *Journal of Infrastructure Systems* 23 (4): 04017034.

Zio, E., and Ferrario, E. 2013. "A framework for the system-of-systems analysis of the risk for a safety-critical plant exposed to external events." *Reliability Engineering & System Safety* 114: 114–125.

Zio, E., and Sansavini, G. 2011. "Modeling interdependent network systems for identifying cascade-safe operating margins." *IEEE Transactions on Reliability* 60 (1): 94–101.

5 International Adaptation and Resilience

DOI: 10.1201/9781003220503-5

SUMMARY

This chapter begins with the transcription of Dr. John Hildebrand's introduction to the keynote speaker, Dr. Lauren Alexander Augustine, Executive Director for the Gulf Research Program at the National Academies. Dr. Alexander Augustine begins by explaining the role of the National Academies in terms of resilience, and then shares her experiences as a member of the National Academies running programs like the Resilient America Program. She concludes by discussing the need to adapt our capacity to build resilience in our society and solve looming challenges, especially in relation to the changing climate. This is followed by a Question and Answer period with the audience.

This is followed by a presentation by Ruth Larbey titled "Building Bridges: Practical Health-Science Diplomacy Tools to Link Public Groups," sharing her experience implementing innovative methods for involving communities (specifically Bristol, U.K.) with decisions about urban development and improving health conditions in those local communities. Lastly is an article that was presented at the conference, and subsequently reviewed by a scientific committee to be included in the proceedings, titled "Understanding problems and finding solutions to achieve 'Good Living' in Upper Gulf of California communities" by Osmar Villalobos-Cristerna and Víctor Ricárdez-García.

INTRODUCTION TO LAUREN ALEXANDER AUGUSTINE

John Hildebrand
Regents Professor, Neuroscience, The University of Arizona

It is a huge privilege and pleasure for me to introduce Lauren Alexander Augustine. She is one of my favorite colleagues from the National Academies in Washington. Lauren received her bachelor's degree in applied math and systems engineering at the University of Virginia and then got a master's degree there in environmental planning. She then went to Harvard, where she got her Ph.D. in a program, although I was on the faculty at Harvard for quite a while and I'd ever heard of this program. It says advanced independent study program in the graduate school of arts and sciences in landscape ecology and physical hydrology. She dealt with not only physical hydrology and ecology but also geomorphology. Lauren has an amazing breadth of training that set her up for a career in science for policy, which is what she's been doing ever since she got her Ph.D. For our good fortune, she's been working for the

National Academies of Sciences, Engineering and Medicine, which used to be called the NRC, in Washington ever since she'd received her crimson robe.

A quick word about some of the things she's been doing there, and the list is way too long to be comprehensive. I'll just hit some high points. When she started out in the old NRC, which is what it was called then, she was senior program officer in the division of earth and life studies, known to us as DELS. That's one of the seven divisions of the NRC, where she's dealt with water science technology as a board director. She went on to another positions in DELS (Division of Earth and Life Studies), where her true colors started to come out because she was director of the disasters resilience program. Now, you'll see there's a theme here because from there she became Senior Board Director in another division, the one where my job is centered, the division Policy and Global Affairs, affectionately known as PGA, where Vaughan Turekian serves as Director. While at PGA, she was the director of the Resilient America Program, meaning disasters. She became known globally as the "dewan of disasters." If ever there was a wonderful match between a person and a personality and disaster, it's this. She has dealt with every kind of pestilence, earthquake, tornado. I mean anything short of nuclear holocaust, Lauren has dealt with and always with a smile and an encouraging word. It's an amazing situation to observe.

In 2018, she was promoted to a very senior position as the executive director of the recently established Gulf Research Program. I'm sure many of you are familiar with Deepwater Horizon, a terrible disaster. So who do you think about when you think about anything connected with the disaster? You think of the "dewan of disasters." After some period of time of establishing the Gulf Research Program, which brought half a billion dollars to the NRC as part of a big settlement, Lauren was promoted to be the executive director of that program and is off to a wonderful start in finding the right ways to shepherd and curate that money so that it is spent properly. With that, I want to turn the podium over to my wonderful friend and colleague, Lauren Alexander Augustine.

USING SCIENCE TO BUILD COMMUNITY RESILIENCE

Lauren Alexander Augustine

Executive Director, Gulf Research Program, National Academies of Science, Engineering and Medicine

Thank you for that very warm welcome and introduction. I want to say thank you to Hassan Vafai and Kevin Lansey, and to all the organizers for inviting me here today to talk about science diplomacy on a slightly different scale than what we've been talking about thus far. It's a real honor to be here and walking among giants. I just want to say that, as John talked about how in early November 2016 the science diplomacy landscape seems to have changed and we're not just diplomats for the United States, but we're also diplomats for science itself, and that is something I just want to emphasize as we go forward.

So let me just start by saying this talk is sort of like a play in three acts. I like to just let people know what they're in for, so there are three acts. First I'll give a little introduction as to what I'm even talking about in terms of resilience and of the

Academy's role here. Then, I'll talk a little bit about the proof of concept. So in addition to being the "queen of disasters," I have always been given the "different" types of assignments at the Academies. If something was a little bit unusual, they would assign it to me, so things like the African Science Academy Development Initiative (ASADI) program, the Resilient America Program, etc. so I will use those as a proof of concept. Finally, I'll talk about building resilience going forward and adapting our capacity.

Introducing Resilience

To begin, let's examine some data on hurricanes. I'm going to go in the order of destruction here in the United States. We all know this first one is Hurricane Katrina in New Orleans, Louisiana in 2015: $259 billion worth of damage and over 1,800 lives loss. This is always an asterisk in the United States history here as extraordinarily devastating. Houston, Texas comes in at number two with Hurricane Harvey just two years ago in 2017 with $125 billion in damages and 86 people killed. New Jersey comes in at number three with Hurricane Sandy in 2012 at $65 billion in damages and 62 killed (NOAA 2020).

Here is a chart of the number of reported disasters around the world by the type of disaster (Figure 5.1).

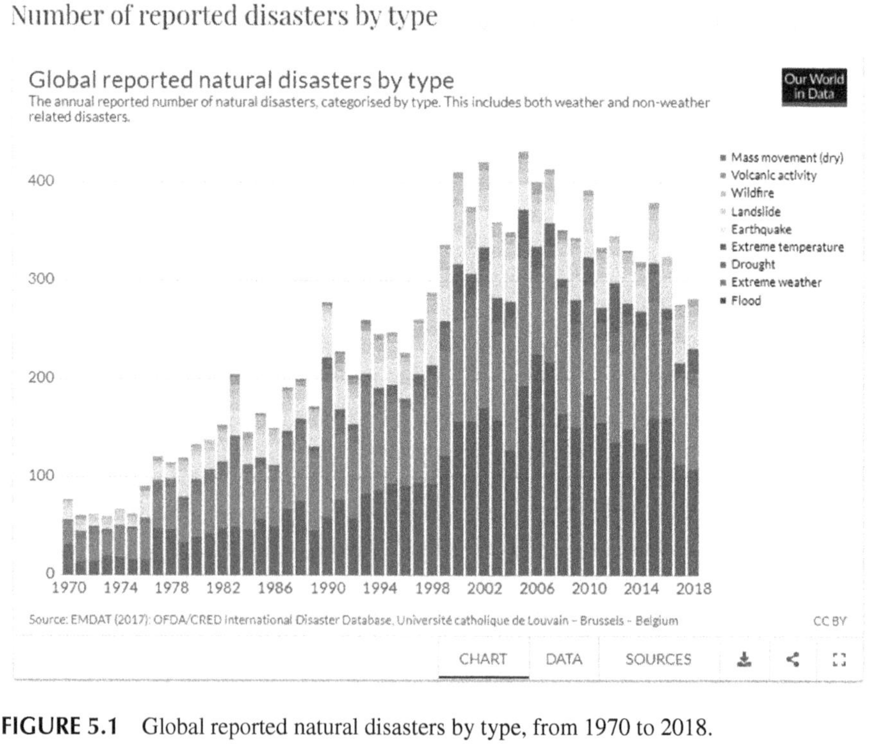

FIGURE 5.1 Global reported natural disasters by type, from 1970 to 2018.

Looking at the figure, we see a pattern. There are two shades of blue here, representing extreme weather and flood, that together make up the majority of reported incidents. What we see in the United States is echoed around the world, that when we talk about disasters, usually it's around weather. And usually it's around wet weather. This is something that is universal and, because I'm a flood person and I'm a hydrologist, it just happens to kind of fit my interest, but also you'll see a theme here. The reason I wanted to point this out is, if we can get our hands around flood resilience, then we can get our hands around a big portion of natural disaster resilience.

While it was raining in Houston in 2017, the exact same days it was raining in Mumbai. There, they actually got *less* rain than Houston, but 1,200 people died and 1.8 million people were displaced (Siddique 2017). Now, I don't even know how to move 2 million people, but they do this every year in the monsoon season in Southeast Asia. So this is just a reminder that even though I'm going to be talking mostly about the United States, this is everywhere and this is where all of our science diplomacy has a huge role to play to save lives.

As was mentioned, I am an engineer. So I always like to be very, very clear about what problem we are trying to solve here. And the problem we're trying to solve here is that the risks and the impacts of these events are evolving. In most cases they're increasing, but in all cases they're changing. And so the problem we're trying to solve is how do we get ready to adapt to these changes. That's what we call resilience.

Introducing the National Academies

So where do the Academies come in? I know many in this audience know the Academies well; I think half of us present are employees for the Academies. But for those who may not know, the National Academies of Sciences, Engineering, and Medicine are not part of the federal government. We are not the National Science Foundation. Under only extreme circumstances do we ever give out money, though we will take money from any source. We are a private nonprofit and we provide independent, high-quality, objective advice around scientific issues with policy implications (Academies 2020). This is what we do. We have an enormous convening power and we bring that to bear to answer thorny questions. Our core belief is that we use science to benefit society, and the beautiful thing about this organization is that we can do it in many, many different ways.

Resilient America Program

Here we are in act two. This is our proof of concept. I always like to ask this question: who in this audience would like to be resilient? Hands up, right? I mean, what's the alternative? "No, I want to be fragile, I want to be weak, I want to be dead." So whenever I ask this, everyone says yes, they'd like to be resilient, and this is always a good start. But there is a second piece. We all want to be resilient, but what does that mean to me, and what does that mean to the people in my community? This is where the question gets a lot harder to answer. And then there's a third part: where do you start?

The story starts for us in 2012. One of the main activities of the National Academies is that we produce these reports and we bring together these blue-ribbon panels of brilliant people around questions. We did that in 2010, and in 2012 we had this report: *Disaster Resilience: A National Imperative* (Cutter et al. 2013). It laid out a definition of resilience, which I'm not here to defend or to promote, but it's good, or at least good enough. President Obama used it in presidential policy directives 21 and 25 (Obama 2013). NATO used it as it amended Article III in 2016 (Shea 2016). So it's good enough; it can be improved, sure, but we're not here to do that today.

When this consensus report came out, there was a lot of excitement around the report and people kept coming to us saying, "Can you help us implement it?" I was like, "No, see, we don't do that. We give you advice and we say good luck." Well, the president at the time, Ralph Cicerone, said "No, no Lauren. I think that there's something here. Why don't you see what you can do? See if you can come up with an implementation program." I thought to myself, "I don't really see an upside to this, because the advice is supposed to work. So if it works, there's no story and if it doesn't work, well, I don't see the upside." It didn't matter. I had my marching orders, and off we went.

We created a new program called Resilient America (Forrest and Milliken 2018). What this was about was to take these four recommendations of the report that basically outlined what must be done if you want to be resilient, and we tried to implement them. So the big things from the report on how to be resilient are:

1. Understand and communicate, so you can better manage your risk.
2. Create partnerships with the community and stakeholders. This is not just a FEMA (Federal Emergency Management Agency) issue; this is a big tent, with a lot of stakeholders, and you have to bring them in.
3. Measure. You have to have some way to track progress or measure your resilience; something to say, which direction you're moving, and how fast and how well.
4. The sharing of information: so share what you know, share data, make sure your data are accurate, those kinds of things.

So we took these recommendations and off we went. We did this in four actual locations around the United States: Seattle, Washington; Tulsa, Oklahoma; Cedar Rapids, Iowa; and Charleston, South Carolina (Academies 2018).

Through these Resilient America pilot communities, we brought science to society and what we learned in this process is we also brought society to science! This was an unexpected, wonderful outcome that we didn't really anticipate. We didn't expect that we would change our ways of how we interacted on scientific issues based on these interactions at the community level. The key takeaways from this were threefold. At the community level, people really do want better information and they want to engage on these issues. Also, relationships and trust are critical to this community engagement. And lastly, specifically for the Academies, there is a really important role we must fill for someone to be a catalyst, a neutral convener, and someone there to start the consensus building and capacity building (Academies 2019). So we learned a lot in those first four and a half years with Resilient America.

Resilience on a Global Scale

Resilience is a global issue though, and I know this is an international, global audience, so I want to point these examples out. This Resilient America program also talked about how, if you can talk about floods, and earthquakes, and climate change in the United States, what does that mean if you had a chemical, biological, radiological, or nuclear (CBRN) event? So with Japan and NATO, we did a big project on CBRN International Resilience with Tokyo 2020 Olympics as a backdrop (National Academies of Sciences 2019). NASA was looking at earth observations to get a better sense of the global scale, and how to use that Earth Observation Data for disaster risk reduction and resilience. Urban flooding, as I already mentioned, is certainly a global issue. We saw Mumbai and Houston flooding at the same time with very, very different results. We have a lot of partnerships in this program, a lot of them being international organizations or internationally focused organizations, including UNEP, UNOCHA, UPS Foundation, World Economic Forum, ASADI partners, NASA/GEO, Zurich Foundation, Willis Reinsurance, NATO, ECHO, and many, many more. This really started to cement in everyone's mind that these are universal issues.

I'll just say one last thing about NATO and Article III. Everyone knows Article V of NATO, where if you pick a fight with one of us, you pick a fight with all of us (NATO 1949). Well Article III sets the conditions for what a country can do to invoke Article V, and in 2016, NATO amended Article III to include "national resilience" based on our report (Shea 2016). So we are very proud of that influence on the global scale.

Resilience on a Local Scale

That said, we still work most intensively at the local scale. What I would say is that we actually all do. Even Glenn Schweitzer, who earlier told us how he has been all over the world to over 200 countries—even held hostage in a couple of them—but the truth is that the whole thing is still an individual conversation. It is still at a local scale even if it is in a different place. So there are some lessons here, in each situation, that I think are quite transferable.

I will just briefly say that in Cedar Rapids and Charleston, two of our Resilient America pilot communities, they wanted to talk about floods. They wanted to make sure that their resilience was around flooding. In Charleston, in particular, I was so excited, being a self-proclaimed "flood person." We had fluvial flooding, pluvial flooding, we had coastal surge! And everyone else was saying, "we just don't want to get wet." I thought, "oh okay, we can work around that." So sometimes it comes down to just a simple statement like that. Seattle took on a climate change and social inequity point of view for their resilience, and Tulsa wanted to talk about money.

I will say in all cases, we went into these communities and once we got comfortable, the first thing they all said was, "We don't want to talk about natural disasters." I was like, "What? What do you want to talk about?" (You'll recall that the title of the book was *Disaster Resilience*.) They said, "we need to talk about vulnerable populations and economic resilience; if we don't take care of our people, and if we don't take care of our livelihoods, there is nothing left for us to talk about." That was a reorienting moment in the program.

So what we learned from this diplomacy at the local level, even in the ASADI program I worked on, and in all of these different places that I've worked, is that these are the things that hold: first, there is no single definition for resilience. I'm not bothered by that, because people have to understand what they are seeking to do in terms that are most important to them. So, be flexible. Second, the more people rallying and the more champions around these thorny issues, the more likely you will have impact, and the more likely you will have someone who can own it and have jurisdiction over—like a local government—the more likely it will be successful. So, someone has to own it; people must participate. The third thing that we learned was that the process was, in some cases, more important than the product. So this is about building trust, this is about building relationships, this is about moving forward together, much more so than actually building a plan or coming up with a strategy. The fourth thing is that resilience is place-based. It has to be here because if it's there, I don't have any control or understanding over it. But if it's here, I can do something. This was a very important piece that as we were working with FEMA, DHS, and some of the federal agencies; they wanted a national "silver bullet" and it didn't really work that way. Finally, we learned that people want this information and they will use it if we, as scientists, do our job and present it in ways and formats that they can understand. This seems, and I'm sure especially to this audience, a bit obvious, but in real life, it took some time to get there, and it is still a work in progress to figure out what those formats are.

So where is the science in all of this? I mean, we are the National Academies of Sciences, not the National Academies of Public Policy (I've been told that once or twice). The translation of data into information is a science. The aspects of moving data and analysis and translating, interpreting, and moving it forward: there's a whole science around that. There's a whole science around community engagement. I just want to explicitly say hats-off to our social scientists on this one. And this last one is the most fun: the science of how you combine different methods of understanding? How do you bridge someone who's extremely quantitative, who does maps and figures, to someone who's qualitative and doesn't, and bring that together to help answer a set of questions? This is a different kind of scientific process. In my abstract, I said there's an art and a science to it. This is the mix, but the science is the backbone.

Adapting Capacity and Enabling Resilience

So the "bad news" is, there is no single way to do this, no silver bullet, no recipe, no cookie cutter, and, I want to emphasize this, there is no scorecard or checklist for resilience (Snel et al. 2019). It's okay though, because in our research and in our experience, the communities don't use these tools anyway. A lot of people produce them. All these gadgets and cool things, and then you get there and people think, "I don't know what that means. I don't have a sum system. I can't process that. I don't know." For example, someone earlier in the conference mentioned the importance of probability and statistics, but that alone reaches a very small audience, so a lot of these tools sit dusty in the decision-makers' offices. So there is a lesson there.

Dimensions for Resilience

In the spirit of building on what we know and not reinventing the wheel (or any other overused adage I can come up with) here's what we really learned in this process: that when you think about these issues, there's a physical dimension, there's a social dimension, economic, human, and a political one. We use all of these, or many of them, in getting a complete picture of what resilience is.

Physical and Natural Dimensions

To illustrate what I mean, you've all seen meteorological radar maps (something that makes a lot of us really). Digital maps like this can run a loop and people are really happy that our abilities are this technically sophisticated to share that kind of visual information. It's important to see what it looks like, how it's coming, when it's coming, and it's kind of in real time.

Here are two other ways that data and information tell a story and we use it all the time in the funding world.

These maps both show Chicago, and the body of water is Lake Michigan. Figure 5.2 shows the FEMA flood map. Now it's curious because FEMA's flood maps are all about riverine flooding with a 100-year floodplain. We can talk about a 100-year floodplain later, but that is what this is, so the red areas show where the streams are and where the 100-year floodplain is. The problem in Chicago is that they sunk all the streams to build the city (McShane and Cain 1980). So if there is not a stream, there can't be a floodplain. So this gives a very inaccurate, incomplete picture of where it is going to flood, because it is not the rivers that are flooding in Chicago. It's flooding all the same, just not from that source; it's a different source of flooding. So this is what policy is built on, what your rates of insurance are going to be built on, what your home insurance might be built on.

In Figure 5.3, here we're getting much finer tuned, with a riverine and pluvial flood model. For those of you just needing a reminder, pluvial flooding is when it rains and the rain, with not really anywhere to go, just kind of sits. Fluvial flooding is when a

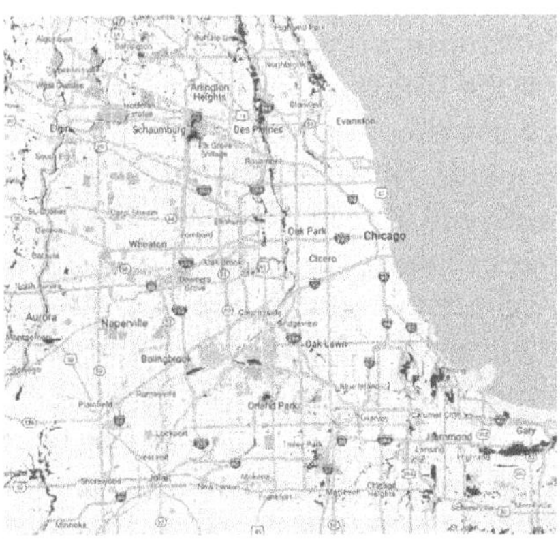

FIGURE 5.2 FEMA riverine flood study. Population exposed: 135,000 (1%).

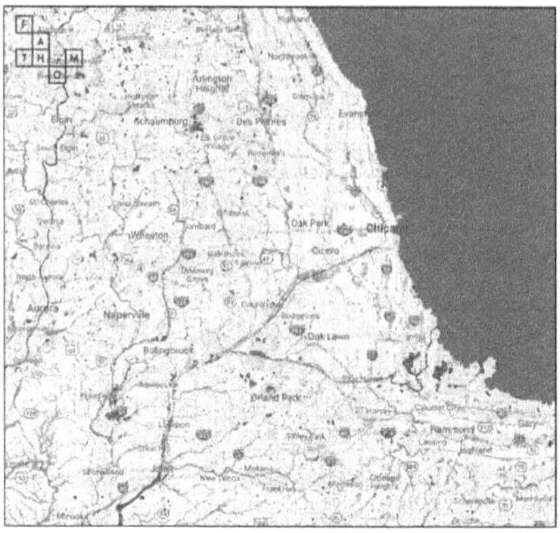

FIGURE 5.3 Riverine & pluvial flood model. Population exposed: 945,000 (10%) (Wing et al. 2018).

river overflows and overspills its banks. So if you look at pluvial flooding in Chicago, which in this figure is all the blue dots, and basically the whole place is blue. This is what Chicago looks like when it rains, despite what FEMA tells us it'll look like when the rivers overflow (Kousky and Golnaraghi 2020). This is where the role of the highly quantitative mapper and the dynamic modelers really comes into play because this is information people can really use.

Social Dimensions

As for the social dimensions it turns out that none of it matters if no one is in the way of the water, and it turns out that it matters a *lot* if there are. When flooding occurs, not every neighborhood floods. One street could flood and the next street over is unharmed. These things don't happen equally. So the scale (and the type of map) matters. How granular can you get this information? Because a little bit of grading, a little bit of elevation, how the road is pitched, will determine whether your house gets wet or not.

Figure 5.4 shows a SoVI map of the Houston area. So in one of our studies, we worked with Eric Tate at the University of Iowa. He took the Social Vulnerability Index, or SoVI, which comes out of the research of Susan Cutter in South Carolina, and thought, we can do a little bit better with this if we utilize some of this flood information (Cutter, Boruff, and Shirley 2003). We started to layer into this some of that physical, topographic, and pluvial information with the SoVI information (Rufat et al. 2015). We started to get this picture of where the vulnerable people around Houston and Harris County are living for flood risk. You can look at the economic dimensions, you can look at NFIP, or the National Flood Insurance Program payouts.

These maps show the amount of payouts due to flooding by county, first of the county (Figure 5.5) and the zoomed in on the Houston area (Figure 5.6). In case you can't tell, the red areas show higher payouts (by county). The darkest red areas are $134 million in payouts, so there is a lot of damage—and that's just insured stuff because the stuff that's not included in that: uninsured losses, loss of work, loss of school time, people,

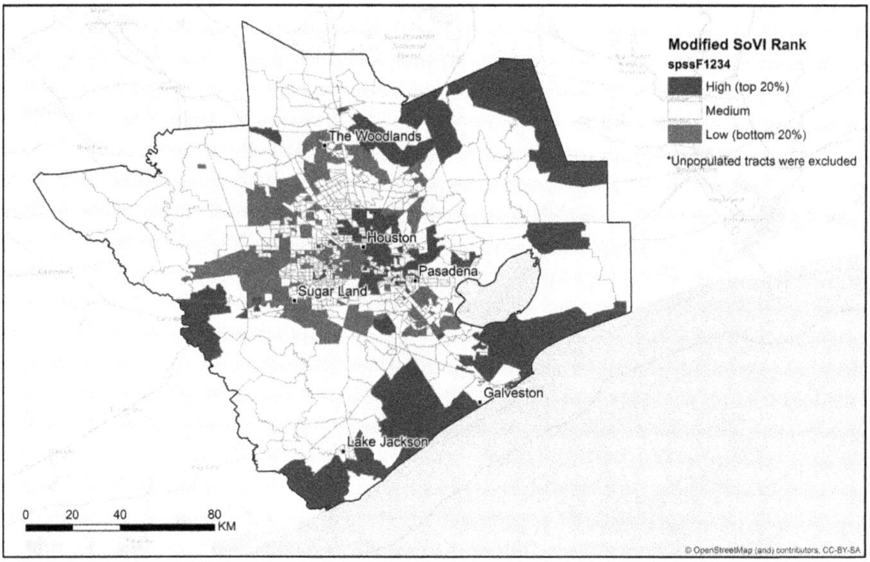

FIGURE 5.4 A modified SoVI map for Houston, courtesy of Dr. Eric Tate, University of Iowa.

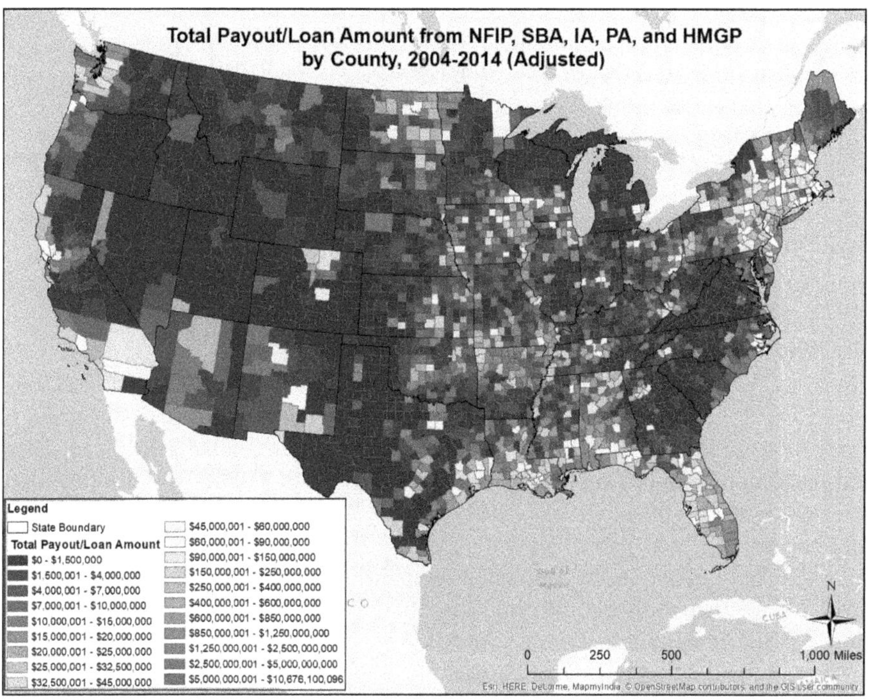

FIGURE 5.5 Total Payout/Loan Amount from NFIP, SBA, IA, PA, and HMGP by County, 2004–2014.

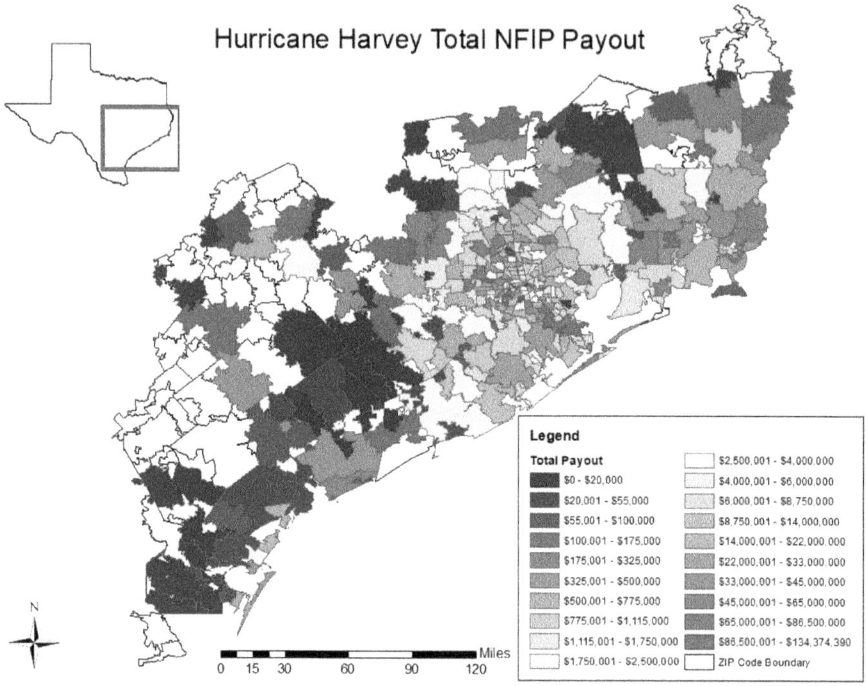

FIGURE 5.6 Hurricane Harvey Total NFIP Payout.

and livelihoods. This third one—there are so many unaccounted losses. People who don't speak English, people who don't want to be found, Black, brown, poor, elderly; those who are the most susceptible to long-term damage are often the least counted. They have the least-heard voice. Then there is the diversity of incomes and financial structures of a community. In short, this is an incomplete picture.

The Role of Science

The role of science here, as we are diplomats for science, because it turns out in some places like Texas and South Carolina, the science is not that popular either, and so you have to be very careful. But here's the role of where it can help: we bring clarity and try to complete this picture of hazards and risks to people who have to make decisions about it, and that turns out to be a pretty nonpartisan issue. So there is a role for everyone. We talked about a big tent, and we do bring these people in all the way from the "feds" down to the individual researchers, NGOs, philanthropists—all are invited. Sometimes it is like herding cats, but that's okay! That's what we do at the National Academies all the time, so we are okay with that.

That work in Resilient America was incredibly hard. I mean, just never being home, always on a plane, always in some community meeting with someone who didn't want to hear what you had to say. And yet it was extremely rewarding. So would I do it again? I'm going to go ahead and say sure, why not?

Gulf Research Program

For those of you who may not know, in 2010, the Deepwater Horizon oil drilling rig exploded, killed 11 people, and spewed oil into the Gulf of Mexico for 87 days (Drilling 2011). As part of the criminal settlement from this spill, the National Academies were endowed with $500 million for 30 years, from 2013 to 2043 (EPA 2017). So, like Dr. Hildebrand said, the Academies Gulf Research Program is my new position where I have been entrusted to make sure that the Academies is a very good and responsible steward for a $500 million endowment, all of which, plus interest, has to be spent by 2043. We have a mandate that involves offshore oil production—since this is out of Deepwater Horizon, we want to make sure we can either reduce the likelihood of such an event happening again, or the impacts of it if it does. So there's a big piece on offshore production. The second of the three pieces is on environmental protection; how can we protect the environment, just generally but in particular to an oil hazard. The third one is around the people, human health, and community resilience. Eleven people died from the incident, and this endowment that comes to the Academies comes with it that heavy responsibility. So we feel very responsible to do good things with this (Figure 5.7).

So given this history, take a look at this map, which was published in *The New York Times* a couple of years ago (Chinoy 2018). This is the whole country, but look where all the red is. This is where you have $5 million or more per year in the zip codes due to natural disasters. The whole Gulf Coast is red, except for the Western panhandle of

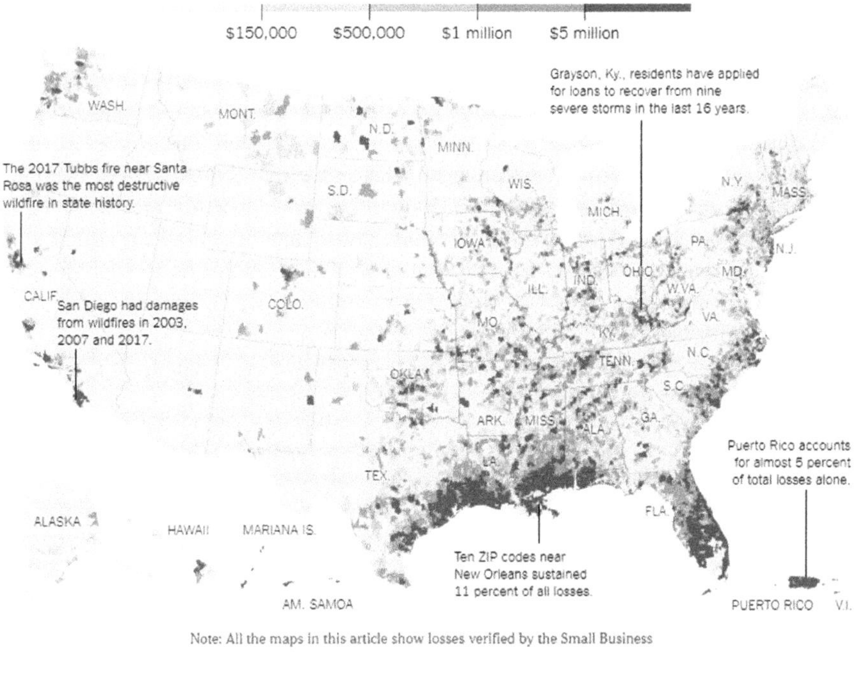

FIGURE 5.7 Map of losses in each ZIP code from major natural disasters, 2002–2017.

Florida which is all occupied by the Navy, so they don't get the federal relief funds, but everyone else does and this is what it looks like. So, this area is really ripe for a lot of these resilience questions. Not to mention, we have a lot of historical inequities in this region, we have social disparities, we have the lowest health outcomes for what they call the North Gulf (Louisiana, Mississippi, Alabama), the lowest health outcomes, the lowest education outcomes; the largest disparities of those who have and those who have not. So we have a lot of issues here in the Gulf.

As I mentioned, part of our charter is for human health and community resilience. So we have a clever new nickname for a program that has not yet begun, called ENCORE, Enhancing Community Resilience. So we're going to take four easy steps that are not easy at all. These steps engender buy-in around the resilience priorities, goals, and leadership. It's going to be a mix of a bottom-up and top-down program, and will include multiple dimensions of resilience, and it will keep measures. We have the advantage of having money *and* time, whereas a lot of these projects don't have enough of either, honestly. But here we are able to start things and we can start to track them over time. So there's a big data piece, there's a big longitudinal, social data platform piece; how do we know we're making change? So we're going to institute a longitudinal data program as well. And the last one is where I really want to get to, where investments that we make result in multiple benefits. So with ENCORE, we want to create a major initiative so that the people of the Gulf feel some benefit from the money that was given to us for the injury they sustained in 2010. This is a very important set of strings to make sure we knit together in a good way. We have the money, and we have the time, so we want to make sure that we get into a space where people can continue to learn from each other, including the Academy learning from the Gulf and vice versa, and then the communities learning from each other in resilience learning collaboratives, which might sound familiar to our ASADI work on learning collaboratives. This includes the longitudinal research program. And here's the one that I think is super important so that we don't continue to make the same mistakes twice: to connect the recovery money—$26 billion for East Texas on Harvey alone, $250 billion New Orleans on Katrina alone (NOAA 2020)—connect those massive resources that can start to get at public assistance for infrastructure and things like that with looking ahead—that anticipatory and modeling strength that the academicians bring—and say let's link this. We have resources here, we have some estimates here, and let's see if we can get this together so we can get the anticipatory mitigation work tied with some of the recovery efforts.

So, in closing, we are on our way. This is where we are headed, on the road to resilience: we convene, we collaborate, and we make an impact.

Questions and Answers

Question

Audience Member: I was working with a couple of colleagues on climate change and the commons. I lived in New Orleans at the time of pre-Katrina. I wanted to say that the real folly of the floodplain maps is that there is that one family that was paid for the same house six times for flooding for $18 million. And I think your work just showed that, how this whole program needs to be thrown out and we really need to do some "real" science, not just this arbitrary drawing of lines. New Orleans was actually very affected by that over the years. We lived in a house that was 4.5 feet below sea level, and we were considered high for flood insurance. So is something that should be stressed more?

Dr. Alexander Augustine: Yeah, it is moving in that direction. So the FEMA flood maps are, in fact, flood insurance rate maps. And just to be fair to FEMA, they were never intended to be flood hazard maps or flood risk maps. It's just a proportion, a relative line. On this side of the line you pay more, on that side you pay less. But to your point, they are the only maps we have for the whole country, and so everyone uses them for much different interpretations. But I think it's moving in that direction. There's a lot of work to get to flood risk and flood impact, which is really what you want.

Question

Audience member: Thank you so much for your presentation. It was really, very interesting. I was curious, with all the engagement efforts that you did, was that mostly engaging with local decision makers and did you do any forums or working directly with community members, or was it mostly using the local decision makers as a proxy?

Dr. Alexander Augustine: It was all of that, truly. I mean, we had the Academies. So, the Academies is a soft money organization, and I just need to say that because this work was funded by a collection of sponsors from the federal agencies and philanthropies and some foundations. I say that just to know where the money's coming from, because that matters. To answer your question, we met at all levels. We met in churches, town halls, hotels. We spent four or five years in each city as we implemented these programs, and so these partnerships are now friendships, and they're strong. We composed what we called ground teams in each place, knowing that the messenger matters and the message matters. So in Charleston, we had a collection of all that list I had from folks in Charleston, so they owned it. A success story out of Charleston is that that ground team that we brought together ended up formalizing itself and calling itself the Charleston Resilience Network, which is still in existence today (CRN 2020). So we really did. The other piece of it, the reason I brought up the federal sponsors, was that there was a great interest to make sure that the local needs and priorities could be made aware to the federal agencies that have a lot of resources. That connection was very hard for both sides to make and it was really natural for us to make. So in many ways we were kind of like a zipper. It really did help connect the laws and the big money to the local needs. So it was intense.

Bill Colglazier: Thank you, first of all, for a wonderful talk. I was really struck by all the insights you've learned about the science community engaging with local communities, listening to the needs, sort of helping the communities, figure out what could actually be done, to do something about the issue. I'm involved with the UN on this role of science, technology for the Sustainable Development Goals. Where that is right now, particularly with developing countries, is focused exactly on the same issue. How can the science community in these countries go and engage with their stakeholders at the local level, sub-national level, the national level, to listen better at what the needs are? To tell them what is known from science, and how together they can sort of begin to implement actions that might address those issues. I'll be talking to you, trying to learn more from your insights to take to this work that's going on now in a number of countries trying to do the same thing that you've done with Resilient America program.

Question

Stephanie Buechler: That was a very interesting talk, thank you. I noticed that your new initiative will include a focus on gender, which I was really happy to see. I work on more slow-moving disasters associated with climate change in agricultural communities and urban, peri-urban and rural areas around the world. One thing that I've really learned from the people that have taught me so much in the communities is that we can learn, you know, across borders, across regions of the world. I was wondering for Resilient America, whether you had any plans to include people from Indonesia, for example, where most of the people that died were women? Or from Central America with Hurricane Mitch, and that experience, where they've developed very innovative responses to disaster and to prevent the kind of loss of life, especially among more-vulnerable groups that include women?

Dr. Alexander Augustine: No plans for Indonesia, specifically, and I think my answer may be unsatisfying to you, but it is based in practicality. That is, the next round of discussions has been with our African Academy counterparts. So, part of my career at the National Academies was that I was part of the African Science Academy Development Initiative (ASADI) and we worked with 12 Academies across Sub-Saharan Africa. Because of those relationships, a lot of these issues, like I said, are universal, and in a lot of ways are being socialized into that community. We just don't have strong ties in Indonesia yet, but I really do think that the key is that wherever we work, that it can't seem like we are parachuting in and telling people what to do. I mean, that's a rub that people just can't get over. So if there is an Indonesian effort, the entrée has to be offered by an Indonesian person or someone with strong Indonesian ties. But on the African side we have those connections in place.

Question

Christopher Scott: My question is actually similar, but maybe a little bit different. You mentioned the Mumbai case. I think a lot of what we think about in resilience is after the event, you know, recovery is one word for it, but maybe preparation for the next time. What do we understand about adaptive capacity in this, and partly that could be contingency planning and different and better infrastructure, but also people in communities' own innate ability to be resilient during and after an event. Are there any lessons to be drawn from that interesting contrast you drew between Houston and Mumbai? Kind of in the same moment and a similar risk event produced the different outcomes. How do we understand that. It's not so much how do we reach Mumbai, but actually, how do we get Mumbai's lessons into our program?

Dr. Alexander Augustine: Yeah, that has a lot of layers. I will say that one of the huge disadvantages, it's an advantage because we're in the United States, so it's good for us. But, you know, the United States is one of the wealthier countries, right? I didn't show in this talk the number of billion dollar events, and that curve of annual billion dollar natural disasters is exponential. It used to be flat and it just keeps going up and up. So what we do here in the U.S. is we throw money at it *after* an event. We do it all the time. Floods alone are $10 billion. It's just how we operate, but not many other places in the world can afford to do that. And so, your question around the individual capacity, the community capacity, the social cohesion, is critical. It is absolutely critical. And as some members of Congress tire of throwing money after events to extract some neighbors in New Orleans with the $18 million six-time rebuild, people get tired of that, that does happen, and it's totally legal, by the way. We might start shifting away from throwing money at it. This is where we can learn from other places that never had that option in the first place, and there is something to be learned, both ways. There's something to be learned about a centralized emergency management agency, if you can afford it. There's absolutely something to be learned from the National Incident Command Center, if you can afford it. And there's something to be learned about not always having to call upon that, just because you have it. So it is two-way learning for sure, I would say.

Question

Audience Member: One of the things that you've clearly been doing, and I think very well, is figuring out how to talk about science to a community that is not comprised of scientists. Could you step back and think about for this or any other example. What advice, in concrete terms, could you give to new scientists who haven't been doing this, who haven't been trained on this, about how to say what they're doing in a way that's understandable and evocative to the community?

Answer

Dr. Alexander Augustine: So the simple, short answer is: work your very best to meet everyone where they are. You know, not where you would think they should be or where you'd like them to be, and certainly not where you are, you know. But really, your best understanding of where they are, and usually people will tell you. I'll tell a very quick story along these lines. We were working in Charleston, South Carolina on flood issues during the time that a Black man got shot in the back by a white officer and the AME's church got shot up and nine people were murdered (Schmidt and Apuzzo 2015; Horowitz et al. 2015). So two racially charged events in Charleston, and we're talking about flood resilience. As sort of outsiders with the volunteers and staff that we brought to the area, we didn't know what to do. We knew we had to address this; everyone in Charleston, I mean, we all felt it. It was very raw, and it was very painful. These experiences were very painful for Charleston. Then the head of the South Carolina Aquarium said "let's remember the sea turtles!" I thought, "what? This is not sounding helpful." But what he was saying was that we can use this "doorway" of flooding to start to get at some of these social issues, because everyone loves sea turtles. Everyone loves sea turtles, and the sea turtles' habitat is shared by the lowest income, the darkest skin, and those with the least aptitude to speak English, and to have high health outcomes. So the poor, the Black, the brown, the immigrant populations in Charleston, are living on this membrane of wet and dry, just like the turtles are. He was saying let's use that to build resilience, for the people, and it worked brilliantly. So there's different ways to get at these issues.

Question

Audience Member: If you were to map out the disaster diplomacy ecosystem, what part of it would represent the desert?

Answer

Dr. Alexander Augustine: The desert meaning like drought? Rural areas. This is actually an enormous divide for those of us with limited resources in time. We are going to go for the biggest bang for our buck, and that's where the people live (so urban). But there are a lot of people at risk in the rural areas, and that is absolutely the research desert.

Audience Member: I think it's important to mention that The University of Arizona has been a leader in this area abroad. Ten years ago, they had a project on resilience in a new suburb of Tehran. They recently had a team over there to talk about all the resilience dimensions, particularly water, and as a result, the local university established a resilience center there based on a two-week visit by the University of Arizona people to Tehran.

Dr. Alexander Augustine: That's awesome, thank you for that.

REFERENCES

Academies, U.S. National. 2018. Community Pilot Partnership Program, 2014–2018. edited by ResilientAmerica Program: The National Academies of Sciences, Engineering, and Medicine.

Academies, U.S. National. 2019. Community Pilot Partnership Program: Successes, Challenges and What We Learned. edited by ResilientAmerica Program: The National Academies of Sciences, Engineering, and Medicine.

Academies, U.S. National. 2020. "The National Academies: An Overview." accessed February 17, 2020. www.nationalacademies.org/brochure/index.html?_ga=2.146899760.648719787.1497270255-141596223.1486144204.

Chinoy, Sahil. 2018. "The Places in the U.S. Where Disaster Strikes Again and Again." *The New York Times*, May 24, 2018. www.nytimes.com/interactive/2018/05/24/us/disasters-hurricanes-wildfires-storms.html.

(CRN), Charleston Resilience Network. 2020. "Charleston Resilience Network (CRN): Who We Are." Charleston Resilience Network (CRN). www.charlestonresilience.net/about/who-we-are/.

Cutter, Susan L., Bryan J. Boruff, and W. Lynn Shirley. 2003. "Social vulnerability to environmental hazards*." *Social Science Quarterly* 84(2):242–261. doi: 10.1111/1540-6237.8402002.

Cutter, Susan L., Joseph A. Ahearn, Bernard Amadei, Patrick Crawford, Elizabeth A. Eide, Gerald E. Galloway, Michael F. Goodchild, Howard C. Kunreuther, Meredith Li-Vollmer, Monica Schoch-Spana, Susan C. Scrimshaw, Ellis M. Stanley, Gene Whitney, and Mary Lou Zoback. 2013. "Disaster resilience: a national imperative." *Environment: Science and Policy for Sustainable Development* 55 (2):25–29. doi: 10.1080/00139157.2013.768076.

Drilling, National Commission on the BP/Deepwater Horizon Oil Spill and Offshore. 2011. Deep Water: The Gulf Oil Disaster and the Future of Offshore Drilling. Report to the President.

EPA. 2017. "Deepwater Horizon – BP Gulf of Mexico Oil Spill." United States Environmental Protection Agency, accessed September 9. www.epa.gov/enforcement/deepwater-horizon-bp-gulf-mexico-oil-spill.

Forrest, Sherrie, and Charlene Milliken. 2018. "Building resilience to disaster: from advice to action." *European Review* 27:1–10. doi: 10.1017/S1062798718000522.

Horowitz, Jason, Nick Corasaniti and Ashley Southall. 2015. *Nine Killed in Shooting at Black Church in Charleston. The New York Times.*

Kousky, Carolyn, and Maryam Golnaraghi. 2020. *Flood Risk Management in the United States: Building Flood Resilience in a Changing Climate.* Zurich: The Geneva Association—International Association for the Study of Insurance Economics.

McShane, Clay, and Louis Cain. 1980. "Sanitation strategy for a Lakefront Metropolis: the case of Chicago." *The Journal of American History* 66:948. doi: 10.2307/1887695.

National Academies of Sciences, Engineering, and Medicine. 2019. *Improving International Resilience and Response to Chemical, Biological, Radiological, and Nuclear Events: Proceedings of a Workshop–in Brief.* Edited by Tom Arrison. Washington, DC: The National Academies Press.

NATO. 1949. The North Atlantic Treaty. Washington, DC.

NOAA. 2020. "Fast Facts: Hurricane Costs." NOAA Office for Coastal Management, accessed August 11. https://coast.noaa.gov/states/fast-facts/hurricane-costs.html.

Obama, Barack. 2013. Presidential Policy Directive – Critical Infrastructure Security and Resilience. edited by Office of the Press Secretary The White House. Washington, DC.

Rufat, Samuel, Eric Tate, Christopher G. Burton, and Abu Sayeed Maroof. 2015. "Social vulnerability to floods: review of case studies and implications for measurement." *International Journal of Disaster Risk Reduction* 14:470–486. doi: https://doi.org/10.1016/j.ijdrr.2015.09.013.

Schmidt, Michael S. and Matt Apuzzo. 2015. *South Carolina Officer Is Charged with Murder of Walter Scott. The New York Times.* Accessed August 11, 2020.

Shea, Jamie. 2016. Resilience: A Core Element of Collective Defence. *NATO Review*. Accessed August 11, 2020.

Siddique, Haroon. 2017. "South Asia Floods Kill 1,200 and Shut 1.8 Million Children Out of School." *The Guardian*, August 30, 2017. www.theguardian.com/world/2017/aug/30/mumbai-paralysed-by-floods-as-india-and-region-hit-by-worst-monsoon-rains-in-years.

Snel, Karin A. W., Patrick A. Witte, Thomas Hartmann, and Stan C. M. Geertman. 2019. "More than a one-size-fits-all approach – tailoring flood risk communication to plural residents' perspectives." *Water International* 44 (5):554–570. doi: 10.1080/02508060.2019.1663825.

BUILDING BRIDGES: PRACTICAL HEALTH-SCIENCE DIPLOMACY TOOLS TO LINK PUBLIC GROUPS

Ruth Larbey
Managing Editor, Science for Environment Policy, University of the West of England

The built environment has a large-scale impact on human health. Despite a growing interest in public involvement in health policy, a lack of evidence means it is difficult to draw firm conclusions about appropriate and effective public involvement activities that can inform healthy decisions in urban development. Ruth Larbey, part of the science communication unit at UWE, reported on some projects in which she has been involved over recent years to address health disparities and promote health-science diplomacy (Science for Environment Policy 2018). She is the managing editor for *Science for Environment Policy*, which is a news and information service run on behalf of the European Commission.

Larbey's recent project is part of the "Our City, Our Health" Program, titled "Shape Our City." Shape Our City is based on a sister project, UPSTREAM, which focused on moving health "upstream" in urban development decision-making, meaning it was all about how to communicate to those in control of urban development in the U.K. and about how they can make human health a priority in their decision-making process (Sardo, Larbey and Laggen 2019). This includes urban planners, developers, construction companies, local authorities, national authorities, and others. The project focused on estimating the scale of the impact that the built environment has on human health in monetary terms in order to have a more tangible conversation about it, and then exploring with the developers about how to minimize the health costs.

BACKGROUND: UPSTREAM

For some background, UPSTREAM (short for "Moving Health Upstream in Urban Development") was a three-year pilot program funded by the Wellcome Trust, a local philanthropic organization in the U.K. (Black et al. 2017, 2019). The program began in 2016 and had four main objectives: (1) demonstrate hidden costs of poor quality urban development, (2) test what impacts these valuations may have on their actions, (3) identify barriers and opportunities, and (4) co-produce the resulting recommendations. The team utilized three different research strategies, including systematic reviews of evidence linking urban environment to health, conducting an economic valuation of the costs and benefits from that data, along with a two-year follow-up program consisting of interviews and workshops with agencies in charge of urban development in the U.K. (Smith et al. 2018, 5). The range of partners included people from the development side, the planning side, and local council and government.

NEXT STEP: SHAPE OUR CITY

While UPSTREAM worked to create feedback loops between researchers, planners, policymakers, and urban developers to work better together, the sister project Shape

Our City focused more on communities and how they should interact with this dialogue. Using the systematic review from UPSTREAM, they came up with five different themes to engage the public of Bristol to create a range of consultation methods: integration, aesthetic, mobility, food, and green space (Bracewell et al. 2019). It was about trying to build trust in expertise, trying to make the valuations more accessible to the public, and to see whether there was some strategy that would work from the top down or from the bottom up, or more ideally, in both directions at the same time (ShapeOurCity n.d.).

They had to adjust their feedback loop, because it seemed that the same voices that provided feedback for UPSTREAM were participating in this sister project, either because they have the comfort to do so or are already engaged in their communities, but that does not accurately reflect the community as a whole; they really wanted to know what would work for the system as a whole, which certainly included more actors. Thus, the researchers had to innovate new approaches to get participation and feedback from a broader range of community members.

Innovative Feedback Methods

They created an online research-based game which gave the participants a choice between six categories: food quality, neighborhood design, building quality, green space, active travel, and road safety. Based on those systematic review categories, the players receive a limited budget, as if the players were in the shoes of policymakers. They performed beta testing on the game with several community groups and got their feedback on user experience and design (Shipman and Laggan 2019).

Another element they created was a research-based board game where participants had to guess their correct health savings from various improvements in the five themes, similar to a puzzle. The board game was used as a conversation starter and demonstration at events throughout Bristol to share information and to have people pledge how they were going to help make a healthy change for their city. It also gave the public the opportunity to ask about the methods and further understand the information.

They also collaborated with a graffiti artist, Luke Jerram, to produce a commissioned sculpture of a diesel soot particle at a scale of three million (BBC News 2018; Jerram 2018). It was placed at the Bristol City Center outside the city hall, right in the path of municipal offices, and served as a conversation invigorator to discuss the health effects of diesel air pollution. Many government officials would then come to talk to us about the sculpture and its significance.

The researchers also gathered peoples' ideas using a Wishing Tree, where community members could add their desires for their neighborhoods on tags and hang them up in the tree. The tags were categorized with a sticker so that the researchers could gauge community-wide preferences.

They also held various focused events, such as food-based events, utilized community organizers, held banner walks where local residents could help map out where health improvements could be made in their local areas. They also taught presentations at different schools around the community, and organized community "Action Days" where UWE planning and architecture students were invited to come in and help carry out conservation plans, but also to implement advice that had been received from engaging with the community.

Results and Findings

Based on the data retrieved from all these various methods of engagement, the researchers were able to extrapolate some of the communities' preferences, even breaking it down by postcode. From the online game, they learned that road safety was least prioritized for most people, and that improving greenspace was the most prioritized. The highest priority from the health savings impact game was the same, though the lowest priority was neighborhood design. As for the art installation piece, most observers found it quite startling, but intriguing and eye-opening as well, once they learned more about the particle and its harmful effects. It was important that someone was there to offer some explanation and education, but many said that they were more instigated to act as a result of the sculpture's shock factor. Simple analysis from the Wishing Tree revealed that "green" was the most-quoted word, followed by "food" and "car."

The Shape Our City efforts have been picked up by national and local media, including the BBC, the BMJ, and others. The researchers found that visuals (both images and visual representations) significantly helped in sharing the content. Community implementation was one the most important results for the researchers, so all the data that was collected was reported back to the community organizations via a conference, as well as individually. The Bristol City Council also got in touch to express interest in conducting some more in-depth food-based consultation events. Two other local councils have also expressed their desire to continue a multi-channel consultation program. The sculpture now hangs in Bristol's local science museum, continuing to spark conversation about air pollution. Since concluding these efforts, there has been increased attention from the mayor of Bristol about further issues regarding the built environment and particularly about air pollution. Another Science Communication Unit project, ClairCity, addresses air pollution specifically and is carried out in several cities across Europe. The public engagement from Shape Our City provided, for the first time really, data on the community's preferences and input, while also increasing awareness and problem-solving the challenges facing urban development.

REFERENCES

BBC News. 2018. "Giant Diesel Soot Particle Unveiled in Bristol." *BBC News*, June 8, 2018. www.bbc.com/news/uk-england-bristol-44411263

Black, D., Ige, J., Pilkington, P., Williams, B., Prestwood, E., Hunt, A., ...Ogilvie, F. 2017. "Moving health upstream in urban development decision-making." Presented at The Sustainable Transport & Health Summit, Bristol, February 2017. Available from https://uwe-repository.worktribe.com/output/898101

Black, D., Scally, G., Orme, J., Hunt, A., Pilkington, P., Lawrence, R., Ebi, K. 2019. "Moving health upstream in urban development: reflections on the operationalization of a transdisciplinary case study." *Global Challenges* 3, 1700103. https://doi.org/10.1002/gch2.201700103

Bracewell, H., Ward, S., Scarpa, L., Gwan, F., Siang, K., and Soobarah, S. 2019. "Shape Our City—Seeding Change: Towards Healthy Urban Development (THUD)." https://urban-health-upstream.info/wp-content/uploads/2019/08/Shape-Our-City-citizens-toolkit.pdf

Jerram, Luke. 2018. *Inhale*. Art installation sculpture commissioned by UWE Bristol's Our City Our Health. www.lukejerram.com/inhale/

Sardo, M., Larbey, R., and Laggan, S. 2019. "Making the Invisible Visible. Using Art to Engage People in Healthy Urban Development." Presented at SciComm South West 2019, Bristol, United Kingdom, June 2019. Available from https://uwe-repository.worktribe.com/output/1492594

Science for Environment Policy. 2018. *Indicators for sustainable cities*. In-depth Report 12. Produced for the European Commission DG Environment by the Science Communication Unit, UWE, Bristol. Available at: http://ec.europa.eu/science-environment-policy.

ShapeOurCity. n.d. Interactive Game Developed by UWE. https://urban-health-upstream.info/shapeourcity/.

Shipman, Eleanor and Laggan, Sophie. 2019. "Shape Our City—Creative Interventions: Creatively Exploring Healthy Urban Planning with Bristol Residents." https://urban-health-upstream.info/wp-content/uploads/2019/08/Shape-Our-City-Newspaper.pdf

Smith, Neil, Barnes, Yolanda, Ferguson, George, Cox, Ian, and Ruairí, Barra Mac. 2018. UPSTREAM Project Report. Bristol: Wellcome Trust. https://urban-health-upstream.info/wp-content/uploads/2018/12/Upstream_Brochure_WEB_single.pdf

UNDERSTANDING PROBLEMS AND FINDING SOLUTIONS TO ACHIEVE "GOOD LIVING" IN UPPER GULF OF CALIFORNIA COMMUNITIES

Osmar Villalobos-Cristerna

Research Associate, CEDO Intercultural | Master's student in arid ecosystem management at the Autonomous University of Baja California

Víctor Ricárdez-García

Research Associate, CEDO Intercultural | Doctoral student in environment and development at the Autonomous University of Baja California

Introduction

The Gulf of California is recognized for its great biological diversity and exceptionally high primary productivity. It is also the most important fishing region in Mexico, contributing more than half of the total annual production (Cisneros-Mata 2010). In the northern portion, the Upper Gulf of California, artisanal fishing origins date back to the early 20th century, when fishermen from southern Sonora, Sinaloa, and Nayarit arrived to meet the demand for the swim bladder of a large Sciaenid fish, the totoaba (*Totoaba macdonaldi*). The totoaba fishery led to the founding and development of the communities of Puerto Peñasco and El Golfo de Santa Clara, Sonora and San Felipe, Baja California, as well as other smaller towns (Cudney-Bueno and Turk-Boyer 1998).

In the historical development of the Upper Gulf of California communities, artisanal fishing has played a key role as a primary source of employment, food, and welfare (Morzaria-Luna, Turk-Boyer and Moreno-Baez 2014). Around this activity, coastal communities have developed specific ways and behaviors in their everyday life, and have acquired deep ecological knowledge, skills, and technologies that are part of their cultural heritage. This interrelationship with the environment through artisanal fishing has also led coastal communities to engage in space appropriation and place attachment processes (Valdez-Gardea 2010). Today, however, a complex environmental, social, and economic crisis is confronting the Upper Gulf of California. This crisis is thought to result from development of an anarchic national fisheries system (Bretón and López 1989), decades of *laissez-faire* fisheries resource management (Cisneros-Montemayor et al. 2013), the prevalence of a conservationist perspective over that of sustainable development (Vázquez-León 2020), the high black market value of totoaba swim bladder (Alvarado and Martínez 2018), and widespread corruption in key government institutions for resource management (Rojas-Bracho et al. 2019).

This article describes some aspects of social organization of artisanal fishing in the Upper Gulf that, if considered in the context of sustainable development strategies, could help to achieve social and ecological sustainability for the region. This diagnosis was done during the Mexican government's third year of implementation of an Integrated Strategy for the Recovery of the Vaquita (ISRV) in San Felipe, Baja California, and El Golfo de Santa Clara, Sonora. The ISRV was designed to respond to the imminent threat to extinction of the endangered vaquita porpoise (*Phocoena sinus*), endemic to this region (DOF 2015). The strategy included an emergency ban to gillnet fishing, financial compensation to fishermen affected by this ban, and increased law enforcement, led by the Mexican Navy and other activities. In practice, as the strategy was implemented, the conflicts and inequalities in the fishing sector, as described previously in other anthropological works (Buitrago-Tello and Valdez-Gardea 2013; Delgado 2013; Valdez-Gardea 2013a, 2013b; Vásquez-León 1999, 2012) became even more evident and were exacerbated. Based on these observations the authors offer some reflections and propose possible pathways for achieving "Good Living"[1] in the Upper Gulf of California.

"Good Living" offers a framework on how to achieve sustainability by reducing environmental degradation and addressing the basic needs of local communities. This framework of political and cultural proposals from the Latin American context includes community participation in design of their own future and pays special attention to their sociocultural dimensions, traditional forms of production, and welfare needs (Altmann 2019; Gudynas 2009; Villalba 2013). For coastal communities, coastal and marine spatial planning of economic activities can offer a framework for strengthening property rights, social organization, and sustainability as has been shown in some Upper Gulf communities (Morzaria-Luna and Turk-Boyer et al. 2020; Turk-Boyer and Valdivia-Jimenez 2019).

1 Good Living comes from the Spanish term "Buen Vivir" or the Kichwa "Sumak Kawsay" proposed by several social movements and scholars as a set of alternatives to conventional development paradigms, seeking social justice and community-based environmental management.

Materials and Methods

This article incorporates the findings and reflections of an anthropological field study carried out during 2018 for the Intercultural Center for the Study of Deserts and Oceans (CEDO Intercultural). Through semi-structured interviews and workshops, artisanal fishermen and women were involved in an evaluation of environmental policies implemented in the context of vaquita conservation efforts. The intention was to encourage stakeholder participation in the design and implementation of future strategies and to provide tools to help them assume greater responsibilities in decision-making (Ricárdez-García 2019). A review of official documents and academic literature was also carried out, and interviews were conducted with focus groups of key social actors involved in the design and implementation of environmental policy and fisheries management.

Authors focus on social organization aspects of artisanal fishing as well as some contradictions or conflicts between groups that became evident and were exacerbated during the implementation of the ISRV. The main purpose is to provide anthropological knowledge to design coherent solutions for a multidimensional crisis taking place in the Upper Gulf of California, an area of prime ecological and economic importance, where the conflicting values of conservation and productive enterprise have collided at the expense of endangered, endemic species and the well-being and cultural integrity of coastal communities.

Governmental Conservation and Sustainable Fishing Efforts in the Upper Gulf of California

Since the 1990s the Upper Gulf marine ecosystem has received a lot of attention from non-government and academic groups focused on conservation. This is due primarily to its diverse, representative ecosystems, great biodiversity and high primary productivity, as well as the existence of endangered, endemic species such as the vaquita porpoise (*Phocoena sinus*) and the large fish, totoaba (*Totoaba macdonaldi*). Given that fishing has been a critical factor causing the population decline of both species, conservation efforts have focused on regulating and reducing the impact of this activity (Bobadilla et al. 2011). In general, these measures have been implemented from top to bottom, directly conflicting with the economic and well-being interests of artisanal fishermen and their families (Turk-Boyer and Valdivia-Jiménez 2019).

For most of the 20th century, the Upper Gulf was primarily a fishing area. Efforts to promote sustainable fishing and conserve marine biodiversity in this region date back to 1940, when a temporary closure was established for the totoaba fishery, which would become permanent in 1975 (DOF 1975). In 1955, the area was recognized as a nursery ground for commercial species (Cudney-Bueno & Turk-Boyer 1998). These and other conservation efforts (Buitrago-Tello and Valdez-Gardea 2013; Valdez-Gardea 2013a, 2013b; Vazquez-Leon 2012) became more relevant with the 1993 creation of the Upper Gulf of California and Colorado River Delta Biosphere Reserve (DOF 1993). The objectives of this Biosphere Reserve, modeled after the United Nations' Man and the Biosphere Program, are to conserve and protect representative ecosystems, biodiversity, and endangered species. It proposes regulations to productive activities to make them compatible with conservation objectives as well as raising the quality of life of communities within a framework of sustainability (SEMARNAT/CONANP 2007). Biosphere Reserves were proposed as areas of the world of special ecological value, where we should focus attention and resources to balance sustainable use and conservation (Bridgewater et al. 1996).

In 1997, the International Committee for the Recovery of the Vaquita (CIRVA) was formed to design a recovery plan for this porpoise using the best available scientific evidence and considering the socioeconomic impacts on coastal communities, with proposals to mitigate or compensate fishers through fishing, economic, or social alternatives (SEMARNAT/CONANP 2008). In 2005, following the advice of CIRVA, a vaquita refuge area was established, expanding the boundaries of the Reserve to include the prime area of distribution of the species (DOF 2005). Three years later, the Vaquita Conservation Action Program (PACE-Vaquita) began operating. Through this program the Mexican government granted a subsidy to fishermen in exchange for their permits and fishing equipment to undertake productive projects in other areas or to implement the use of alternative fishing gear. The three buyout subsidy options included: (a) productive reconversion and diversification; (b) technological reconversion and best fishing practices; and (c) conservation activities (SEMARNAT/CONANP 2008).

Despite these efforts, between 2011 and 2014 (CIRVA 2015), coincident with a boom in totoaba poaching, the vaquita population declined by 67%. In 2013 and 2014, a new integrated strategy (ISRV) was designed by the government in response. The strategy, designed to operate in the Upper Gulf as an emergency measure for a maximum period of two years, began in April 2015 (DOF 2015), and included: (a) an expansion of the vaquita protection area; (b) a complete ban of gillnet fishing; (c) a compensation program for fishermen and others in the industry; and (d) strengthening inspection and surveillance actions, under the charge of the Navy. The Welfare Secretariat (SEDESOL) set the income amounts for the compensation program to mitigate the negative economic impacts of these actions, in accordance with the last five years of fisheries landings in each community (see Table 5.1).

TABLE 5.1
Social Compensation Program types and amounts of support (Data from SEMARNAT/CONANP 2018)

		Support		
	Type	Amount (Mexican Pesos)		
Monthly	Fishers	8,000		
	Productive chain	4,000 to 8,000		
Annually	San Felipe permit holders	**Shrimp permit**	**Fishes permit**	**Shark permit**
		287,441.44	153,213.90	28,160.00
	Santa Clara permit holders	**Shrimp permit**	**Fishes permit**	**Shark permit**
		265,558.20	133,538.69	4,266.00

According to CIRVA (2016a), these temporary measures would only be successful if adequate conditions for artisanal fishermen were generated for developing alternative livelihoods, sustainable technology, and management practices. These conditions were not fully met and the boom in illegal totoaba fishing continued its course, seriously threatening the survival of the vaquita and the quality of life of coastal communities (Ricárdez-García 2019). Given the existence of less than 30 individuals of vaquita (Thomas et al. 2017), CIRVA (2017) recommended the permanent suspension of gillnets and in 2018 the government announced that the strategy would continue indefinitely (DOF 2018).

Social Organization of Artisanal Fishing

In the following sections, some aspects of the social organization of artisanal fishing and contradictions and conflicts identified during the implementation of the ISRV in San Felipe and El Golfo de Santa Clara are presented: (a) production organization; (b) division of labor; (c) inequalities; (d) mutual aid; and (e) labor immobility. A general description of the totoaba poaching problem is also presented. In addition, the authors propose some pathways for moving towards ecological and social sustainability, focused on greater self-sufficiency of local communities.

Production Organization

In Mexico, the waters of territorial seas, inland marine waters, and their fishing resources are national property. As part of its sovereignty, the Mexican federal government, through the competent authority in fisheries management,[2] is empowered to transfer rights for the use of these resources. Through fishing permits, CONAPESCA establishes who has access to a fishery, the level of fishing effort allowed, where and when one can fish, what fishing gear can be used, and the volume of catch allowed. What usually happens in practice, however, is that users define rules and rights at the local level to regulate access to resources (Marín 2007a).

In general, the privileged economic agent designated by the Mexican government to exploit fisheries resources is the fishing cooperative. Cooperatives were established in the 1920s and 1930s as part of the post-revolutionary Mexican government policy of nationalization of all productive sectors (Breton and López 1989). Cooperatives are given access to credits to acquire means of production such as boats, engines, and fishing gear, which has supported this activity of great importance for local populations (Ramírez and Verduzco 2016).

Ideally, cooperatives are self-sufficient organizations whose social function is the economic and social improvement of their members, characterized by collective ownership of production means, free association, collaborative relationships, and democratic self-management. Members' rights in decision-making and the distribution of economic benefits are equitable, since they are based on labor contributions (Alegret 1999).

2 Competent authorities in fisheries management are Secretariat of Agriculture and Rural Development (SADER) and the National Aquaculture and Fisheries Commission (CONAPESCA).

In the Upper Gulf, as in many coastal communities, fishing cooperatives have been the main form of work organization, commonly formed between friends and family networks. This production model is based on the social economy (López-Torres et al. 2018). In the Upper Gulf more than 4,000 associates and permit holders, as well as an undetermined number of "free" or unorganized fishermen are grouped into fishing cooperatives (Rodríguez-Quiroz and Bracamonte-Sierra 2008). Although, ideally, cooperatives are egalitarian productive organizations, as described in the following section, conflicts often arise from the discrepancies and inequalities between those who hold fishing permits, *pangas*,[3] and fishing gear and those who live from their own labor (Delgado 2013).

Division of Labor

Decades of neoliberal reforms in Mexico have shaped the artisanal fishing sector of the Upper Gulf of California through the economic incentives of global markets and the international seafood trade (Frawley, Finkbeiner, and Crowder 2019). After the structural transformation of Mexican's rural sector that began in the 1990s, marine fisheries moved from business models where cooperatives dominated to private investment and commoditization schemes (Vásquez-León 2012; Buitrago-Tello and Valdez-Gardea 2013). This resulted in a concentration of marine resource access rights by local elites or business managers, while access for ordinary fishermen became increasingly difficult, and many became illegal social actors without incentives to follow the best practices for sustainability (Vásquez-León 1999). This is how deep inequalities have been structured around access to resources in the Upper Gulf artisanal fishing sector, undermining the capacity for local self-sufficiency and self-regulation.

Although fishing cooperatives are still the most widespread form of organization, as Buitrago-Tello and Valdez-Gardea (2013) argue, they are usually characterized for their correct legal constitution, without recognition of their fictitious social function. The dominant cooperative management scheme is like that of a fishing company managed by an individual who concentrates power. Permit holders usually own their means of production, and play the role of boss of their *pangas*, supervising others or operating the vehicles used to move the boats. Each *panga* operates with several fishing permits for different types of resources exploited (finfish, shark, or specific species of crustaceans or mollusks). Free fishermen, on the other hand, do not have permits or any fishing gear, and are hired at any time through informal agreements. Since they have no means of production and cannot produce for themselves, their subsistence and that of their families depends on the payment for a fishing day (Buitrago-Tello and Valdez-Gardea 2013).

The *panga* fishing team usually consists of a captain and one or two fishermen, depending on the fishery. The captain is usually the most experienced and knows how to use each type of fishing gear and understands the seasonality of water and wind, so his role is to make decisions about fishing sites and maneuvers. During the capture, fishermen oversee placing and extracting the nets, untangling the fish and

3 A small flat-bottomed fishing boat with an outboard motor, used in the Upper Gulf of California.

placing them on the bottom of the *panga*. At the end of the day, the sale of the catch is equally distributed between the members of the fishing team and the permit holder, once the operating costs have been covered (Buitrago-Tello and Valdez-Gardea 2013; Delgado 2013).

Since the benefits are linked directly and proportionally to the volume of capture, and are not temporally specific, this system has the potential to induce maximum extraction to obtain the greatest economic benefit, leading to overexploitation (Delgado 2013). This is especially likely considering that the system operates within the framework of dependency relations with external markets and intermediaries. On the other hand, it conceals the discrepancies between capital and labor by inducing a perception of parity in profit sharing. In the absence of formal labor contracts, permit holders can impose their conditions on unorganized fishermen (Buitrago-Tello and Valdez-Gardea 2013).

Inequalities

These discrepancies became more relevant and destructive in the framework of the ISRV. According to free fishermen, before the gillnet ban their income ranged between 15,000 and 30,000 Mexican pesos per month during a fishing season. Under the ISRV compensation program, however, while some received a payment of 8,000 pesos a month, many free fishermen were excluded as beneficiaries. Delays of up to four months in the payment of the compensation also resulted in systematic indebtedness and widespread discontent with authorities and environmental NGOs, which sometimes ended in clashes.

When fishing was banned, the relationship of mutual dependence between permit owners and unorganized fishermen was ruptured. One expression of this was the choice of many permit holders and cooperatives to exclude fishermen who had actively engaged in harvest with them for many years, and instead place deceased people, people who were never engaged in fishing or others from outside the communities on the list of recipients for the compensation. In the absence of formal labor contracts, free fishermen remained invisible and permit holders assumed no legal responsibility for them. It was common for permit holders to retain the cards, whereby the compensation was paid and to charge unorganized fishermen fees for distributing their payment. A total 7.2 million pesos of such irregularities were found in the delivery of compensations by the Superior Audit of the Federation (Méndez 2019a).

This deepened the conditions of poverty and marginalization for unorganized fishermen since they could not perform the activity on which their livelihood depended nor were they being compensated. In contrast, because of the concentration of fishing permits by local businessmen through the process of privatization and commodification (Vázquez-León 2012; Buitrago-Tello and Valdez-Gardea 2013), some permit holders received up to 12 million pesos per year (SEMARNAT/CONANP 2017), consolidating a wealthy, entitled class. This created a deepening inequality not only in terms of distribution of economic power, but also political power. Some permit holders strengthened their political capital by accessing the decision-making spaces and controlling who would be the beneficiaries of the compensation program.

Mutual Aid

Artisanal fishermen in Mexico have developed mutual aid systems to overcome the high levels of marginalization and deficiencies in basic goods and services present in most coastal communities (Marín 2007b; EDF Mexico 2019). Local social security networks have evolved that allow communities to overcome a life without stable employment, savings, or the attention of the welfare state. The fundamentals of these systems are solidarity, reciprocity, and redistribution that give members of the communities' access to goods and services that are necessary for survival (Alcalá 1992).

As described by Graciela Alcalá (1992) for other coastal communities in Mexico, in the Upper Gulf of California these social relations are primarily developed and replicated within the operational space of the *panga*. Since fishing gear cannot be operated individually, a good catch depends on the development of agreements and cooperative relationships between the members of each fishing team. Likewise, risks inherent to the activity force them to provide mutual support to be safe. During a fishing day, the integrity and life of a fishermen are always reciprocally in the hands of the other. Trust and mutual help are fundamental principles for survival and are strengthened in everyday practice.

Fishermen usually get two kinds of catch. One is the target species, backed by a permit and traded according to market values; its sale covers operating costs and is the basis for distributing the profit. The other, incidental catch, does not usually have a value in the market and is, in part, the result of poor selectivity of fishing gear. Fishermen generally distribute this catch for self-consumption or as an exchange currency to acquire other goods and services that they would have difficulty accessing otherwise. Part of this capture also serves to pay those who collaborate on the docks, helping to land or to clean the capture (Alcalá 1992).

Despite its great relevance for coastal communities, these relationships of mutual aid have remained invisible. Furthermore, fishing is an integral and important part of the economy of the communities in which fishermen live. The fishing ban implemented with the ISRV, articulated with previously described inequalities and the lack of alternative fishing gear, caused the sensation in many people in coastal communities to be alienated from their main supply of livelihoods (Ricárdez-García 2019). It also affected the mechanisms of redistribution and reciprocal assistance that communities have built around fishing, and that have fed a population sector that otherwise could barely survive. Fishermen suggest, for example, that those who used to benefit from these exchange networks in the docks have today become thieves or members of organized crime.

Labor Immobility

Two pervasive characteristics of artisanal fishing are the high dependence on subsistence fishing and the few opportunities to be employed in other sectors (Marín 2007b). Precisely, during the ISRV, fishermen hardly managed to find employment in other areas. Those who did get jobs did so in temporary and poorly paid activities. The limitations for employment in coastal communities are a critical reason for this dependence on fishing. Other factors that inhibit the incorporation of artisanal fishermen into formal labor markets are their age, low level of schooling, and inexperience in other productive sectors. The fishing sector is characterized by a widespread resistance to abandoning a life in the sea and by the pleasure for fishing (Ricárdez-García 2019).

Except in some cases, fishermen do not have experience in wage labor schemes. The system of remuneration by parts, linked to the randomness of catches and outdoor work (Delgado 2013), induces in fishermen a different space-time experience from that of a person whose livelihood depends on a well-defined salary, workplace, and working hours. In addition, other productive sectors do not generally require the same level of knowledge and skills that fishermen have (Marín 2007b). Some fishermen, however, have managed to be employed to extract ghost nets, support acoustic monitoring of vaquita, or test new fishing gear. Others, in contrast, have admitted to fishing illegally for totoaba, at least once. In both situations, fishermen put into practice the cultural capital developed in their fishing years, maintaining a relationship with the marine environment and its species, although focused on divergent activities (Ricárdez-García 2019).

Women play a fundamental role in artisanal fishing, although little recognized. They have the responsibility for cooking and serving food, cleaning, and ensuring the welfare of their family members. In addition to these reproductive activities,[4] they carry out other economic activities in the public space: when the *pangas* arrive at the dock, they receive, classify, and clean the capture to be subsequently processed and sold. It is common to keep non-commercial species for home consumption. Also, though not frequently, women in the Upper Gulf participate in harvesting.

During the ISRV suspension of fishing, activities carried out by women were limited, in general, to reproductive labor. In the cases where women got a job, these were activities related to cooking, cleaning, and caring for children or the elderly. The jobs they accessed were of low pay and their productive ventures generally had little added value for the family situation. Some women, however, did organize to create civil organizations such as *Uniendo Fuerzas por Nuestra Reserva*, in order to address public health, education, and social security issues at the local level.

Given the exclusion of a large segment of fishermen from the social compensation program, coupled with delays in the payment of the compensation and the lack of employment, groups of fishermen chose to move into other communities' fishing areas, causing new conflicts over resources (Morzaria-Luna and Turk-Boyer et al. 2020). Others moved with their families to cities along the northern border of Mexico or the United States. In many cases, totoaba fishing appeared as an attractive alternative.

Organized Crime and Dispossession?

Although commercial fishing for totoaba has been banned since 1975, it has experienced a renewed boom due to the high value of its swim bladder in the illegal market. It is assumed that, locally, each swim bladder brings around 1,500 dollars.

4 The term reproductive labor has been proposed by feminist theory to make visible those unpaid activities performed primarily by women in the private space: cleaning, cooking, family caring, among other activities.

This value multiplies as it advances in a global chain that includes markets in Mexico, the United States, China, and Hong Kong, with a price of up to 60,000 dollars paid to the final consumer (EIA 2016). It is also presumed that there is a close interaction between local fishermen, Mexican organized crime linked to drug trafficking, Chinese mafias and authorities that are colluding in extraction, transportation and marketing networks, and sharing in this lucrative activity.

It is important to consider that, in the Mexican context, the growth of criminal organizations linked to drug production and trafficking has its origins in neoliberal deregulations that led the rural sector to socioeconomic and political deterioration, undermining the livelihoods of small shareholders, and allowing a revaluation of rural areas (Valencia 2018). In recent decades, however, criminal organizations in Mexico have undergone a process of industrialization and economic diversification (Galindo 2005). In the spaces in which they operate, they usually trigger processes of accumulation by dispossession (Harvey 2003): processes of alienation and closing access to natural resources to rural communities (Zibechi 2014). It is increasingly common for these criminal organizations to develop governance systems to command and regulate access to resources within well-defined territories (Mbembe 2003; Valencia 2018). For this, they develop negotiation-pressure-cooptation mechanisms or produce scenarios of violence (Merchand 2018). In fact, spaces in Mexico where murders and disappearances systematically occur are often the object of interest for energy and extractive projects in which transnational consortiums, local authorities, and criminal groups work in collusion (Paley 2014).

In this context, it can be understood that a process of accumulation by dispossession at the hands of organized crime is taking place in the Upper Gulf of California. While access to the sea and its resources was restricted for local fishermen for conservation, members of criminal organizations from other regions took their place. They began to establish themselves in the region at least since 2012 and have developed a governance system based on violence to meet the demand for totoaba swim bladder by Chinese elites. Currently, this illegal wildlife trade network is considered the most profitable in the world (Alvarado and Martínez 2018) and constitutes the main threat to the survival of the vaquita because of its by-catch (CIRVA 2016b).

According to local fishermen, it is common for members of this illegal wildlife trade network to enter the sea armed and to intimidate those who extract ghost nets or test alternative fishing gear. They develop clandestine information networks to monitor authorities, local social networks, and NGO members; they use buoys, seagulls, and satellite geolocators to mark the points where gillnets are installed. It is presumed that market intermediaries with connections to the black market use the routes for drug and arms trafficking between the United States and Mexico, and those involved in the illegal fishing of totoaba establish complex relationships with authorities, ranging from direct confrontation to alliances and complicities.

Given the huge profits from totoaba poaching, unorganized fishermen excluded from the compensation program and inexperienced teenagers who drop out of school started illegally fishing this resource. Violence has become a tool that ensures access and control over fishery resources by these groups. Illegal fishing is, for many, the most accessible option not only to overcome marginalization but also to comply with certain norms of masculinity in the Mexican context such as the role of suppliers, economic respectability, and indifference to danger (Valencia 2014). Violence also functions as a mechanism to regulate behaviors in coastal communities and even to control market dynamics.

CONAPESCA has recognized the problem of illegal fishing of totoaba and other species by organized crime, as well as their collection of quotas from local fishermen to permit them to carry out their productive activity. Local fishermen must pay the mafia around 1,000 pesos a week to fish in the Upper Gulf, as well as a percentage of the catch (Méndez 2019b). This expansion of organized crime into artisanal fisheries is now reconfiguring coastal communities socially, culturally, economically, and politically, with serious consequences for ecological and social sustainability.

As part of the ISRV, the Mexican federal government implemented a militarized security strategy to combat illegal totoaba fishing (DOF 2017). With the Navy in charge, this strategy included the Federal Police, the Environmental Gendarmerie, and the Federal Attorney for Environmental Protection (PROFEPA), as well as some environmental NGOs that patrol the area to extract illegal gillnets. The presence of military and police elements was increased, and technology developed by the U.S. military industry was introduced (Maíz 2016). From the local perspective, however, the security strategy that follows the logic of the war on drugs has only increased insecurity, violence, and corruption networks.

It is important to consider research into "green violence," which consists of the overt use of police surveillance and militarization in Natural Protected Areas (Büscher and Ramutsindela 2016; Kelly and Ybarra 2016). Although there are cases in which increased policing has reduced levels of poaching (Hilborn et al. 2006), there are many other cases where it tends to escalate conflict and violence (Duffy et al. 2015). In the Mexican context, since 2006, militarization has contributed to strengthening a social order based on corruption, injustice, and impunity. These approaches are generally abstracted from social, cultural, political, and economic causes that have fueled illegality, as well as the porosity between the state apparatus and organized crime (Piñeyro 2010; Suárez de Garay 2013). Alternatively, it is necessary to counteract the undoubted economic attractiveness of criminality by strengthening the productive and self-sufficient capacity of coastal communities that engage in legal and social and environmentally sustainable activities, so that they may be able to adapt to difficult, complex, and changing situations of the globalized world.

Alternative Development for "Good Living"

Recently, several Latin American scholars and social movements have proposed a series of alternatives to conventional development paradigms (Altmann 2019; Gudynas 2009; Villalba 2013). These proposals from the local seek to reverse poverty and environmental degradation in rural areas. Likewise, using the framework of coastal and marine spatial planning for economic activities, a concerted effort was made to organize six coastal fishing communities on the Sonora side of the Upper Gulf with emphasis on strengthening property rights, social organization, and sustainability in the context of space (Morzaria-Luna et al. 2020; Turk-Boyer and Valdivia-Jimenez 2019). Grouped under the concept of Good Living, and from a biocentric and relational perspective, these alternatives are directed towards social justice and community-based environmental management. They emphasize the need to build, based on continuous processes of community reflection, "new" forms of social organization to meet needs, being consistent with local culture and traditional forms of production and management of common-pool resources.

Based on dialogue with community members and the scenario observed during the project's fieldwork, to achieve "Good Living" in the Upper Gulf an effective strategy of productive diversification that allows greater local self-sufficiency is needed. Instead of a reconversion strategy, a strategy to accompany fisher people in their efforts to strengthen their existing organizations and undertake sustainable productive projects in line with their experiences should be supported. This accompaniment should provide the necessary skills and knowledge to help people consolidate community-based entrepreneurships that allow greater income and must help meet local needs (Barkin 2000; Barkin and Lemus 2012). Fishermen, coastal communities, and their organizations need to be recognized as the main allies for protecting biological and cultural diversity.

Achieving a more equitable distribution of political and economic power within productive organizations is also fundamental. Structurally excluded groups need to be included and engaged in decision-making spaces. This implies facilitating the organization of free, invisible fishermen, the recognition of reproductive labor mostly carried out by women, and the creation of social security systems for the elderly. Development of coastal fishermen must include strengthening their capacities for democratic self-management and inter-cooperation between cooperatives, as they work to develop an integrated set of sustainable productive projects. Mutual aid systems should be recognized as the basis for the development of ecological and solidarity economic schemes for cooperatives (Quiroz-Niño and Murga-Menoyo 2017).

These schemes allow cooperative members to secure their livelihoods, strengthen the social fabric in their community, and protect ecosystems at the same time. It is also important for facilitating the formation of inter-community exchange networks that help pool resources to meet local needs and enforce rights that have not been met satisfactorily by the Mexican state. From this point of view, market relations are understood as links between people, families, groups, and communities that seek to ensure a *good living*, rather than the accumulation of capital (Giarraca 2008).

It is vital to develop or strengthen capacities within cooperatives, so that fishermen take responsibility for limiting negative environmental impacts as part of their production and exchange processes (Quiroz-Niño and Murga-Menoyo 2017). For example, rates of exploitation should not exceed the capacity of fishing stocks to regenerate, closed seasons should be respected, and agreements should be developed to define the level of fishing exploitation and location of recruitment areas. This implies initiating a dialogue, and incorporating local knowledge for the design of selective fishing gear, as some NGOs are already doing. The implementation of new fishing gear and the shortening of value chains can help reduce the carbon footprint generated by fishing and require less energy. Together with democratic self-management, if implemented, these steps can give co-ops the possibility to access fair trade certifications and sustainable markets.

The defense of facilitating sustainable livelihoods in a context marked by the processes of dispossession at the hands of organized crime requires not only strengthening the productive capacity of communities but also changing other types of behaviors to remake the social fabric. Fishermen and communities must resist those masculine behaviors which, obeying neoliberal logics and demands, make violence and capital acquisition a means of empowerment (Valencia 2014). This implies de-learning processes in which men, mainly young men, generate new conditions to build and execute their masculinities. This process of deconstruction of violent masculinity also implies overcoming the historical sexual division of labor, promoting co-responsibility in family care, and the satisfaction of needs as the central axis of the economy (Vega 2016).

Conclusions

In this article, we have highlighted some key social organization aspects of Upper Gulf of California fishing communities. Our purpose is to provide anthropological and participatory knowledge to find solutions to a multidimensional crisis that has been brewing for decades. From the perspective of this article, artisanal fishermen and coastal communities should be considered as the main allies to achieve ecological and social sustainability. To accomplish this, authorities and NGOs should pay more attention to the sociocultural dimension and welfare needs of local communities, whose livelihoods depend on the use of natural resources. More precisely, objectives and work agendas should be defined together with the communities.

People who seek to strengthen their productive organizations and diversify their economic activities need to be accompanied in their process to acquire the necessary knowledge and skills. Community businesses that focus on meeting employment, food, health, and social security needs should be developed. The formation of inter-community fair trade networks should be facilitated. Agreements should be developed between communities to develop consensus on fishing exploitation, areas for recruitment of fisheries species, and technological gear alternatives. To reconstruct the social fabric in a context with strong presence of organized crime, the productive capacity of communities must be improved, but also new non-violent conditions must be created, and the needs of the local community must become the central axis of the local economy.

REFERENCES

Alcalá, G. 1992. La ayuda mutua en las comunidades de pescadores artesanales de México. *Anales de antropología,* 29 (1): 179–203.

Alegret, J. 1999. "La dimensión organizativa del sector pesquero de Cataluña: Las Cofradías de pescadores." In Montes del Castillo, A. (ed.). *Antropología de la pesca. Debates del Mediterráneo,* pp. 157–178. Murcia: University of Murcia.

Altmann, P. 2019. Los pueblos indígenas en el buen vivir global, un concepto como herramienta de inclusión de los excluidos. *Iberoamerican Journal of Development Studies,* 8 (1): 58–80.

Alvarado, I. & Martínez, E. 2018. "Trafficking of Totoaba Maw". In Arroyo-Quiroz, I. & Wyatt, T. (eds.). *Green Crime in Mexico. A Collection of Case Studies,* pp. 149–170. Cham: Palgrave Macmillan.

Barkin, D. 2000. "Superando el paradigma neoliberal: desarrollo popular sustentable". In Giarraca, N. (comp.). *¿Una nueva ruralidad en América Latina?* pp. 81–102. Buenos Aires: CLACSO.

Barkin, D. & Lemus, B. 2012. La economía ecológica y solidaria: una propuesta para nuestra crisis. *Sustentabilidades,* 5, 1–15.

Bobadilla, M., Álvarez-Borrego, S., Ávila-Foucat, S., Lara-Valencia, F. & Espejel, I. 2011. Evolution of environmental policy instruments implemented for the protection of totoaba and the vaquita porpoise in the Upper Gulf of California. *Environmental Science and Policy,* 14, 997–1008.

Bretón, Y. & López, E. 1989. *Ciencias sociales y desarrollo de las pesquerías: modelos y métodos aplicados al caso de México.* Mexico City: Instituto Nacional de Antropología e Historia-Secretaría de Pesca.

Bridgewater, P., Phillips, A., Green, M. & Amos, B. 1996. *Biosphere Reserves and the IUCN System of Protected Area Management Categories.* Canberra: IUCN.

Buitrago-Tello, D., & Valdez-Gardea, C. 2013. "La conservación de la biodiversidad en el Alto Golfo de California y la organización del trabajo en la pesca: Omisiones y sustracciones". In Valdéz-Gardea, C. & Galindo-Bect, M. (coord.). *Pesquerías globalizadas,* pp. 157–179. Hermosillo: El Colegio de Sonora/Universidad Autónoma de Baja California.

Büscher, B. & Ramutsindela, M. 2016. Green violence: rhino poaching and the war to save Southern Africa's Peace Parks. *African Affairs,* 115 (458): 1–22.

Cisneros-Mata, M. 2010. "The importance of fisheries in the Gulf of California and ecosystem-based sustainable co-management for conservation." In Brusca, R. (ed.). *The Gulf of California. Biodiversity and Conservation,* pp. 120–134. Tucson: University of Arizona Press.

Cisneros-Montemayor, A. M., Cisneros-Mata, M. A., Harper, S. and Pauly, D. 2013. Extent and implications of IUU catch in Mexico's marine fisheries. *Marine Policy,* 39:283-288.

Cudney-Bueno, R., & Turk-Boyer, P. 1998. *Pescando entre mareas del Alto Golfo. Una Guía Sobre la Pesca Artesanal, Su Gente y Propuestas de Manejo.* Puerto Peñasco: Intercultural Center for the Study of Deserts and Oceans.

Delgado, C. 2013. "La pesca ribereña en el contexto capitalista. Apuntes sobre una propuesta de análisis para comunidades pesqueras del litoral sonorense". In Valdéz-Gardea, C. & Galindo-Bect, M. (coord.). *Pesquerías globalizadas,* pp. 203–223. Hermosillo: El Colegio de Sonora.

Duffy, R., St. John, F. , Büscher, B., & Brockington, D. 2015. The militarization of anti-poaching: undermining long term goals? *Environmental Conservation,* 42 (4): 345–348.

EDF Mexico. 2019. Impacto social de la pesca ribereña en México: propuestas para impulsar el bienestar social en el sector pesquero. Mexico city: EDF Mexico.

EIA. 2016. *Dual extinction: The Illegal Trade in the Endangered Totoaba and Its Impact on the Critically Endangered Vaquita.* London: Environmental Investigation Agency.

Frawley, T., Finkbeiner, E., & Crowder, L. 2019. Environmental and institutional degradation in the globalized economy: lessons from small-scale fisheries in the Gulf of California. *Ecology and Society,* 24(1): 7.

Galindo, M. 2005. El capitalismo criminal, fase superior del imperialismo. *Mundo Siglo XXI,* 2 (6): 45–49.

Giarraca, N. 2008. "Producción y mercados para la vida: una posibilidad emancipadora para el siglo XXI". In Giarraca, N. & Massuh, G. (comp). *El trabajo por venir. Autogestión y emancipación social,* pp. 36–41. Buenos Aires: Editorial Antropofagia.

Gudynas, E. 2009. La dimensión ecológica del Buen Vivir: entre el fantasma de la modernidad y el desafío biocéntrico. *Revista Obets,* 4, 49–53.

Harvey, D. 2003. *The New Imperialism.* Oxford: Oxford University Press.

Hilborn, R., Arcese, P., Borner, M., Hando, J., Hopcraft, G., Loibooki, M., ... Sinclair, A. 2006. Effective enforcement in a conservation area. *Science*, 314, 1266.

Kelly, A. & Ybarra, M. 2016. Introduction to themed issue: "Green security in protected areas". *Geoforum,* 69, 171–175.

López-Torres, V., Moreno-Moreno, L., Pérez-Rivas, A., & González-Rosales, V. 2018. Impacto de las estrategias de conservación en una comunidad local de México. *Revista Venezolana de Gerencia*, 23 (83): 719–139.

Maíz, J. 2016. La Armada de México incorpora el avión sin piloto Arcturus T-20. *Defensa*, August 12, 2016. Retrieved from www.defensa.com/mexico/armada-mexico-incorpora-avion-sin-piloto-arcturus-t-20

Marín, G. 2007a. Pesca artesanal, comunidad y administración de recursos pesqueros. Experiencias en la costa de Michoacán, México. *Gazeta de Antropología*, *23*, 1–16.

Marín, G. 2007b. *Vidas a contramarea: Pesca artesanal, desarrollo y cultura en la costa de Michoacán.* Ciudad de México/Zamora: CIESAS/COLMICH.

Mbembe, A. 2003. Necropolitics. *Public Culture*, 15 (1): 11–40.

Méndez, E. 2019a. Cobran derecho de mar a pescadores; Conapesca reconoce problema. *Excelsior*, November 14, 2019. Obtenido de www.excelsior.com.mx/nacional/cobran-derecho-de-mar-a-pescadores-conapesca-reconoce-problema/1347663.

Méndez, E. 2019b. Desvían 7.2 millones de pesos destinados a protección de vaquita marina. *Excelsior*, February 22, 2019. Retrieved from www.excelsior.com.mx/nacional/desvian-72-millones-de-pesos-destinados-a-proteccion-de-vaquita-marina/1298025.

Merchand, R. 2018. Extractivismo, acumulación por despojo y organizaciones criminales en México. *Revista Kavilando*, 10 (1): 11–30.

Morzaria-Luna, H., Turk-Boyer, P., & Moreno-Baez, M. 2014. Social indicators of vulnerability for fishing communities in the Northern Gulf of California, Mexico: implications for climate change. *Marine Policy*, 45, 182–193.

Morzaria-Luna, H., Turk-Boyer, P., Dorantes-Hernández, J., Polanco-Mizquez, E., Downton-Hoffmannan, C., Cruz-Piñón, G., Carrillo-Lammens, T., Loaiza-Villanueva, R., Valdivia-Jiménez, P., Sánchez-Cruz, A., Peña-Mendoza, V., López-Ortíz, A., Koch, V., Vázquez-Vera, L., Arreola-Lizárraga, J., Amador-Castro, I. Suárez, A. & Munguia-Vega, A. 2020. Fisheries management tools to support coastal and marine spatial planning: a case study from the Northern Gulf of California, Mexico. *MethodsX* 7, 1–19. DOI: 10.1016/j.mex.2020.101108

Official Journal of the Federation [DOF]. 1975. ACUERDO que establece veda para la especie Totoaba, Cynoscion MacDonaldi, en aguas del Golfo de California, desde la desembocadura del Río Colorado hasta el Río Fuerte, Sinaloa en la costa oriental, y del Río Colorado a Bahía Concepción, Baja California, en la costa occidental. August 1, 1975. Retrieved from www.dof.gob.mx/nota_detalle.php?codigo=4786520&fecha=01/08/1975#:~:text=ARTICULO%202o.,California%2C%20en%20la%20costa%20occidental.

Official Journal of the Federation [DOF]. 1993. DECRETO por el que se declara área natural protegida con el carácter de Reserva de la Biosfera, la región conocida como Alto Golfo de California y Delta del Río Colorado, ubicada en aguas del Golfo de California y los municipios de Mexicali, B.C., de Puerto Peñasco y San Luis Río Colorado, Son. June 10, 1993. Retrieved from http://dof.gob.mx/nota_detalle.php?codigo=4744610&fecha=10/06/1993#:~:text=DOF%3A%2010%2F06%2F1993,San%20Luis%20R%C3%ADo%20Colorado%2C%20Son.

Official Journal of the Federation [DOF]. 2005. Acuerdo mediante el cual se establece el área de refugio para la protección de la vaquita (Phocoena sinus). September 8, 2005. http://extwprlegs1.fao.org/docs/pdf/mex176172.pdf.

Official Journal of the Federation [DOF]. 2015. ACUERDO por el que se suspende temporalmente la pesca comercial mediante el uso de redes de enmalle, cimbra y/o palangres operadas con embarcaciones menores, en el Norte del Golfo de California. April 10, 2015. Retrieved from www.dof.gob.mx/nota_detalle.php?codigo=5388486&fecha=10/04/2015.

Official Journal of the Federation [DOF]. 2017. ACUERDO por el que se prohíben artes, sistemas, métodos, técnicas y horarios para la realización de actividades de pesca con embarcaciones menores en aguas marinas de jurisdicción federal de los Estados Unidos Mexicanos en el Norte del Golfo de California. June 30, 2017. Retrieved from www.dof.gob.mx/nota_detalle.php?codigo=5488674&fecha=30/06/2017.

Official Journal of the Federation [DOF]. 2018. ACUERDO por el que se suspende temporalmente la pesca comercial mediante el uso de cimbras operadas con embarcaciones menores, en el Norte del Golfo de California. March 14, 2018. Retrieved from www.dof.gob.mx/nota_to_imagen_fs.php?codnota=5516131&fecha=14/03/2018&cod_diario=280744.

Paley, D. 2014. *Drug War Capitalism*. Oakland: AK Press.

Piñeyro, J. 2010. "Las fuerzas armadas mexicanas en la seguridad pública y la seguridad nacional". In Alvarado, A. y Serrano, M. (coords.). *Los grandes problemas de México XV. Seguridad nacional y seguridad interior,* pp. 155–190. Mexico City: El Colegio de México.

Quiroz-Niño, C. & Murga-Menoyo, M. 2017. Social and solidarity economy, sustainable development goals, and community development: the mission of adult education and training. *Sustainability*, 9 (2164): 1–16.

Ramírez, A. & Verduzco, F. 2016. El papel del Estado en la configuración de la realidad actual de las sociedades pesqueras en México. *Antropologías del Sur,* 5, 125–139.

Ricárdez-García, V. 2019. *Una estrategia transdisciplinaria para el desarrollo sustentable de comunidades pesqueras del Alto Golfo de California (Master's thesis).* Ensenada: Autonomous University of Baja California.

Rodríguez-Quiroz, G. & Bracamonte-Sierra, Á. 2008. Pertinencia de las ANP como política de conservación y mejoramiento de la calidad de vida. Análisis de percepción en la Reserva de la Biosfera del Alto Golfo de California del Delta del Río Colorado. *Estudios sociales*, 16 (32): 141–176.

Rojas-Bracho, L., Brusca, R., Álvarez-Borrego, S., Brownell, R., Chamacho-Ibar, V., Ceballos, G., De la Cueva, H., García-Hernández, J., Hastings, P., Cárdenas-Hinojosa, G., Jaramillo-Legorreta, A., Medellín, R., Mesnick, S., Nieto-García, E., Urbán, J., Velarde, E., Vidal, O., Findley, L. & Taylor, B. 2019. Unsubstantiated claims can lead to tragic conservation outcomes. *BioScience*, 69 (1): 12–14.

SEMARNAT/CONANP. 2008. *Action Program for the Conservation of the Species: Vaquita (Phocoena sinus).* Retrieved from www.gob.mx/cms/uploads/attachment/file/350199/PACE_Vaquita_en_Ingles.pdf.

SEMARNAT/CONANP. (2007). *Programa de Conservación y Manejo Reserva de la Biosfera Alto Golfo de California y Delta del Río Colorado.* Retrieved from https://simec.conanp.gob.mx/pdf_libro_pm/2_libro_pm.pdf.

SEMARNAT/CONANP. 2017. Programa de Recuperación y Repoblación de Especies en Riesgo (PROCER). Componente de compensación social para contribuir a la conservación de la vaquita marina: padrón de beneficiarios. Retrieved from www.conanp.gob.mx/pdf_procer/PadronBeneficiariosPROCER2017.pdf.

SEMARNAT/CONANP. 2018. Lineamientos para la ejecución del Programa de Recuperación y Repoblación de Especies en Riesgo. Componente de compensación social para contribuir a la conservación de la vaquita marina. Retrieved from www.conanp.gob.mx/procer/documentos/procer_2018/LineamientosVaquitaMarina2018.pdf

Suárez de Garay, M. 2013. "Violencias policiales en México: de las rutinas, las prácticas, los hechos extraordinarios y el gobierno directo de los ciudadanos." In Arteaga-Botello, N. (ed.). *Violencia en México,* pp. 131–160. Madrid: Catarata.

Thomas, L., Jaramillo-Legorreta, A., Cárdenas-Hinojosa, G., Nieto-García, E., Rojas-Bracho, L., Ver Hoef, J.M., Moore, J., Taylor, B., Barlow, J., & Tregenza, N. 2017. Last call: passive acoustic monitoring shows continued rapid decline of critically endangered vaquita. *The Journal of the Acoustical Society of America Express Letters*, 142 (5): EL512–EL517.

Turk-Boyer, P. & Valdivia-Jiménez, P. 2019. *Un Nuevo Rumbo hacia el Manejo Ecosistémico y Pesquero en México: Planeación Espacial Marino-Costera y Manejo Integral del Corredor Biológico y Pesquero Puerto Peñasco – Puerto Lobos, Sonora.* Puerto Peñasco: Intercultural Center for the Study of Deserts and Oceans.

Valdez-Gardea, G. 2010. Pesquerías globalizadas: Revisitando a la comunidad marítima en el Alto Golfo de California. *Estudios sociales*, 18 (35): 135–163.

Valdez-Gardea, G. 2013a. …Los pescadores seguimos donde mismo. Reconversión productiva en el Alto Golfo de California. In G. Valdez-Gardea, & S. Galindo-Bect, *Pesquerías globalizadas,* pp. 109–130. Hermosillo: El Colegio de Sonora/Universidad Autónoma de Baja California.

Valdez-Gardea, G. 2013b. Especies en peligro de extinción y vidas precarias en el Alto Golfo de California . In G. Valdez-Gardea, & M. Galindo-Bect, *Pesquerías globalizadas,* pp. 47–64. Hermosillo: El Colegio de Sonora.

Valencia, S. 2014. Teoría transfeminista para el análisis de la violencia machista y la reconstrucción no-violenta del tejido social en el México contemporáneo. *Universitas humanistica, 78* (78): 65–88.

Valencia, S. 2018. *Gore Capitalism.* Pasadena: Semiotext(e)/Intervention Series.

Vázquez-León, C. 2020. La crisis pesquera en el Alto Golfo de California; protagonismos, deficiencias y escenarios. *La Jornada.* September 23, 2020. Retrieved from https://jornadabc.mx/opinion/23-09-2020/la-crisis-pesquera-en-el-alto-golfo-de-california-protagonismos-deficiencias-y.

Vásquez-León, M. 1999. Neoliberalism, environmentalism, and scientific knowledge: redefining use rights in the Gulf of California. In Heyman, J. (ed.). *States and Illegal Practices*, pp. 233–260. Berg: Oxford.

Vásquez-León, M. 2012. "Policies of conservation and sustainable development: fishing communities in the Gulf of California, Mexico." In Greenberg, J., Weaver, T., Alexander, W. & Browning-Aiken, A. (eds.). *Neoliberalism and Commodity Production in Mexico.* pp. 165–185. Boulder: University Press of Colorado.

Vega, S. 2016. La sostenibilidad de la vida como eje para convergencias teóricas y políticas. *Dossieres Economistas Sin Fronteras*, 23, 34–37.

Villalba, U. 2013. Buen vivir vs development: a paradigm shift in the Andes? *Third World Quarterly,* 34 (8): 1427–1442.

Vaquita Conservation International Recovery Committee [CIRVA]. 2015. *Final report. Sixth Meeting of the Comité Internacional para la Recuperación de la Vaquita (CIRVA-6).* Retrieved from: www.iucn-csg.org/wp-content/uploads/2010/03/CIRVA-6-Report-Rev-19-July-2015.pdf

Vaquita Conservation International Recovery Committee [CIRVA]. 2016a. *Eighth Meeting of the Comité Internacional para la Recuperación de la Vaquita (CIRVA-8).* Retrieved

from www.biologicaldiversity.org/species/mammals/vaquita/pdfs/CIRVA_8_Report_2_1_17.pdf.
Vaquita Conservation International Recovery Committee [CIRVA]. 2016b. *Seventh Meeting of the Comité Internacional para la Recuperación de la Vaquita.* Retrieved from www.mmc.gov/wp-content/uploads/CIRVA-7-Final-Report.pdf.
Vaquita Conservation International Recovery Committee [CIRVA]. 2017. *Ninth Meeting of the Comité Internacional para la Recuperación de la Vaquita (CIRVA-9).* La Jolla: CIRVA. Retrieved from www.iucn-csg.org/wp-content/uploads/2010/03/CIRVA-9-Final-Report-May-11-2017.pdf.
Zibechi, R. 2014. "El estado de excepción como paradigma político del extractivismo". In Composto, C. & Navarro, M. (comp.). *Territorios en disputa. Despojo capitalista, luchas en defensa de los bienes comunes naturales y alternativas emancipatorias para América Latina,* pp. 76–88. Mexico City: Bajo Tierra Ediciones.

6 Hydrodiplomacy

DOI: 10.1201/9781003220503-6

SUMMARY

This chapter begins with the paper "U.S.-Mexico Hydrodiplomacy: Foundations, Change, and Future Challenges" by Margaret O. Wilder, Robert G. Varady, Stephen P. Mumme, Andrea K. Gerlak, Nicolás Pineda Pablos, and Christopher A. Scott, which was presented at the conference by some of the contributing authors as a panel discussion. They offer some background regarding the water governance situation at the U.S.-Mexico border, discuss the evolving hydrodiplomatic relationship, and explain the challenges yet to face. This is followed by a summary of the panel discussion "Hydrodiplomacy in *Ambos Nogales*: Lessons Learned from Binational Water Management," which was chaired by Adriana Zuñiga Terán, with panelists Francisco Lara-Valencia, Laura Norman, Joaquin Murrieta, and Stephanie Buechler. They discuss the unique challenges facing this region, and explain the efforts throughout the last several years to find a way to cooperatively manage the binational area and resources, offering potential solutions for this area and for others in similar circumstances.

Next are two articles that were presented at the conference and subsequently reviewed by a scientific committee to be included in the proceedings: "Forging Sister Wetlands in El Paso, Texas and Ciudad Juárez, Mexico: Urban Transboundary Conservation and Restoration on the U.S-Mexico Border" by Ana Cordova, Adriana Zuñiga Terán, and John Sproul, and "A Model for Transboundary Water Cooperation: Perspectives from a Science Diplomat" by Katherine E. Himes.

U.S.-MEXICO HYDRODIPLOMACY: FOUNDATIONS, CHANGE, AND FUTURE CHALLENGES

Margaret O. Wilder
Associate Professor, School of Geography, Development and Environment, The University of Arizona

Robert G. Varady
Research Professor, Udall Center for Studies in Public Policy, The University of Arizona

Stephen P. Mumme
Professor, Department of Political Science, Colorado State University

Andrea K. Gerlak
Professor, School of Geography, Development and Environment; and Research Professor, Udall Center for Studies in Public Policy, The University of Arizona

Nicolás Pineda Pablos
Professor, Public Policy Studies, El Colegio de Sonora

Christopher A. Scott
Director, Udall Center for Studies in Public Policy; and Professor, School of Geography, Development and Environment, The University of Arizona

Note: This article was presented at the conference and was subsequently published in the online journal *Science & Diplomacy* in December 2019.

(Reprinted with permission from E. William Colglazier (ed.), *Science & Diplomacy*)

The year 2019 marked a significant milestone in U.S.-Mexico hydrodiplomacy: 75 years since the two countries adopted the Mexican Water Treaty of 1944, which apportioned between them the waters of the Rio Grande, Colorado River, and Tijuana River. Although the treaty is the countries' most notable and enduring act of hydrodiplomacy, numerous cooperative agreements have followed over the past seven-plus decades (Mumme 2019). These arrangements have transformed transboundary water management in the region, in an arc away from conflict and toward greater cooperation and collaboration.

Water scarcity along the U.S.-Mexico border has intensified due to various effects of climate change, economic development, population growth, and increasing water demand, resulting in a complex array of water-governance challenges (Garfin et al. 2013; Wilder et al. 2013). But paradoxically, transboundary binational institutions between the United States and Mexico have become more adept—not less so—at devising innovative adaptations in water management. To some extent, this experience accords with study findings that show nations tend toward cooperation rather than conflict over water and that complexity need not be a deal breaker if water management institutions are robust (Wolf et al. 2005). This raises a crucial question to be addressed in this essay: how precisely have transboundary U.S.-Mexico relations evolved toward cooperation and collaboration?

Hydrodiplomacy in Border Regions

The European Commission defines science diplomacy as "the use of science to prevent conflicts and crises, underpin policy making, and improve international relations in conflict areas where the universal language of science can open new channels of communication and build trust" (Müller and Bona 2018). It can likewise be thought of as "the use of scientific expertise to advance diplomacy" (Colglazier 2018). In essence, science diplomacy corresponds with the idea that "working together on research projects can build trust and networks between countries, improving their overall relations" (Matthews 2019). Specifically, hydrodiplomacy can be understood as science diplomacy applied to transboundary waters. According to E. William Colglazier's 2018 review, in this journal, of the state of science diplomacy, transboundary diplomacy over water may be one of the most consequential forms of future international scientific and political collaboration. Colglazier, citing a "future world" scenario of global warming challenges (Busby 2018)—most of which revolve around water—notes an anticipated "rise in international tensions and conflict, especially in developing countries over water" amid global warming conditions (Colglazier 2018).

Hydrodiplomacy generally refers to international cooperation around water resources shared by two or more countries and varies according to the particular sets of power relations between those countries. Hydrodiplomatic initiatives may explicitly utilize scientific expertise to different degrees in various contexts, ranging, for example, from descriptive data to hydrological and water-resource models to decision-making support for alternative future scenarios. Lawrence Susskind and Shafiqul Islam note that successful hydrodiplomacy rests on moving away from zero-sum thinking that conceives of water as a fixed resource to more flexible understandings that allow space for cooperation between countries (Susskind and Islam 2012).

We contend that effective hydrodiplomacy must be based on adaptive and collaborative transboundary water governance, defined as a form of collective decision-making that is "formal, consensus-oriented, and deliberative," where responsibility is shared among state and nonstate stakeholders working together to seek consensus (Ansell and Gash 2008; Cosens and Fremier 2015). Yet an important distinction in the U.S.-Mexico context is that not all stakeholders are coequals, given that the two countries' sections of the binational boundary water commission, as state actors, retain sovereign decision-making authority in carrying out their responsibilities under the 1944 treaty.

Adapting terms from organizational management, we offer straightforward working definitions of these key concepts aimed at clarifying important distinctions (Seapoint Center for Collaborative Leadership n.d.). *Conflict* in water governance is when two or more parties have different goals they view as mutually exclusive. *Cooperation* is when the parties share resources and information to help each achieve its own goal. And *collaboration* is when the parties work together toward achieving a shared goal. Yet this framework risks understating the complexity of these concepts in transboundary settings. Mark Zeitoun and Naho Mirumachi view conflict and cooperation as two essential faces of the same coin; rather than occupying opposite ends of a spectrum, they coexist and are in constant interaction with each other (Zeitoun

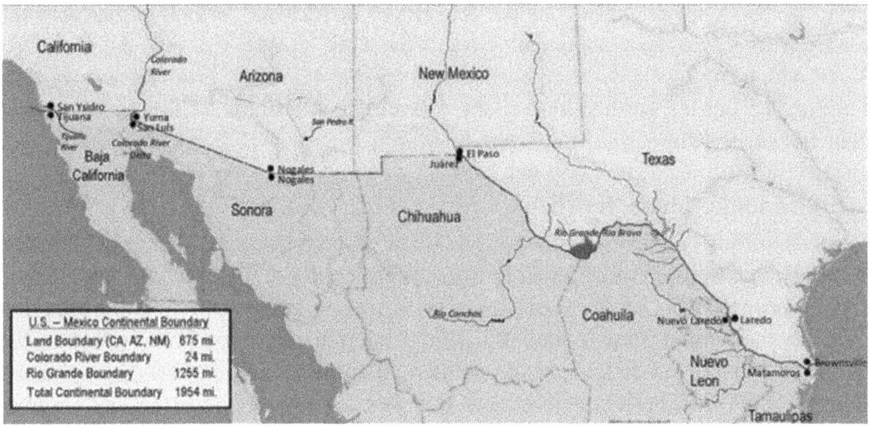

FIGURE 6.1 U.S.-Mexico border, border states, and selected rivers (Source: Adapted by the authors from the International Boundary and Water Commission (IBWC) n.d.)

and Mirumachi 2008). Claudia W. Sadoff and David Grey, in contrast, espouse a continuum theory that ranges from conflict to coordination to collaboration to joint action. They assert that "collaboration results when national plans are adapted either to secure gains or to mitigate harm in another riparian country," with joint action going even further and occurring when "riparians act as partners in the design, investment, and implementation" of the development of shared rivers (Sadoff and Grey 2009). In the U.S.-Mexico border region, intermittent episodes of conflict have largely been outweighed by an increasing spirit of cooperation over recent decades; thus, we argue that U.S.-Mexico hydrodiplomacy in general has moved away from conflict and toward cooperation and collaboration.

The U.S.-Mexico Border Region

Recognized as a "hot spot" both for climate change and socioeconomic vulnerability, the U.S.-Mexico border region presents an important test case for current and future hydrodiplomacy (Climate Central 2018; Garfin et al. 2013). The 1,956-mile (3,218 km) border between the two countries encompasses parts of ten states, four in the United States and six in Mexico (see Figure 6.1).

Physical Geography

The border's climate ranges from hyperarid to semiarid to temperate, and its landscape is at once rugged and fragile. The border features a number of important protected areas, migratory bird corridors, and rich wildlife and vegetation in diverse ecological regions. By midcentury, average temperatures are projected to increase significantly in the region, with widespread drying, drought, and rising water stress (Overpeck et al. 2013, 1–20; Wilder et al. 2013). Mean annual rainfall is highly variable across the region, but climate models project decreasing rainfall by midcentury (Scott 2011). There are a dozen transboundary rivers in the border region, including two major ones: the Rio Grande (known as the Río Bravo in Mexico), which demarcates the border from Texas and Chihuahua east to the Gulf of Mexico, and the Colorado River, which crosses the western border. Several transboundary groundwater aquifers underlie the border, further linking the hydrological systems of the two countries (Sanchez et al. 2016).

Socioeconomic Characteristics

Fourteen million people live in the region—in large, paired border cities, including San Diego/Tijuana and El Paso/Ciudad Juárez, and in smaller towns and rural settlements. Population growth is rapid, intensifying urbanization and encroachment on natural areas. In the United States, Sunbelt-oriented internal migration has resulted in the spread of housing subdivisions and commercial sprawl, while in Mexico unplanned colonies often lack basic services such as water, sanitation, and electricity (Ingram et al. 1994).

Asymmetries of wealth, power, and influence are hallmarks of this border, where an advanced industrialized economy comes face-to-face with a developing economy. The two countries' economies, cultures, and political systems all manifest unevenness; yet the past has seen cycles of conflict and accommodation over immigration, labor, trade, disaster management, narco-trafficking control, and—sometimes—water (Ganster and Lorey 2008; García-Acevedo and Ingram 2004; Sánchez-Rodriguez and Mumme 2010).

The first decade of the 21st century witnessed two significant conflicts over water; yet the region's robust transboundary institutions and relationships helped overcome them. Background for incident one dates to the 1990s, when severe precipitation deficits caused Mexico to declare an "extreme drought"—a term used but not defined in the 1944 treaty (Hanson 1991). This led to Mexico's curtailment of water deliveries from its tributary rivers in the eastern Río Bravo to the United States, resulting in a massive water debt to Texas farmers. In 1999–2002, the issue became an international incident involving both countries' presidents, resolved when wetter years allowed deliveries to resume (Texas House of Representatives 2002; Viña 2005). The situation was also assuaged due to a mechanism that allows five-year average deliveries to meet Mexico's treaty obligation. Some analysis indicates that the water shortage resulted from regional conflicts and mistrust between upstream and downstream irrigators, but that these groups also made common cause in demanding that their respective government representatives step in to resolve the crisis (Walsh 2004, 43–58). The dispute shed light on the need for bilateral cooperation to define shortage-sharing mechanisms. In 2016, Mexico fully paid off its water debt to the United States, and in 2017 the International Boundary and Water Commission (IBWC; in Spanish, *Comisión Internacional de Límites y Aguas*, CILA) created working groups to study a joint modeling system for the Rio Grande/Bravo. Yet Texas farmers remain concerned that the fundamental issues underlying the water debt in drought periods are not fully resolved.

A second conflict occurred in 2005–2007, when the United States unilaterally decided to line with concrete a 23-mile section of the earthen All-American Canal (AAC), over objections and court challenges by a coalition of environmental organizations, communities, and the Mexican government (Imperial Irrigation

District n.d.; Cortez-Lara, Donovan, and Whiteford 2009). The AAC is an 82-mile (128 km) aqueduct in southeastern California that conveys Colorado River water west to Imperial Valley farms and nine cities. Considerable political pressure in favor of the canal lining was exerted by the U.S. cities' water utility corporations and the rural irrigation districts. Plans for the lining extended back to 1983, when the United States, again unilaterally, declared ownership of the seepage water from the earthen canal; in 1988, the U.S. Congress directed the Department of the Interior to develop specific plans for the lining. Nearly two decades later, in 2005, eight claims were brought against the canal-lining project in the U.S. Ninth Circuit Court of Appeals. But the United States prevailed in these court cases, and subsequent appeals in 2006 and 2007, based on the claim that the lining achieved loss prevention for water perceived as belonging to the U.S. allocation. Mexico, for its part, viewed the lining as essentially a "taking" of cross-border seepage water used by farmers in the Mexicali area for more than 60 years (Cortez-Lara, Donovan, and Whiteford 2009). The canal lining was completed in 2009 (Imperial Irrigation District n.d.). Scholars note that once the conflict became a legal matter, the IBWC/CILA—as well as U.S. and Mexican federal agencies—was sidelined and thereafter had limited involvement in resolving the case (Cortez-Lara, Donovan, and Whiteford 2009; de la Parra Rentería 2006). In 2010, an earthquake damaged Mexicali and prevented Mexico from accepting Colorado River water deliveries, engendering a humanitarian response by the United States (U.S. Department of the Interior 2010). This led to Minute 319 (discussed later) and set the stage for a new era of binational collaboration.

Evolution of U.S.-Mexico Hydrodiplomacy

Despite periods of conflict, with remnants persisting today, the United States and Mexico, as already noted, have tended to cooperate over the management of transboundary water resources, an impulse that has only gained strength in the last three decades. This is the case even though the two neighbors have sometimes clashed over immigration, narco-trafficking, trade, and other issues of geopolitical significance.

Bilateral efforts to develop institutions to manage the border date back 130 years, to 1889, when the countries entered into the U.S.-Mexico Convention, whose signal creation was the joint International Boundary Commission. In 1944, this commission was renamed the International Boundary and Water Commission (Comisión Internacional de Límites y Aguas), reformulated under a mandate to manage shared water resources. The Treaty of 1944, monitored and enforced by the IBWC/CILA, allocated the waters of border rivers—the Colorado, Rio Grande/Río Bravo, and Tijuana—and established a method of "minutes" (i.e. treaty extensions or applications) that would permit modifications to the treaty over time.

Thus, in 1973, the IBWC/CILA adopted Minute 242 to address water quality in response to Mexico's demands to reduce the salinity of Colorado River water delivered by the United States (Mumme 2017, 133–183). Minute 242 guaranteed a minimum level of water quality and vindicated a national U.S. interest over the narrower interests of the seven U.S. Colorado River basin states (Arizona, California, Colorado, Nevada, New Mexico, Utah, and Wyoming).

In 1983, the La Paz Agreement became the first bilateral framework agreement to address the environment. Mexico and the United States agreed to "protect, conserve, and improve the environment" of the border region, and committed to a procedure for discussing border environmental issues and sharing related data and information (U.S. Department of State 1983). La Paz led to official recognition of a broad group of stakeholders, including states, *municipios* (in Mexico), and county and local administrations (in the United States), as well as environmental nongovernmental organizations (Varady 2009; Mumme and Collins 2014, 303–330).

The 1994 North American Free Trade Agreement (NAFTA), signed by Mexico, the United States, and Canada, was accompanied by NGO and interest-group concerns about the potential environmental impacts the agreement would wreak in terms of pollution and toxic waste dumping, especially in Mexico. NAFTA thus became a unique trade deal that pioneered an environmental side agreement resulting in the creation of the trinational Commission for Environmental Cooperation (CEC), headquartered in Montreal. Separately, the U.S.-Mexico Border Environment Cooperation Commission (BECC) and the North American Development Bank (NADB) were established at the same time. In 2018, the BECC was subsumed by the NADB, which along with the CEC marked 25 years in 2019.

The 2006 establishment of the Transboundary Aquifer Assessment Program (TAAP) committed the United States—via its network of U.S. Geological Survey offices and university-based water resources research centers—to hydrogeologic assessment and data collection on selected transboundary groundwater aquifers in Arizona, New Mexico, and Texas. Mexico, although not a signatory, has participated in scientific research and information sharing via its national water authority, the Comisión Nacional del Agua (CONAGUA), CILA, state governments, and regional universities. Until the creation of TAAP, the IBWC had paid little to no attention to transboundary groundwater resources, and the program helped highlight the importance of data sharing around transboundary aquifers.

Major Turning Points

The 1944 treaty and the subsequent Minutes 319 and 323 represent significant inflection points in the evolution of U.S.-Mexico hydrodiplomacy. The components of the first major turning point, the treaty and creation of the IBWC/CILA, have proven durable yet flexible foundations for the subsequent 75 years of hydrodiplomacy. The treaty's complexity, coupled with its political malleability through the "minutes" process for operationalizing and extending provisions, contributes to its durability. With specific agreements concerning the management of three named rivers, as well as a mandate to manage transboundary sanitation and administer all existing bilateral boundary agreements, the treaty serves numerous and varied stakeholders, helping widen its utility and political support.

The principal challenge to the treaty's legitimacy, a dispute over U.S. responsibility to control the salinity of the Colorado River, was resolved by Minute 242 in 1973, further consolidating binational support for the agreement. With Minute 242, the United States effectively conceded to Mexico's argument that the treaty guaranteed a minimum threshold of water quality to co-riparians. Six years later, in 1979,

Minute 261 extended the treaty's application to transboundary sanitation, enabling the IBWC/CILA to address water-quality problems.

Over the past two decades, the treaty's capacity to resolve water-shortage disputes and address ecological conservation has been tested on both the Rio Grande/Bravo and the Colorado River. On the Rio Grande/Bravo, Article 4's shortage-sharing provisions have led to binational investments in water conservation infrastructure in the watershed of Mexico's Río Conchos (a Río Bravo tributary), though binational differences over treaty interpretation persist. In contrast, the Colorado River, thought by the treaty's negotiators to present the most vexing shortage problems, has seen unprecedented advances in shortage sharing and ecosystem conservation (Minutes 319, 323). Two institutional features contributing to the functionality of the IBWC/CILA are (1) its operation outside the political-electoral arena in both countries and (2) the professionalization of its commissioners and personnel. At least in the case of Mexico, where most public offices are granted to insiders and political appointees whose service overlaps with presidential terms, the CILA commissioners have stayed in office much longer than the six-year presidential terms, including one commissioner who remained 31 years in office. A U.S. commissioner held the post for 20 years. The commissioners, engineers by profession with a few exceptions, have traditionally kept a low political profile, playing a mostly technical role.

A second inflection point came in 2012 with Minute 319, which authorized environmental flows of water to the Colorado River Delta (located entirely within Mexico), marking the first time the two nations had delivered water to the delta to promote ecosystem restoration (International Boundary and Water Commission 2014). Minute 319 is one of the "most significant agreements" signed to date, given the diversity of issues and cooperation measures embraced and its long-term collaboration on Colorado River basin management (Sánchez and Cortez-Lara 2015). Following issuance of the minute, parties successfully negotiated in March 2014 a pulse-flow release meant to mimic the ecologically vital spring floods that historically inundated the delta. For the first time in 50 years, the pulse-flow release reconnected the Colorado River with the Sea of Cortez. Environmental NGOs helped raise the necessary funding to meet the water and restoration commitments. In the larger picture, Minute 319 helped launch a binational stakeholder process under the IBWC/CILA that brings together agency representatives, university researchers, and environmental NGO representatives to conduct research and monitoring (Flessa, Kendy, and Schlatter 2016). Building from this experience, the successor, Minute 323 (2017), required both countries to provide water and funding for habitat restoration and scientific work in the Colorado River Delta for the subsequent decade, set forth voluntary cutbacks in times of drought, and allowed Mexico to continue to store water in Lake Mead, located upstream on the Colorado River between the states of Arizona and Nevada (International Boundary and Water Commission 2017).

Minutes 319 and 323, and the successful negotiation of a pulse-flow release to begin restoration of the delta's wetlands, demonstrate the significance of the Treaty of 1944 and the IBWC/CILA as bedrock institutions, and underscore the argument that strong binational institutions are critical for managing complexity in transboundary basins (Wolf et al. 2005). Furthermore, the case illuminates the emerging role of environmental networks led by natural and social scientists, policy scholars, and environmental NGOs that has altered the hydrodiplomacy landscape and helped reshape IBWC/CILA activities (Gerlak 2015).

A weakness in the 1944 treaty, however, is its inclusion only of the United States and Mexico as official parties and the lack of a requirement for consultation with affected sovereign tribal nations (Lewis 2019, 255). Despite Minute 319 being generally regarded as a success, legal scholar Madeleine J. Lewis notes that the Cocopah Indian Tribe and other delta stakeholders "have not received due representation during the negotiations of past environmental flow programs." She argues that the absence of a requirement for input from the Cocopah tribe "is a major failure of the entire Minute process to date" (Lewis 2019, 256). Nevertheless, Lewis also points out that Cocopah tribal members and other delta stakeholders were among those most overjoyed to witness the water flowing during the 2014 release, with the majority observing this for the first time in living memory.

Hydrodiplomacy has benefited from sustained scientific leadership on one hand, and strong political and institutional foundations on the other hand. Leadership by scientists in the United States and Mexico to devise and implement innovative solutions (e.g. pulse-flow release) and monitor project impacts (e.g. ecological recovery and societal benefits), as well as internal scientific leadership within the binational network, has lent enduring stability and credibility to the longtime efforts to protect the delta's delicate wetlands (Pitt and Kendy 2017). Political leadership demonstrated by IBWC/CILA authorities has been crucial to hydrodiplomatic successes, just as environmental NGO representatives have been instrumental in providing scientific support and financial resources, alongside challenging the IBWC/CILA to provide a more open and participatory stakeholder process (Gerlak 2017). Finally, the treaty is the terra firma upon which the rest of the transboundary cooperative apparatus rests. Lewis, like Stephen P. Mumme, the political scientist, argues for the essential importance of the treaty as an orchestrator of all transboundary water relationships (Lewis 2019, 231–257; Mumme 2003). Without a treaty, "the likelihood of contriving an equitable and sustainable water-sharing agreement between the two nations was improbable" (Lewis 2019, 234).

Anticipating Hydrodiplomacy Challenges at Midcentury

The evolution of transboundary relations over shared water resources has been dynamic and responsive to newly arising challenges. Although the U.S.-Mexico relationship has experienced significant instances of conflict, precipitated by drought or infrastructure problems, the overarching direction has been toward cooperation and even collaboration. The Treaty of 1944 has been a stabilizing influence on bilateral water relations, cushioning political oscillations. Indeed, for the past 75 years, the two countries' water relations have been remarkably insulated from geopolitical nuances and differences, responding more directly to local and watershed concerns and to the perpetual need to accommodate and respond to development asymmetries along and beyond the international boundary.

Minutes 319 and 323, which allowed water flows into Mexico's Colorado River Delta, demonstrated fresh capacities for innovation, cooperation, and collaboration. After decades of relationship building among scientists, environmental NGO and government agency representatives, and IBWC/CILA officials, the two countries defined a shared goal of restoring some environmental flows to the delta's stressed wetlands. The United States and Mexico collaborated to allow the peak flow release and an ongoing mechanism for identifying and purchasing other water allocations, followed by sustained scientific monitoring of the restoration effort. One key participant indicated that the "binational collaboration, bolstered by the heavy advocacy by nongovernmental organizations from both countries, was a necessary force in negotiating commitments to delta flows" (Lewis 2019, 239). As part of the same agreement, the two countries cooperated to achieve Mexico's goal of securing water storage capacity, when needed, in U.S. reservoirs and the United States' goal of identifying a formal means of sharing Colorado River shortages with Mexico. These actions fulfill Sadoff and Grey's concept of collaboration and even joint action.

Lewis emphasizes on both the durability of working relationships over long periods and the role of successive treaty minutes in building layer upon layer of institutional architecture that ultimately resulted in the "remarkable" Minute 323 (Lewis 2019, 237). Amendments stretching back to Minute 306—which was adopted in 2000 and dedicated the two countries to a "conceptual framework" to protect the delta's wetlands—attest to the "IBWC's growing but hesitant commitment to the environmental health of the Colorado River, and of the delta in particular" (Lewis 2019, 237). Such institutional partnerships are not created overnight. The monumental strides in binational collaboration to secure dedicated environmental flows in this water-stressed landscape were the work of decades, not mere months or years, and of a strong coalition—with an engaged IBWC/CILA at its center—that likewise developed over many years and remained solidly guided by both science and a commitment to restoring essential, environmental flows. At the same time, in working toward Minutes 319 and 323, both the United States and Mexico were pursuing their own national goals and interests.

As we look toward the future, many potential obstacles lie in the path of successful U.S.-Mexico hydrodiplomacy. In the last decade, scientists have recognized that the assumption of stationarity—that "natural systems fluctuate within an unchanging envelope of variability"—no longer holds for water resources, given that "anthropogenic change of Earth's climate is altering the means and extremes of precipitation, evapotranspiration, and rates of discharge of rivers" (Milly et al. 2008). Hydrological nonstationarity requires that the region rely on continued binational commitment and robust transboundary governance—an institutional stationarity of sorts—to build resilience against future challenges.

Much of the challenge transcends climate change and fluctuation. Along the border, as noted earlier, the sociopolitical environment creates conditions of increasing complexity—including uneven economic development, rapid population growth, and chronic water scarcity. These changes are placing increased demands on the treaty and IBWC/CILA for ever-more nimble institutional responses. Moving toward midcentury, hydrodiplomacy initiatives must continue to engage these challenges creatively and, most important, binationally. While little evidence suggests that shifting geopolitical relations have affected binational environmental relations, they could at some point hamper, halt, or even set back hydrodiplomacy, should national interests on either side favor such a course. This is arguably a somewhat more acute concern under the Trump administration, which has given dangerous credence to antiscientific rhetoric—namely, denying the science behind the anthropogenic causes of climate change—with the potential to affect U.S.-Mexico geopolitical relations. In essence, the U.S. leadership has sparked the opposite of science diplomacy in its desire to declare a "crisis" at the U.S.-Mexico border. Such an approach is at odds with the cooperative hydrodiplomatic initiatives in the transboundary region for three-quarters of a century.

The history of effective U.S.-Mexico hydrodiplomacy carries implications for other global transboundary regions under stress, pointing to the fundamental value of strength, flexibility, and, above all, trust in the foundational instruments and management institutions governing the relationship. Dynamic transboundary regions require adept institutions and scientific and political leadership to face the challenges that inevitably confront them—not as instances for prevailing in conflict but as opportunities to further cooperation and collaboration toward serving the individual interests of each country while also advancing their shared goals.

Acknowledgments

The authors gratefully acknowledge the contributions of Adriana A. Zuñiga, Sharon B. Megdal, and Karl W. Flessa to a 2018 conference presentation on which this essay was partially based: R. G. Varady, S. Mumme, A. K. Gerlak, S. B. Megdal, M. O. Wilder, N. Pineda Pablos, C. A. Scott, K. W. Flessa, and A. Zuñiga, "Seventy-Five Years of Hydrodiplomacy on the U.S.-Mexico Border," presented at the "Empowering Hydro-diplomacy Conference," The Hague, Netherlands, October 5, 2018.

The work that resulted in this essay was undertaken as part of the International Water Security Network, a project funded by Lloyd's Register Foundation, an organization that helps protect life and property by supporting engineering-related education, public engagement, and the application of research. We further acknowledge the Inter-American Institute for Global Change Research for Project CRN3056, which is supported by the U.S. National Science Foundation (grant no. GEO-1128040) and NSF grant no. DEB-1010495. The paper also benefited from support from the Morris K. Udall and Stewart L. Udall Foundation, in Tucson, Arizona.

REFERENCES

Ansell, Chris, and Gash, Alison. 2008. "Collaborative Governance in Theory and Practice." *Journal of Public Administration Research and Theory* 18, no. 4 (October 2008): 543–572.

Busby, Joshua. 2018. "Warming World." *Foreign Affairs* (July/August 2018): 49–55. www.foreignaffairs.com/articles/2018-06-14/warming-world.

Climate Central. 2018. "American Warming: The Fastest-Warming Cities and States in the U.S." www.climatecentral.org/news/report-american-warming-us-heats-up-earth-day.

Colglazier, E. William. 2018. "Science Diplomacy and Future Worlds," *Science & Diplomacy* 7, no. 3 (September 2018). www.sciencediplomacy.org/editorial/2018/science-diplomacy-and-future-worlds.

Cortez-Lara, Alfonso A., Donovan, Megan K., and Whiteford, Scott. 2009. "The All-American Canal Lining Dispute: An American Resolution over Mexican Groundwater Rights?" *Frontera Norte* 21, no. 41 (June 2009). www.scielo.org.mx/scielo.php?script=sci_arttext&pid=S0187-73722009000100006.

Cosens, Barbara, and Fremier, Alexander. 2015. "Assessing System Resilience and Ecosystems Services in Large River Basins: A Case Study of the Columbia River Basin." *Idaho Law Review* 51, no. 1. https://digitalcommons.law.uidaho.edu/idaho-law-review/vol51/iss1/3.

de la Parra Rentería, Carlos. 2006. "The All-American Canal: An Issue of Contexts." In *The U.S.-Mexican Border Environment: Lining the All-American Canal; Competition or Cooperation for Water in the U.S.-Mexican Border?* Edited by V. Sánchez Munguía, SCERP Monograph Series 13. San Diego: SDSU/SCERP/El Colef.

Flessa, Karl, Kendy, Eloise, and Schlatter, Karen, eds. 2016. *Minute 319: Colorado River Limitrophe and Delta; Environmental Flows Monitoring*, Interim Report. El Paso: International Boundary and Water Commission. www.ibwc.gov/Files/Minutes%20 319/2016_EFM_InterimReport_Min319.pdf.

Ganster, Paul, and Lorey, David. 2008. *The U.S.-Mexican Border into the 21st Century*, 2nd ed. New York: Rowman & Littlefield.

García-Acevedo, María, and Ingram, Helen M. 2004. "Conflict in the Borderlands: Report on Water," *NACLA Report on the Americas* 38, no. 1 (July/August 2004): 19–41. https://nacla.org/article/conflict-borderlands.

Garfin, Gregg, et al. eds. 2013. *Assessment of Climate Change in the Southwest United States: A Report Prepared for the National Climate Assessment.* Washington, DC: Island Press. www.swcarr.arizona.edu/content/about-report.

Gerlak, Andrea K. 2015. "Resistance and Reform: Transboundary Water Governance in the Colorado River Delta." *Review of Policy Research* 32, no. 1 (2015): 100–123.

Gerlak, Andrea K. 2017. "Regional Water Institutions and Participation in Water Governance: The Colorado River Delta as an Exception to the Rule?" *Journal of the Southwest* 59, no. 1–2 (2017): 184–203.

Imperial Irrigation District. n.d. "All-American Canal." www.iid.com/water/water-transportation-system/colorado-river-facilities/all-american-canal.

International Boundary and Water Commission. 2014. "Colorado River Reconnects with the Sea following Historic Release of Water for the Environment." May 16, 2014. www.ibwc.gov/Files/Press_Release_051614.pdf.

International Boundary and Water Commission: United States and Mexico. 2017. "Minute No. 323: Extension of Cooperative Measures and Adoption of a Binational Water Scarcity Contingency Plan in the Colorado River Basin." September 21, 2017. www.ibwc.gov/Files/Minutes/Min323.pdf.

Lewis, Madeleine J. 2019. "Minute by Minute: An Assessment of the Environmental Flows Program for Restoration of the Colorado River Delta." *Wyoming Law Review* 19, no. 1 (2019): 255, ft. 206.

Matthews, David. 2019. "Can Science Diplomacy Work?" *Inside Higher Education*, June 6, 2019. www.insidehighered.com/news/2019/06/06/advocates-science-diplomacy-see-great-need-era-world-divisions.

Milly, P. C. D., et al. 2008. "Stationarity Is Dead: Whither Water Management?" *Science* 319, no. 5863 (2008): 573–574. doi:10.1126/science.1151915.

Müller, Jan Marco, and Bona, Maurizio. 2018. "Past, Present, and Future of Science Diplomacy in Europe." *Science & Diplomacy* 7, no. 3 (September 2018). http://sciencediplomacy.org/perspective/2018/past-present-and-future-science-diplomacy-in-europe.

Mumme, Stephen P. 2003. "Revising the 1944 Water Treaty: Reflections on the Rio Grande Drought Crises and Other Matters." *Journal of the Southwest* 45, no. 4 (Winter 2003): 649–671.

Mumme, Stephen P. 2017. "Constructing Equity during the Salinity Crisis, 1961–1973: Mexico, the United States, and the Politics of Sharing on the Colorado River," *Journal of the Southwest* 59, no. 1–2: 133–183. https://doi.org/10.1353/jsw.2017.0009.

Mumme, Stephen P. 2019. *The 1944 Water Treaty as a Constitutional Document.* Houston: Rice University's Baker Institute for Public Policy. www.bakerinstitute.org/media/files/files/fb5668a4/mex-pubwatertreaty-031419.pdf.

Mumme, Stephen P., and Collins, Kimberly. 2014. "The La Paz Agreement 30 Years On." *Journal of Environment & Development* 23, no. 3 (2014): 303–330.

Overpeck, Jonathan, et al. 2013. "Summary for Decision Makers," in *Assessment of Climate Change in the Southwest United States*, edited by Gregg M. Garfin et al. Washington, DC: Island Press. pp. 1–20. www.swcarr.arizona.edu/chapter/one.

Pitt, Jennifer, and Kendy, Eloise. 2017. "Shaping the 2014 Colorado River Delta Pulse Flow: Rapid Environmental Flow Design for Ecological Outcomes and Scientific Learning." *Ecological Engineering* 106, Part B (September 2017): 704–714. https://doi.org/10.1016/j.ecoleng.2016.12.002.

Sadoff, Claudia W., and Grey, David. 2009. "Cooperation on International Rivers: A Continuum for Securing and Sharing Benefits." International Water Resources Association. *Water International* 30, no. 4 (2009): 420–427. https://doi.org/10.1080/02508060508691886.

Sanchez, Rosario, Lopez, Victoria, and Eckstein, Gabriel. 2016. "Identifying and Characterizing Transboundary Aquifers along the Mexico-U.S. Border: An Initial Assessment." *Journal of Hydrology* 535 (2016): 101–119.

Sánchez, Vicente, and Cortez-Lara, Alfonso A. 2015. "Minute 319 of the International Boundary and Water Commission between the U.S. and Mexico: Colorado River Binational Water Management Implications." *International Journal of Water Resources Development* 31, no. 1 (2015): 17–27. https://doi.org/10.1080/07900627.2014.922465.

Sánchez-Rodriguez, Roberto, and Mumme, Stephen P. 2010. "Environmental Protection and Natural Resources." In *Mexico and the United States: Confronting the Twenty-First Century*, Working Paper Series. San Diego: University of California. https://scholar.google.com/scholar?hl=en&as_sdt=0%2C3&q=sanchez-rodriguez+and+mumme&btnG=

Scott, Christopher A. 2011. "The Water-Energy-Climate Nexus: Resources and Policy Outlook for Aquifers in Mexico." *Water Resources Research* 47. https://doi:10.1029/2011WR010805.

Seapoint Center for Collaborative Leadership. (n.d.), https://seapointcenter.com/cooperation-teamwork-and-collaboration/.

Susskind, Lawrence, and Islam, Shafiqul. 2012. "Water Diplomacy: Creating Value and Building Trust in Transboundary Water Negotiations." *Science & Diplomacy* 1, no. 3 (September 2012). www.sciencediplomacy.org/perspective/2012/water-diplomacy.

Texas House of Representatives. 2002. "Mexico's Water Debt: Behind the U.S.-Mexico Water Treaty Dispute." House Research Organization 77–7 (April 30, 2002).

U.S. Department of State. "Agreement between the United States of America and Mexico Signed at La Paz, August 14, 1983." Washington, DC: U.S. Government Printing Office.

U.S. Department of the Interior. 2010. "Salazar, Elvira Announce Water Agreement to Support Response to Mexicali Valley Earthquake," December 20, 2010. www.doi.gov/news/pressreleases/Salazar-Elvira-Announce-Water-Agreement-to-Support-Response-to-Mexicali-Valley-Earthquake.

Varady, Robert G. 2009. "The North American Free Trade Agreement (NAFTA) and Environment in the U.S.-Mexico Border Region." *Encyclopedia of Environment and Society* vol. 5, edited by P. Robbins. Thousand Oaks, CA: Sage Publications. pp. 1261.

Viña, Stephen R. 2005. *The United States–Mexico Dispute over the Waters of the Lower Rio Grande River.* Washington, DC: Congressional Research Service. www.everycrsreport.com/files/20050321_RS22085_90ee6592ed31594967978e6d8af28409d5d13b16.pdf.

Walsh, Casey. 2004. "*Aguas Broncas*: The Regional Political Ecology of Water Conflict in the Mexico-U.S. Borderlands." *Journal of Political Ecology* 11: 43–58.
Wilder, Margaret, et al. 2013. "Climate Change and U.S.-Mexico Border Communities." In *Assessment of Climate Change in the Southwest United States*, edited by Gregg Garfin et al. Washington, DC: Island Press. pp. 340–384. www.swcarr.arizona.edu/chapter/16.
Wolf, Aaron T., et al. 2005. "Managing Water Conflict and Cooperation," chap. 5 in *State of the World 2005: Redefining Global Security*, edited by Erik Assadourian et al. Washington, DC: Worldwatch Institute. pp. 80–208.
Zeitoun, Mark and Mirumachi, Naho. 2008. "Transboundary Water Interaction I: Reconsidering Conflict and Cooperation." *International Environmental Agreements* 8 (2008): 297–316. https://doi.org/10.1007/s10784-008-9083-5.

HYDRODIPLOMACY IN *AMBOS NOGALES*: LESSONS LEARNED FROM BINATIONAL WATER MANAGEMENT

Panel Discussion

Chair:
Adriana Zúñiga-Terán
Assistant Research Scientist, School of Landscape Architecture and Planning and Udall Center for Studies in Public Policy, The University of Arizona

Panelists:

Francisco Lara-Valencia
Associate Professor, School of Transborder Studies, Arizona State University

Laura Norman
Supervisory Research Physical Scientist, United States Geological Survey

Joaquin Murrieta
Cultural Ecologist, Watershed Management Group

Stephanie Buechler
Associate Research Professor, Agricultural Sciences Global, The Pennsylvania State University

In this session, we heard from experts from academia (UA and ASU), a federal agency (USGS), and a local environmental NGO (Watershed Management Group), who have been conducting studies and projects in these twin cities on these particular issues. They gave brief presentations about (1) governance, cooperation requirements, and challenges for a binational solution including the role of scientists and universities amidst the current political environment; (2) efforts to model runoff at the watershed scale regardless of political boundaries and the practical use of these outputs; (3) the implementation of green infrastructure in Nogales, Sonora, as a mitigation strategy for flooding and erosion and its influence in the development of the first green infrastructure regulation in Mexico; and (4) potential sustainable solutions that can reduce the burden on both sides of the border while enhancing environmental protection.

Background

Ambos Nogales, the sister cities of Nogales that sit on both sides of the U.S.-Mexico border, are located along the tributary of the Santa Cruz River near Nogales Wash. About 42% of the Nogales Wash watershed is in the United States and about 58% is in Mexico. The key feature in this fascinating case study is that water flows north from Mexico into the United States, for better or worse. The urban settlements span both sides of the border, following along the Nogales creek.

Flooding issues have been a problem in these sister cities for a long time. There was a major flood in 1930 that was particularly devasting to both cities, and which prompted the construction of the earliest storm water management infrastructures with the construction of the Nogales Wash canal in *Ambos Nogales*, a shared infrastructure. Once the Water Treaty of 1944 was signed, there was finally a mechanism in place to address these issues of water at the binational scale (Utilization of Waters of the Colorado and Tijuana Rivers and of the Rio Grande 1944). In 1951, the two commissions, the International Boundary and Water Commission (IBWC) and the Mexican counterpart, Comisión Internacional de Límites y Aguas (CILA), authorized the construction of a binational sewage system in *Ambos Nogales*. This consisted of a sewage system on the Mexican side and a wastewater treatment on the U.S. side, which when you think about it makes a lot of sense because water flows north. The commission also funded the Nogales International Wastewater Treatment Plant (NIWTP) in Nogales, AZ. In 1972 the binational treatment plan was relocated from Nogales to Rio Rico, AZ—just nine miles north of Nogales. The International Outfall Interceptor, or IOI, is a tunnel that conveys Mexican sewage from the border to the plant. The problem is that it desperately needs repairs, though no one claims ownership due to the massive costs of the necessary repairs (García et al. 2019).

Another big piece of the puzzle is the design of the streets in Nogales, Sonora. The streets were built directly on creeks (not beside them) with sewage pipe systems below. This means that when it rains, the streets become rivers again, so the main streets of Nogales, Sonora end up being a river, a street, and a sewage system all existing in the same space. After the September 11, 2001 attack on the World Trade Center, concerns over border security resulted in the construction of a less-permeable border wall, which blocks stormwater flow on both sides of the border. Now stormwater can no longer flow north freely. Therefore, Nogales, Sonora, experiences severe flood events, like the one in 2008 which caused massive amounts of damage to sewage system infrastructure (McCombs 2008). We continue to see severe flooding events on the Mexican side of border, and as they repair damages, they continue aligning the streets and sewer systems, not fully addressing the larger problem. The Mexican population in Nogales gets desperate when these catastrophic events occur because functionality greatly decreases as the city becomes paralyzed by the flooding, so people open up the manholes of the sewage system to relieve the flooding. The big problem is that this results in the stormwater, along with debris, sediment, and trash from the streets, are all combined with the raw sewage, exceeding projected volumes of Mexican sewage conveyed to the NIWTP in Rio Rico, and overflowing onto the streets. The damaged IOI then leaks all this debris and raw sewage on the U.S. side of the border, which violates several regulations of water quality, making it a serious public health concern.

Another important complexity factor here is the financial aspect of the wastewater treatment plant operations. Mexico pays for the water that is being treated in the NIWTP, but at an agreed volume. When these storm events happen, the volume exceeds the set volume, and Mexico is charged at a much higher rate than what was agreed upon, which places a greater burden on the City of Nogales, Sonora. Then the operations of the plant are in jeopardy because they are working in deficit and are not actually able to pay for the operations of the treatment plant. To reduce the Mexican volume of sewage conveyed to the NIWTP, a second wastewater treatment plant, Los Alisos, was built south of the Nogales Sonora watershed. The problem with that one is that Nogales has to pump the sewage up the hill to get it across the next watershed, which is really costly and very high energy intensive.

Lastly, another big issue they are dealing with is what happens with the effluent discharged from the NIWTP in Rio Rico, because this water cannot be used by either country. It technically belongs to Mexico, but there is currently no way to bring it back to Mexico for them to use it, because it would require a huge amount of energy to pump it back, not to mention the lack of consistent land tenure and other land challenges that are difficult to meet. So Mexico cannot physically use that water, but the United States cannot use it either because it belongs to Mexico. The United States is not allowed to allocate it for use in agriculture or development. As a result, Mexican effluent becomes an environmental flow discharged into the Santa Cruz River. We have seen a lot of ecosystem services derived from this environmental flow. Some endangered species have returned such as the Gila topminnow, an endangered fish. A lot of coalitions of people have been formed to defend the environmental flow, including Friends of the Santa Cruz River, birdwatchers and biodiversity advocates, who have seen a lot of greening in these riparian areas. However, none of these benefits are enjoyed by the people in *Ambos Nogales*.

All these binational issues have resulted in some efforts to address these problems, which the panelists expand upon in their presentations.

Introduction to Panel Members

Francisco Lara-Valencia is an associate professor in the School of Transborder Studies and director of the program for transborder communities at Arizona State University. His current work focuses on understanding the role of structure and institutional change on cooperation and sustainable urban governance in border settings. Francisco is also interested in the potential for green infrastructure approaches as a strategy to mitigate and adapt to climate change in border cities. At the Transborder Studies, Francisco teaches classes on transborder theory and U.S.-Mexico border development. He is also affiliated faculty of the Centro de Investigación en Alimentación y Desarrollo and the Global Institute of Sustainability.

Laura Norman is a Supervisory Research Physical Scientist at the USGS since 1998. Her Ph.D. is in watershed management from The University of Arizona, minoring in Remote Sensing and Spatial Analysis. Her research uses complex hydrologic and hydraulic models to predict the fate of non-point source pollutants and identify their sources, consider potential growth scenarios, and identify watershed processes. Her current research is to improve restoration, site selection and design, assess efficacy, and predict impacts of change.

Joaquin Murrietta is a cultural ecologist who has been working at Watershed Management Group, or WMG, since 2011. WMG is a local nonprofit organization that develops community-based solutions to ensure the long-term prosperity of people and health on the environment. Joaquin is responsible for managing a variety of community-based projects, with an emphasis on the interface of water, people, and culture. Specifically, Dr. Murrietta looks at the role of urban environments in the overall management of watersheds. Primary responsibilities of the job include managing the schoolyard program, international and U.S.-Mexico border region, expanding and coordinating activities with Native cultures of the Sonoran Desert, and leading green program project implementation.

Stephanie Buechler has a background in public policy, sociology, and geography. Her work focuses on the intersections of water resources (including wastewater), environmental change, agriculture, sustainability planning, poverty, youth, and gender. Stephanie has worked with binational applied research teams as well as nonprofits and grass-roots organizations and government agencies on these issues. Recently, her work has focused on urban agriculture, on the water scarcity in Tucson's low-income neighborhoods, youth's decision-making in the context of environmental change in rural Sonora, Mexico, women's innovation in water and climate change adaptation in Sonora, Mexico, and women's use of renewable energy in Zacatecas, Mexico and Arizona. Her recent work is very relevant for this panel as it includes cross-border perception of the international wastewater treatment plant (WWTP) in Nogales.

Governance, Cooperation Requirements, and Challenges for a Binational Solution to Watershed Management

In order to give a broad perspective of water management in *Ambos Nogales*, Dr. Lara-Valencia looks at the past, especially his prior experience and research of water management in the area, and also looks at the issues from the perspective of a planner.

The background just given lays out some of the complexity of the water management issues in *Ambos Nogales*. Many of the water management problems in the area are linked to sprawling urbanization, economic growth, and institutional fragmentation. Since their establishment, these two cities have been dealing with these problems of flooding and management. The city was established at the lower valley of the Nogales Wash which runs up north. The terrain of this region is very complex and very steep, and nearly all the water that falls in the area ends up in downtown Nogales, Sonora and Nogales, Arizona.

Growing urbanization has caused problems to the local ecology, causing it to lose its innate functionality. The precipitation partners created and intensified these problems with greater substitution of natural elements. Water resources are under a lot of pressure due to population and economic expansion. Water quality has also become a very serious issue, further exacerbated by severe rainfall events and ongoing drought and climate change patterns.

These challenges are not different from those identified 30 years ago (Ingram et al. 1994). There was a group of researchers from The University of Arizona and El Colegio de la Frontera Norte (COLEF) working together to create baseline data on water demand and supply on both sides of the border. The group, which was led by Helen Ingram, Robert Varady, and Roberto Sanchez of COLEF, was also interested in identifying institutional disparities in water quality both locally and across the border. Among the issues identified in that report from the early 1990s was contamination from pollutants and spills. It was estimated that about 25% of the people in Nogales were exposed to raw sewage in various ways throughout the city, had indirect exposure due to spills, because the quality was compromised due to issues with pollution (Ingram and White 1993). The concern of that time was not only of exposure to raw sewage, but also with the infiltration of untreated sewage into the water supply system in Nogales, and obviously the contamination of groundwater in the area. Another main concern was the frequent occurrence of flooding in Nogales and its subsequent economic disruption, including urban destruction and social, environmental, and fiscal costs for both sides of the border because major flooding would occur multiple times every single year (Ingram et al. 1995). These issues, though pinpointed over 30 years ago, are still the major concerns of the area, as we see in the Arizona-Sonora Environmental Strategic Plan for 2017–2021 that their goals are still addressing these same issues (Arizona Department of Environmental Quality 2017).

This is not to say that there has been no change or improvement in the last three decades. There has been important and positive change in many areas, including the expansion of the sanitation and water supply in Nogales, Sonora, thanks to a lot of investments and grant money from the NAD Bank, EPA, and others through IBWC/CILA. There has been substantial growth in the number of highly qualified and motivated water professionals on both sides of the border. The last decades have also seen a significant increase in the number and quality of binational collaboration between agencies along the border. The mechanisms put in place over the years have created a space for open and flexible communication between the sister cities. The level of trust and consensus that can be observed in meetings among technicians, administrators, and policymakers on both sides of the border has greatly increased when contrasted with the high levels of distrust and limited communication experienced several years ago in *Ambos Nogales*. Another important development is the formation of cross-border knowledge and practice communities around water issues where you can see water scientists and water managers working together on solutions using actionable knowledge supporting place-based solutions and innovation. Laura Norman and Adriana Zuniga provide a good example of this.

With all these developments, you can see a transformation occurring, which provides the basis for a more effective form of collaboration and governance for sustainable water management in *Ambos Nogales*. With each transformation, it develops stronger and more sustainable governance.

As a university professor, Dr. Lara-Valencia looks at how academic institutions can contribute to solving the problems at *Ambos Nogales*. There is still a need to "bridge" the various aspects of these problems instead of looking at them one dimensionally. For example, connecting water management to public health, water quality and transportation planning, and waste management and residential development gives a more complete picture of the actual issues at hand. The importance of the ecological functionality provided by the vanishing natural landscape of *Ambos Nogales* has been clear as climate and precipitation patterns change in the area. Water management and water governance must understand and be involved in a multitude of urban environmental issues, not only water.

Another idea is to address the physical fragmentation that the border has created. From an environmental point of view, the border is an artificial imposition upon the landscape and has an effect not only on the ecological functioning of the watershed but also on social and professional practices and values toward this shared ecosystem. Unfortunately, the "we" and "them" narratives created by the border have hindered any approaches to collaborative governance, and it is very difficult to convince urban planners of the importance of recognizing the binational impacts of resource management.

Finally, one encouraging and relatively new development is the emergence of new modes of collaboration between scientists, public institutions and communities along the border. When the binational water study was launched in the early 1990s, there was an effort to connect with local stakeholders, but it was not very successful, especially on the Mexican side. Looking back, this was due to limited local capacity, distrust between Mexican and U.S. water agencies, and because collaboration was often framed as a zero-sum process. What is seen now is quite different, because the interaction between science and water communities along the border is increasing. Now, knowledge and expertise are sought after on both sides because we can see mutual benefits. We should continue to encourage this in *Ambos Nogales* and try to create programs at universities that facilitate and support binational scientific collaboration and environmental management. The local capacity to perform water management planning has increased substantially on both sides of the border. The amount of planning from many agencies and the access they have to information is impressive. Those in academia provide a natural step in better understanding the problems in *Ambos Nogales* and how to deal with them. This connection should be seen more as equal partners, as academics are creating knowledge and developing ideas, others (local governments and NGOs) are implementing those ideas and putting them into practice; both are equally important in resolving the issues and developing new plans to address them.

Question

Audience Member: From your perspective, has there been a positive change over the last thirty years, particularly on the Mexican side of the border, to address the issue? Are there more active NGO's on the Mexican side today than there were then?

Answer

Francisco Lara-Valencia: Right now, it can be difficult to differentiate between NGO's that are on one side or the other, which could be seen as a positive thing. Many of the civic organizations in Mexico are binational in nature. They don't see their work as merely for one side or the other, but rather they see Ambos Nogales as an entity. When we did the work in the 1990s, that was not the case. That period spurred the creation of a lot of NGO's, but they were new, did not have a lot of resources, and they didn't have an understanding of the border as a binational space. So there is certainly more collaborative work in recent years compared to previous engagements.

Watershed Modeling in *Ambos Nogales*

Watershed modeling is not a typical topic discussed at diplomacy conferences, though the research and application of effective and sustainable watershed models are something that can help cities throughout the world adapt to the effects of changing precipitation patterns. Laura Norman, a supervisory research physical scientist at the United States Geological Survey (USGS), defines a watershed as an area that topographically contributes to the drainage to a point of interest. Runoff precipitation collects in catchments, which collectively form watersheds that eventually feed into rivers or replenish groundwater basins.

The Santa Cruz Watershed flows north from Mexico into the United States, so Tucson residents are the recipients of watershed activity upstream from San Rafael Valley and *Ambos Nogales*. The watershed covers about 235 km^2, just under half of which is in Mexico. Besides the thousands of residents in *Ambos Nogales* (about 20,000 in Nogales, AZ and about 220,000 in Nogales, SON), there are also many agencies involved in governing the watershed, meaning there are a lot of stakeholders concerned with what happens here. Watershed planning and management provide a great mechanism for hydrodiplomacy to occur in a binational setting. If we can use watershed modeling across borders to span the whole region and make research-based planning decisions, we can build our capacity for drought and flood resilience more effectively than by attempting to address each challenge piecemeal privately.

Dr. Norman's research in Ambos Nogales began in 1998 looking at heavy metals and distribution routes using geospatial technology (Brady et al. 2002). That research was later used for municipal planning, because it was the first time anyone had ever collected that geospatial data before. In 2003, she began a HUD-funded project with "*colonias*," housing settlements within 150 miles of the border that lack adequate housing or infrastructure such as sewage or potable water (Norman, Parcher, and Lam 2004; Norman 2010).

As the other presenters suggested, the rapid growth in Nogales has been driving a lot of the environmental changes they have experienced. Working with city officials on both sides of the border and based on geospatial data previously collected by Varady and Mack (1995), she developed an online GIS database though the Colonias Monitoring Program to enhance binational planning and to provide tools that may assist in the acquisition of funds set aside for these borderland communities (Norman et al. 2006). She modeled how these problems become exacerbated by increasing development encroaching hillslopes over time (Norman et al 2006; Norman, Guertin, and Feller 2008). An increase in impervious surfaces and accumulation of nonorganic matter that accompanies urban growth has significantly changed the ground's absorption capacity and led to increased erosion. And as the urban sprawl continues to grow and the land increasingly erodes, there is more and more sediment yield (Norman, Feller, and Guertin 2009). This means more deposition in the waterways and increased potential to act as carriers of other pollutants, including heavy metals (Norman, Guertin, and Feller 2008).

So what is the potential for sustainable development? Dr. Norman's thesis was to identify areas that are at-risk or "hotspots" for erosion, secure those areas and identify them for revegetation, and then identify locations that are not as at-risk for erosion to expand development if necessary. Her modeling found that this system of regenerating targeted areas would actually decrease the rate of erosion by over 11,000 tons per year by 2030 (Norman 2005, 2007).

In 2004, the USGS launched the Border Environmental Health Initiative (BEHI), a major project encompassing the entire U.S.-Mexico border region. In 2009, a study of the Santa Cruz River Watershed (SCW) was initiated as part of the BEHI, with support from the Mendenhall Program, as an interdisciplinary effort to track sources of contamination over land, through water, plants, and animals (Norman, Callegary, van Riper, and Gray 2010). From this research they learned just how big of an issue flooding is for *Ambos Nogales*.

The flooding in *Ambos Nogales* is now the number one priority for the Arizona Department of Emergency and Military Affairs (DEMA). Flooding is a major economic and socio-cultural problem, leading to multiple deaths on the Mexican side of the border every year. Even with "normal" rainfall, there is the potential of realizing abnormal runoff, which augments loss of life and property through flooding, soil-loss and watershed-capacity for revegetation, and nonpoint-source pollution of shared drainages and infrastructure (sediment and solid waste), which puts public health and the environment at risk (Norman 2007; Norman, Huth, et al. 2010). Using geospatial monitoring and rainfall-runoff data, Dr. Norman and her team determined areas throughout the watershed that were more at-risk for flood runoff (Norman 2010; Norman, Huth, et al. 2010). These at-risk areas span across the border, making them a binational issue.

The IBWC/CILA used this research to further develop models in order to determine how different structures, such as gabions, would impact flooding scenarios (Norman, Levick, et al. 2010). Since then, there have been numerous studies and projects branching off of some of these earlier undertakings to examine socio-environmental health in the colonias and disparities or environmental justice of access to ecosystem services (Norman, Caldeira, et al. 2012; Norman, Villarreal, et al. 2012; 2013). Thus, the partnerships in the watershed have created opportunities to advance the environmental health and safety for the watershed. Most recently, the IBWC commissioners Ed Drusina and Roberto Salmón Castelo said in 2017 that they would create a new Minute to strategize a framework for binational cooperation on transboundary stormwater control in the Nogales Watershed. Once approved by both Governments, the Minutes enter into force as binding obligations of the United States and Mexican Governments.

In summary, Dr. Norman concluded that, managing water resource programs on a watershed basis (rather than a municipal or national basis) makes sense—environmentally, financially, and socially. Watershed modeling is especially valuable to be used as a planning tool for watersheds that span administrative boundaries; models can portray how actions throughout the watershed might impact different communities. Products of transboundary watershed analyses allow stakeholders to envision scenarios of management, visualize mutual goals for healthy watersheds, and develop comprehensive approach strategies (Norman 2019).

The Role of Green Infrastructure in Watershed Management

There are so many studies and so many analyses, but how do you actually get stuff done? Sometimes you present findings to city, county, state, or other local government, but that's as far as it gets. So how can you cause real action and change? The Watershed Management Group (WMG) was founded as a nonprofit to do just that; Joaquin Murrieta-Saldivar, a cultural ecologist at WMG, and his team look at the latest research and scientific data to help determine how it will actually get carried out on the ground.

WMG wrote a proposal based on Dr. Lara-Valencia's research and acquired $55,000 from BECC (border environmental cooperation commission) USD to conduct watershed management work in *Ambos Nogales*. There is an abundance of science and innovation, but a lot of times the problem is not coming up with the solution, but rather *implementing* the solution. Scientists have found solutions, but convincing officials and gaining access to implement change is another story. WMG uses the scientific data to benefit the practical aspects of watershed management.

Currently, Nogales, Sonora spends between five and seven million USD every year to manage the fallout of flooding and accumulation of sediment in the watershed. Dr. Murrieta proposed that they try to save some of that money by addressing the infrastructural problems causing the flooding. Though it would require some upfront investment, he projected that they would be able to save about 50% of that annual budget by predicting and preventing damage and losses. The project began in 2017 and was completed in 2018. As urbanization increases, with its numerous impervious surfaces, issues of erosion increase, leading to increased sediment in runoff water, it is the large amount of sediment that facilitates extreme flooding in Nogales, Sonora. Thus, there is a need to work on the urban environment in order to alleviate the amount of sediment arriving through the watershed. Highly developed urban areas, which are characterized by 75%–100% impervious surfaces, cause about 55% of precipitation to runoff as opposed to a rural or natural environment, in which up to 50% of precipitation infiltrates the ground (EPA 2014, 7). This contributes to higher temperatures in urban areas, as well as more drastic flood events. So the more urban expansion in Nogales, the more surface runoff there will be, and the more flooding problems they will experience, even with normal precipitation.

So what can be done about this? How can you maintain urban growth *and* prevent devastating flood events? One of the answers could be is Green Stormwater Infrastructure (GSI) planning and implementation. WMG defines Green Stormwater Infrastructure as "constructed features that use living, natural systems to provide environmental services, such as capturing, cleaning and infiltrating stormwater; creating wildlife habitat; shading and cooling streets and buildings; and calming traffic" (Watershed Management Group n.d.).

As WMG works to implement research-based, sustainable change, they try to do so in a way that increases the capacity of local management. This is the case for their work with Nogales, Sonora. WMG held a series of training events and conferences to better inform the community about GSI, and the majority of the green infrastructure development—the digging, trenching, planting, and building of swales and gabions—was performed by the community. They built a series of basins to decrease the speed of water and amount of erosional sediment moving into the city from the surrounding hillsides. The results inspired the initiation of a new green infrastructure law by the Congress of the State of Sonora: *Ley 254 de Ordenamiento Territorial y Desarrollo Urbano* (Law 254 of Land Use and Urban Development) and *Ley del Equilibrio Ecológico y Protección al Medio Ambiente* (Law of Ecological Balance and Environmental Protection). This is a big step in the right direction, because these laws create the legal framework for the municipalities to not only communicate and plan more effectively but also modify the corresponding related regulations to adapt to change, making *Ambos Nogales* that much more resilient.

Question

Audience Member: To what extent did IBWC/CILA help with the implementation of Green Stormwater Infrastructure?

Answer

Joaquin Murrieta: The IBWC has helped with other projects, though not with this particular project. Arizona and Sonora are in a unique position because there are several government mechanisms that help facilitate communication, including IBWC/CILA and the Arizona-Sonora Commission, which is something no other state has implemented. In addition to these, we also have a lot of scientific collaboration, which I think is somewhat unique to our region.

Potential Sustainable Solutions to Water Resources Management Challenges

Stephanie Buechler, an associate research professor at the School of Geography and Development and at the Udall Center for Studies in Public Policy at The University of Arizona, along with a binational team of professors from The University of Arizona until June 2021, now Associate Research Professor in Agricultural Sciences Global at the Pennsylvania State University, El Colegio de la Frontera Norte (COLEF) Nogales, COLEF Ciudad Juarez, and El Colegio de Chihuahua, recently completed a study evaluating the transborder impact of sanitation projects in *Ambos Nogales*, (García et al. 2019). The study ran from May 2018 to May 2019 and was included the application of surveys on both sides of the border with populations in Nogales, Sonora and Nogales, Arizona, as well as key informant interviews and an overall literature review of data pertaining to the NIWTP. As mentioned previously, the treated wastewater from the NIWTP in Rio Rico discharges into the Santa Cruz River, creating an economic boom for Arizona particularly with the revenue made from birdwatching due to the ecological restoration. Unfortunately, the Mexican side does not receive any economic bonuses because of the financial and logistical constraints required to pump the water back into Mexico after treatment.

The results from the binational survey found that only a little over half the population in Nogales, Sonora perceived environmental changes due to NIWTP were beneficial, whereas nearly 80% of Nogales, Arizona's population perceived beneficial impacts (García et al. 2019). So people on the U.S. side saw positive impacts, whereas their Mexican counterparts did not see the same benefits. However, in both cities, international cooperation was nearly unanimously considered necessary to protect the quantity and quality of water resources. The study also revealed that most respondents were not even aware that the binational WWTP existed prior to the study. The researchers conducting the survey recommended that education about the binational effort be emphasized in order to continue sustained collaborations.

The key informants interviewed for the project were individuals with a deep knowledge of the binational nature and border of the sanitation system of *Ambos Nogales* and the impacts generated. These key informants thus had deep knowledge of the situation before and after the beginning of the operation of this system, and included staff from governmental agencies, nonprofit organizations. This was necessary to obtain additional perspectives that be compared with the survey results that gauged public opinion. All informants thought that the sanitation infrastructure has generated positive impacts in the region. The interviews also uncovered issues that needed to be addressed:

- Improve communication between both countries.
- Mexico has not been able to meet financial obligations associated with operating and maintaining the NIWTP. (This seems to be a global problem with paying for operation after construction of the treatment plant is completed, as there is no willingness to continue to support the infrastructure once it has been built.)
- Improve the system of measuring and controlling local contaminants.
- Further investigate a way to get the treated water back to Nogales, Sonora so that they may receive the economic benefits.
- Consider further uses for treated wastewater in diverse sectors and/or for parks and gardens.

Based on their results, the researchers recommended a few ways to reduce costs and increase sustainability of the binational storm and wastewater management, including adding renewable energy to increase sustainability of energy sources and to lower treatment costs, as well as to develop constructed wetlands to better utilize stormwater.

In 2012, Nogales, Sonora opened the Los Alisos Wastewater Treatment Plant to try to stem some of the flow into the United States so that Sonora could use some of that water, with a capacity to treat five million gallons per day (OOMAPAS de Nogales 2014). Funded by NADBANK loans and grants, 902 kW solar panels were installed for its sole electricity source, making it the first WWTP in Latin America—and one of the first in the world—to be powered entirely by solar panels.

So what lessons have been learned from Los Alisos? Dr. Buechler found that 60% of costs to run the NIWTP are just for electricity. Thus, there is a very viable potential to install solar panels for the treatment plant to reduce costs and improve its economic performance, which is important because both sides of the border are behind in their payments for the NIWTP.

Other potential projects that were recommended by Dr. Buechler and Dr. Zuniga-Teran include creating constructed wetlands and hybrid by-ways. Constructed wetlands are made by creating a confined area with a barrier to prevent pollutants from infiltrating into the groundwater, and then using the treated wastewater to mimic natural wetlands. Aspects of this include incorporating green infrastructure, using gravel and plants to aid in filtration and uptake heavy metals, and it creates habitats for wildlife while restoring the local ecology. Dr. Zuniga-Teran suggested that in order to prevent flooding of the streets (that had once been streams but were paved over or diverted), hybrid by-ways, or dedicating one to two lanes to serving what they once were, as water runoff streams, complete with GSI traffic techniques and vegetation to decrease runoff, could be very effective. Utilizing these strategies creates economic and socio-cultural improvements merely by exploiting the natural courses of water that existed prior to the development of the city.

Question

Audience Member: Are people in Nogales paying any special taxes for the project?

Answer

Stephanie Buechler: Funding is through sewage fees, but no additional taxes are in place currently. There is the possibility for newer housing developments to have a tax due to area greening and increased property values.

Question

Audience Member: When recommending that the NIWTP implement solar panels, did you look at the upfront costs for installation and pay off?

Answer

Buechler: Absolutely, we did. We found that solar panels would pay for themselves relatively quickly, especially because they are now 20 times more efficient than they were in 2009, making the payoff come that much sooner. And the efficiency of panels, for example, is continually increasing with technological developments. NADBANK seems to be receptive to this recommendation, especially because they paid for the panels at the Los Alisos Wastewater Treatment Plant.

Question

Audience Member: Why did constructed wetlands not gain more traction? The idea has been around for over twenty years, and it seems like such a great idea, why has it not been implemented more?

Answer

Buechler: Labor is a major deterrent, particularly due to the cost to maintain the constructed wetlands. Landscape terrain can also be a deterrent, since the location cannot be too steep or in too close proximity to certain things. A key thing to remember is that we don't have to implement every one of these ideas, but that progress is made by even just doing one or two small projects gradually over time, so that you all at once, are decentralizing wastewater treatment somewhat, and improving the lives of residents on both sides of the border.

REFERENCES

Arizona Department of Environmental Quality. 2017. Arizona-Sonora Environmental Strategic Plan 2017–2021. https://static.azdeq.gov/pub/obep_strat_plan.pdf

Brady, L. M., Gray, F., Castaneda, M., Bultman, M., & Bolm, K. S. 2002. Preliminary United States – Mexico Border Watershed Analysis, Twin Cities Area of Nogales, Arizona And Nogales, Sonora. *U.S. Geological Survey Open-File Report, OF 02-112*. Retrieved from http://geopubs.wr.usgs.gov/open-file/of02-112/

EPA. 2014. *Reducing Urban Heat Islands: Compendium of Strategies – Urban Heat Island Basics*. www.epa.gov/sites/production/files/2014-06/documents/basicscompendium.pdf

García, R., C. Scott, A. Zúñiga, L. Cervera, S. Buechler, S. Peña, J. Muñan, and K. Martínez. 2019. Transboundary Impact Assessment of Wastewater Infrastructure Projects in Ambos Nogales (*Evaluación del Impacto Transfronterizo de Proyectos de Infraestructura de Saneamiento en Ambos Nogales*). Mexico: North American Development Bank.

Hanson, R.L. 1991. "Evapotranspiration and Droughts." In *National Water Summary 1988–89—Hydrologic Events and Floods and Droughts: U.S. Geological Survey Water-Supply Paper 2375*, compiled by Paulson, R.W., Chase, E.B., Roberts, R.S., and Moody, D.W. 99–104. Retrieved from https://geochange.er.usgs.gov/sw/changes/natural/et/.

Ingram, H., N. Laney, and D. Gillilan. 1995. *Divided Waters: Bridging the U.S.–Mexico Border*. Tucson: University of Arizona Press.

Ingram, H., Milich, L., and Varady, R.G. 1994. Managing Transboundary Resources: Lessons From Ambos Nogales. *Environment* 36(4). May 1994. www.researchgate.net/profile/Robert_Varady/publication/320741102_Environment_1994/links/59f8fd5d458515547c26aae1/Environment-1994.pdf

Ingram, H., and White, D. 1993. International Boundary and Water Commission: An Institutional Mismatch for Resolving Transboundary Water Problems. *Natural Resources Journal* 33(1): 153–175. Retrieved December 22, 2020, from www.jstor.org/stable/24884684.

McCombs, Brady. 2008. "Mexico Ties Flooding in Nogales to U.S. Border Patrol-Built Wall." *Arizona Daily Star*. https://tucson.com/news/local/border/mexico-ties-flooding-in-nogales-to-u-s-border-patrol-built-wall/article_a11265b0-17c3-5bed-b50a-72e5d17e1369.html

Norman, L. M., Caldeira, F., Callegary, J., Gray, F., O' Rourke, M. K., Meranza, V., & Van Rijn, S. 2012. "Socio-Environmental Health Analysis in Nogales, Sonora, Mexico." *Water Quality, Exposure, and Health* 4(2): 79–91. https://doi.org/10.1007/s12403-012-0067-x

Norman, L. M., Callegary, J., van Riper, C., & Gray, F. 2010. "The Border Environmental Health Initiative; Investigating the transboundary Santa Cruz watershed." *U.S. Geological Survey Fact Sheet, 2010-3097*, 2.

Norman, L. M. 2005. *Modeling Land Use Change and Associate Water Quality Impacts in the Ambos Nogales Watershed, United States-Mexico Border*. The University of Arizona. Retrieved from http://phdtree.org/pdf/25545994-modeling-land-use-change-and-associate-water-quality-impacts-in-the-ambos-nogales-watershed-united-states-mexico-border/

Norman, L., Guertin, D., & Feller, M. 2008. A Coupled Model Approach to Reduce Nonpoint-Source Pollution Resulting from Predicted Urban Growth: A Case Study in the *Ambos Nogales* Watershed. *Urban Geography*, 29(5), 496–516. https://doi.org/10.2747/0272-3638.29.5.496

Norman, L. M. 2007. United States-Mexican Border Watershed Assessment: Modeling Nonpoint Source Pollution in Ambos Nogales. *Journal of Borderlands Studies* 22(1): 79–97. https://doi.org/10.1080/08865655.2007.9695670

Norman, Laura M. 2010. "Flash-Flood Forecasting at Ambos Nogales." In the *Border Climate Summary*. Climate Assessment for the Southwest (CLIMAS) program, The University of Arizona, Tucson., https://climas.arizona.edu/sites/default/files/BorderClimateSummary_Jun10.pdf

Norman, L. M., Huth, H., Levick, L., Shea Burns, I., Phillip Guertin, D., Lara-Valencia, F., & Semmens, D. 2010. "Flood Hazard Awareness and Hydrologic Modelling at Ambos Nogales, United States–Mexico Border." *Journal of Flood Risk Management* 3(2): 151–165. http://doi.org/10.1111/j.1753-318X.2010.01066.x

Norman, L. M., Feller, M., & Phillip Guertin, D. 2009. Forecasting Urban Growth across the United States–Mexico Border. *Computers, Environment and Urban Systems 33*(2): 150–159. https://doi.org/10.1016/j.compenvurbsys.2008.10.003

Norman, Laura, Villarreal, Miguel, Niraula, Rewati, Meixner, Thomas, Frisvold, George, and Labiosa, William. 2013. Framing Scenarios of Binational Water Policy with a Tool to Visualize, Quantify and Valuate Changes in Ecosystem Services. *Water* 5: 852–874. DOI:10.3390/w5030852.

Norman, L. M. (2019). Watershed Modeling, Sediment and Flood Flows at Ambos Nogales. Science, Health, Eng Policy & Diplomacy Conference Sustainable Development Conference: Session "Hydrodiplomacy in Ambos Nogales: Lessons Learned from Binational Water Management," Tucson, Ariz.

Norman, L. M., Levick, L. R., Guertin, D. P., Callegary, J. B., Quintanar Guadarrama, J., Zulema Gil Anaya, C., Prichard, A., Gray, F., Castellanos, E., Tepezano, E., Huth, H., Vandervoet, P., Rodriguez, S., Nunez, J., Atwood, D., Patricio Olivero Granillo, G., & Octavio Gastelum Ceballos, F. (2010). Nogales flood detention study. U.S. Geological Survey Open-File Report, 2010–1262, 112. https://doi.org/10.3133/ofr20101262

Norman, L. M., Villarreal, M. L., Lara-Valencia, F., Yuan, Y., Nie, W., Wilson, S., Amaya, G., & Sleeter, R. (2012). Mapping socio-environmentally vulnerable populations access and exposure to ecosystem services at the U.S. -Mexico borderlands. *Applied Geography*, 34, 413–424. https://doi.org/10.1016/j.apgeog.2012.01.006

OOMAPAS de Nogales. 2014. *"Los Alisos" Wastewater Treatment and Conveyance Project Close-Out Fact Sheet*. Nogales, Sonora: NADBANK. Retrieved from www.nadb.org/uploads/files/601_nogales_los_alisos_eng.pdf.

U.S. Fish and Wildlife Service. 2011. "Chapter 10 Santa Cruz River Watershed." *Biological Assessment of the Arizona Game and Fish Department's Statewide and Urban Fisheries Stocking Program*. January 2011. www.fws.gov/southwest/federal_assistance/pdfs/chapter%2010%20santa%20cruz%20river%20watershed.pdf

Utilization of Waters of the Colorado and Tijuana Rivers and of the Rio Grande. U.S.-Mex. February 3, 1944. Treaty Series 994. Treaty Between the United States of America and Mexico. Washington: 1946. www.ibwc.gov/Files/1944Treaty.pdf

Watershed Management Group. n.d. Scaling up Green Stormwater Infrastructure (GSI). https://watershedmg.org/article/scaling-green-stormwater-infrastructure-gsi#:~:text=Green%20stormwater%20infrastructure%20(GSI)%20refers,and%20buildings%3B%20and%20calming%20traffic.

FORGING SISTER WETLANDS IN EL PASO, TEXAS AND CIUDAD JUÁREZ, MEXICO: URBAN TRANSBOUNDARY CONSERVATION AND RESTORATION ON THE U.S-MEXICO BORDER

Ana Córdova
Research Professor, Department of Urban and Environmental Studies, El Colegio de la Frontera Norte

Adriana Zúñiga-Terán
Assistant Research Scientist, School of Landscape Architecture and Planning and Udall Center for Studies in Public Policy, The University of Arizona

John Sproul
Program Coordinator/Manager, Center for Environmental Resource Management, University of Texas at El Paso

The Ciudad Juárez-El Paso metropolitan area is home to a human population of nearly two million. With the IBWC-CILA Rio Grande Rectification Project in the 1930s, the river was channelized in the El Paso-Juárez Valley, and its meanders and natural riparian areas were eliminated. With this engineering project, the ecosystem services provided by riparian areas practically vanished, while runoff peaks and sediment volumes increased. The Rio Bosque Wetlands Park (RBWP) in El Paso, Texas, with 372 acres, aims to "establish meaningful examples of native wetland and riparian habitats historically found along the Rio Grande in the El Paso region." Across the river, in Mexico, is one of the few remaining large open spaces in Ciudad Juárez, which presents a unique opportunity. Scientists in Ciudad Juárez and El Paso have been collaborating to create a sister wetland directly across the border from RBWP which could double the wetland/wildlife refuge surface area and boost many of the ecosystem services through ecological synergies that can benefit both communities.

In this paper we describe the collaboration that has taken place to date, current activities and steps forward, and how science can provide a basis for diplomacy in the international policy and political environment of the US-Mexico border, where it is sorely needed.

Introduction

Major challenges nowadays are transboundary in nature, and science can help address these challenges; therefore, there is a need to integrate science cooperation into international relations in what is known as "science diplomacy" (Turekian 2018). Turekian (2018, 5) defines science diplomacy as "the effort to leverage science engagement and exchange in support of broader objectives beyond science discovery", and states that this term implies both a new field of study and a formalized set of operations. Science diplomacy is usually driven by institutions, where a link to an institution is required (e.g. governments, private sector, universities, civil society) and the goals of science diplomacy necessarily fit into the objectives of the institutions involved. In this process, individual scientists are key protagonists, whether they are diplomatic or not (Turekian 2018). Similarly, Kaltofen and Acuto (2018b) define science diplomacy as twofold—(1) as "a set of relations between two or more actors that identify and/or are identified as representing distinct legitimate political entities" and these relations are fundamentally scientific; and (2) "the study of such phenomena" (Kaltofen & Acuto 2018b, 8). This way, science diplomacy is comparable to the field of study of international relations—"both practice and scholarship that unpacks that practice and where both inextricably intertwine but without agreeing what is and isn't part of the study" (Kaltofen and Acuto 2018b, 9).

Even though nowadays science diplomacy is at the core of important global challenges, there is limited understanding on the complexity of the relationships (Kaltofen & Acuto 2018a). Science diplomacy may not be enough to provide accurate perspectives of world politics; therefore, it is necessary to increase our understanding of the dynamics of negotiations (Kaltofen & Acuto 2018a). The purpose of this paper is to increase our understanding of the dynamics of negotiations of science diplomacy through the analysis of a case study in Ciudad Juárez–El Paso metropolitan area located on the U.S.-Mexico border.

Landscape Transformations in the Paso del Norte Region

The Ciudad Juárez–El Paso metropolitan area has experienced dramatic landscape transformations over time. These landscape transformations have been a consequence of permanent farming and irrigation structures practiced and built during the European settlement, which followed the establishment of the Guadalupe Mission in 1659. Since then, this area grew in size and importance because its strategic location served as a stop for resource supply for travelers along the Camino Real (which ran south–north and connected Mexico City with Chihuahua City and Santa Fe) and the travelers following the Gold Rush of 1849 (east–west). During that time, large parties of travelers, such as one reported to include 1,100 people, 300 wagons, 4,000 cattle heads, and 300 mules (Stotz 2000), stopped regularly in this area to replenish their supplies. In addition, the 20th century saw intense urbanization in this region, and the 1930s U.S.-Mexico river rectification project further dramatically altered the landscape.

As a result of a lack of instrumental data dating from the 17th, 18th, and 19th centuries, historians rely on written accounts of travelers that passed through the Guadalupe Mission and El Paso del Norte to have an idea of the landscape. According to Stotz (2000), Villagra (1933, 126–127) quotes a description of this area from 1598 as—"the pleasant shade of the wide spreading trees which grew along the river banks." Bartlett (1854) describes the type and density of trees that existed along the Rio Grande in a quote from 1845 that reads "thickly timbered with cottonwood… large size… cottonwood and the roots of the mesquite constitute the fuel of the country." Ruxton (1973, 168) provides an idea of the width of this riparian bosque (forest) in his 1846 account of the cottonwoods lining the river extending "a few hundred yards on each side of the banks." We know about the wetlands of this area from Gallegos (Hammond and Rey 1927, 23), as he describes the 1581 Rodriguez Expedition encountering, in the valley below modern El Paso, "a valley of swamps which extends over eight leagues," or about 21 miles (Stotz 2000). Finally, we know of the existence of the "Rio Bosque" (forested river) from the map by Augustus Kock from 1886 that shows a bird's eye view of El Paso.

Even though an international border has divided the northern and southern banks of the Rio Grande since 1848, the communities on both sides have developed a continuous urban area, with residents crossing the border freely at the beginning (Flores Simental et al. 2010), and with few restrictions and short wait times up until 9/11/2001. In 1875, the settlement north of the river took the name of El Paso and in 1888, the settlement south of the river, previously named Villa Paso del Norte since the 17th century, took the name of Ciudad Juárez after the Mexican president, Benito Juárez, who took refuge in this city during the French occupation of Mexico in 1865 and 1866 (Flores Simental et al. 2010). Today, this binational metropolitan area is home to roughly two million people.

Wetlands along the Rio Grande

Landscape transformations along the Rio Grande in Ciudad Juárez–El Paso have resulted in a destruction of natural ecosystems, as the river has been dammed, rectified, and channelized. The completion of Elephant Butte Dam upriver in New Mexico in 1916 effectively eliminated the high flows following spring snowmelt that were critical to sustaining historic wetland and riparian habitats in the Ciudad Juárez–El Paso region. The Rio Grande Rectification Project shortened the river from 155 miles to 86 miles between El Paso and Fort Quitman as a way to maintain the location of U.S.-Mexico border along an otherwise meandering and changing riverbed. This binational rectification project was authorized in 1933 and completed in 1938.

In 1973, the U.S. Department of the Interior conveyed 276.92 acres located along the Rio Grande in southeast El Paso to the City of El Paso under the Federal Lands to Parks Program. This parcel is adjacent to one of the old river meanders cut off by the Rectification Project. With its transfer to the City, it became Rio Bosque Park. The City added another 95 acres in later years to bring the park to its current size of 372 acres.

In 1990, the U.S. Congress authorized the American Canal Extension project, a project to upgrade the existing American Canal in El Paso and extend it 12.1 miles downriver, paralleling the Rio Grande. Congress placed responsibility for building this project with the U.S. Department of State, through the U.S. Section of the International Boundary and Water Commission (USIBWC). Subsequent environmental studies determined that the project would create local habitat loss, and in 1993, the USIBWC agreed to mitigate that loss with the creation of 30 acres of shallow-water emergent wetland. The USIBWC and the U.S. Fish and Wildlife Service (USFWS) (2011) had identified Rio Bosque Park as the preferred site for this mitigation project (Figures 6.2 and 6.3).

The prospect of a 30-acre mitigation wetland at Rio Bosque Park stimulated discussion in El Paso about the possibility of making that wetland the foundation for a larger wetland project at the 372-acre park. There was considerable enthusiasm for this idea, and a diverse group of partners coalesced around it. The plan that ultimately emerged included re-building the old river channel where it wound through the park, excavating a series of shallow basins that could be flooded by diverting water from the old river channel, and using treated effluent from a WWTP immediately north of

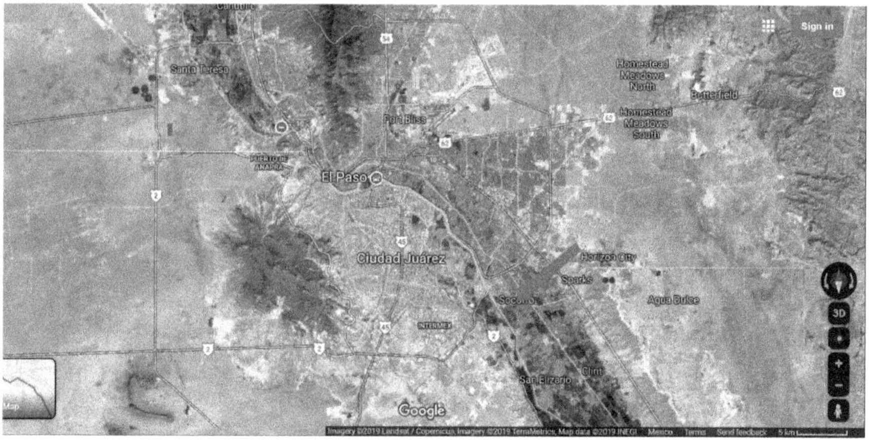

FIGURE 6.2 Location of the preferred site for a restored wetland. (Adapted from Google Maps.)

FIGURE 6.3 Location of Rio Bosque Wetlands Park. (Adapted from Google Maps.)

the park to sustain the wetlands. The University of Texas at El Paso (UTEP) agreed to manage the site.

In 1997, the USIBWC did the earth-moving and other work needed to prepare the site, and the park's name changed to Rio Bosque Wetlands Park (Figure 6.4). The

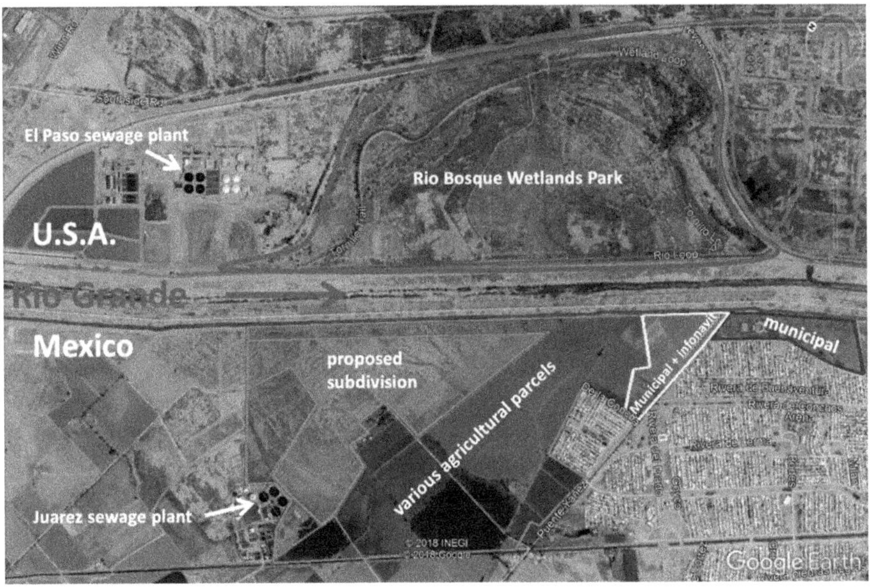

FIGURE 6.4 Location of the potential sister wetlands in El Paso and Ciudad Juárez. (Adapted from Google Maps.)

overarching goal of RBWP is "to recreate the mosaic of habitats characteristic of the Rio Grande and its floodplain in pre-settlement days" (City of El Paso & University of Texas at El Paso 2002, 21), specifically, the river's historic riparian vegetation (cottonwood-willow bosque), native upland habitat, and moist-soil habitat for migrating waterfowl. Overall the park aims to act as a wildlife refuge, provide public space for hiking and biking as well as important opportunities for education and research.

In the years immediately following 1997, water from the WWTP was consistently available to the park only in late fall and early winter. Up to 100 acres could be flooded, attracting large numbers of migrating and wintering water birds. The rest of the year, including throughout the growing season, little or no water was available. During those intervals, all treated effluent from the WWTP was dedicated to agriculture in the El Paso Valley. As a result, for over a decade, progress establishing wetland and riparian vegetation at RBWP was limited.

Since 2009, a number of projects have come to fruition that, in combination, have made some water available to the site throughout the year. The sources include treated effluent from the adjacent WWTP, groundwater (two electric-powered wells, two windmills), and irrigation water from the Rio Grande. The more consistent water availability during the growing season has greatly accelerated the restoration of wetland and riparian habitats at the site.

Exactly across the Rio Grande from RBWP, there is still some open space in Ciudad Juárez, Mexico which could potentially be used for a similar ecological/wetland restoration purpose. This is a unique and time-sensitive opportunity because Ciudad Juárez is very densely populated and continues to urbanize very rapidly. There are few remaining open space parcels that could be transformed into greenspace. In addition, at this site, there is also a Mexican WWTP that could provide effluent to this specific area. At RBWP, the amount of water currently available varies each year but is on the order of 6,500 acre-feet (3,500 acre-feet in the irrigation season, 3,000 acre-feet in the non-irrigation season). This amount represents less than 20% of the Mexican effluent, which is currently not committed to any use. Thus, it would be possible to obtain enough water to sustain ecological restoration in a Mexican sister wetland. However, the land on the Mexican side is not public nor dedicated to restoration. Thus, on the U.S. side, RBWP had publicly owned open space available for a restoration project but securing enough water to sustain wetland and riparian habitats presented a significant challenge, whereas the Mexican side has sufficient water available, but land acquisition presents the most significant challenge (Figure 6.4).

Land tenure on the Mexican side is diverse. We know that there is a proposed housing subdivision and there are various agricultural parcels that still operate as such. One of the agricultural fields is flooded several months of the year, functioning in effect as a small wetland already in place. Finally, land tenure in this area also includes municipal land and some Infonavit (federal housing) land. Securing land in this area would require not only having the funds to purchase the land but negotiating with the landowners, identifying which organization could manage the land, and engaging with neighboring communities.

Stakeholder Engagement

Although the challenges associated with creating a sister wetland in Ciudad Juárez are somewhat different from those that faced the wetland project at RBWP, the RBWP project provides a useful model for achieving a successful outcome. RBWP exists because of the partnerships established between several agencies and organizations. El Paso Water–Public Service Board administers the park and cooperates to provide water; El Paso County Water Improvement District #1 also cooperates to provide water to the park; UTEP manages the park under an agreement with the Public Service Board. The USIBWC built the wetland complex, and the U.S. Bureau of Reclamation was active in the initial planning of the wetlands park and provides ongoing restoration support. Local and national NGOs involved in this project include Ducks Unlimited, who provided the project design, initial management plan, and water-control structures; and the Friends of the Rio Bosque, who have been involved in all aspects of the project.

The experience in El Paso suggests that a network of actors and organizations is needed to make things happen. All organizations that have had a role in the development of the park are interested in some aspect of the project that aligns with the goals of the organization. The level of involvement also corresponds to their capacity. The sum of all efforts is what makes projects like this one possible.

Stakeholder engagement for the sister wetland in Ciudad Juárez is equally needed to make the idea a reality. So far, potential project partners include academic institutions along the U.S.-Mexico border, such as El Colegio de la Frontera Norte (El Colef),

UTEP, The University of Arizona (UA), Western New Mexico University (WNMU), New Mexico State University (NMSU), and Universidad Autónoma de Ciudad Juárez (UACJ). Conservation organizations that could be involved in this project include ProNatura, The Nature Conservancy Mexico, Cuenca Los Ojos, Friends of the Rio Bosque, and Ducks Unlimited. Governmental organizations that may be interested in collaborating include the Municipio of Ciudad Juárez, the State Government of Chihuahua, the local water utility (Junta Municipal de Agua y Saneamiento—JMAS), Comisión Internacional de Límites de Agua (CILA, the Mexican Section of IBWC), Mexico's Federal Protected Areas Agency (Comisión Nacional de Areas Naturales Protegidas—CONANP), and the National Commission for Biodiversity (Comisión Nacional para el Conocimiento y Uso de la Biodiversidad—CONABIO). In addition, because the expansion of restored wetland and other habitat types right next to RBWP will likely result in increases of wildlife and possibly improvements in the local hydrology, we believe some U.S. governmental organizations may be interested in this project, including the USFWS, the USIBWC, and the U.S. Environmental Protection Agency. Finally, we see potential collaboration and interest from private foundations in both countries.

Next Steps

The next steps of this project include raising funds to secure the land in Ciudad Juárez right across from RBWP. Also, there is a need to secure water for the wetlands. In addition to securing water and land, there is a need to identify the organization that would be responsible of managing the park. This organization may or may not be the actual landowner. Further, there is a need to develop a plan for the ecological restoration of the site so that historical habitat is recreated, while it provides recreational, educational, and research opportunities. Finally, next steps include stakeholder engagement so that collaboration between organizations may happen.

Science Diplomacy

There are important research questions that may attract scientists to this project and involve diplomacy between the two countries. The scientific collaboration activities include sharing inventory, research, outreach plans, protocols, and plant material between the RBWP and the new sister wetland, as well as groundwater monitoring as water table depth changes with new sources of water. Potential ecological and water quality research questions include:

1. How fast can the sister wetland establish native habitats with higher quantities of water?
2. Does increasing habitat area increase populations, biodiversity, prospects of species prone to localized extinctions?
3. What additional treatment can the wetlands provide to the water released from WWTPs on either side?

The border between the United States and Mexico, and in particular the presence and continued construction of the border wall by the United States, has been a highly politicized issue and one where diplomacy is sorely needed (De La Parra & Córdova 2012). This wall has wide-ranging impacts on the shared natural environment of both countries (Córdova & De La Parra 2007). In this context, a localized, binational effort to restore the habitat along the Rio Grande, which is an important and critical ecological element in the Chihuahuan Desert (Stotz 2000) and which failed to reach the Gulf of Mexico in 2001 ("Rio Grande—New Mexico, Texas," n.d.), presents an opportunity to strengthen diplomacy while also bolstering ecological restoration and historic reconstruction. Should the project be successful and there be an increase in terrestrial wildlife populations, it is possible to envision negotiations for localized and monitored openings in the border wall for the transboundary movement of small wildlife. And this could represent an opening for other sorts of constructive actions along and surrounding the border wall in general.

Conclusion

The idea of creating a sister wetland in Ciudad Juárez mirroring the one in El Paso comes from the historical landscape where a single bio-social environment has been deeply altered by an array of human activities and marked by a political boundary. This is a unique opportunity to re-establish historical native habitats on both sides of the Rio Grande, directly opposite each other. The uniqueness is not only spatial, but also due to water availability from the effluent of WWTPs to sustain habitats, including wetlands. This project entails binational interests because it would benefit both sides of the border and has the potential to enhance the efforts at the RBWP. Finally, this project is a concrete example of science spearheading transboundary ecological restoration and paving the way for diplomacy.

REFERENCES

Bartlett, J. 1965. *Personal Narrative of Explorations and Incidents in Texas, New Mexico, California, Sonora, and Chihuahua, connected with the United States Mexican Boundary Commission, During the Years 1850, 1851, 1852, and 1853* (Vol. 1). New York: D. Appleton & Company.

City of El Paso, & University of Texas at El Paso. 2002. *A Biological Management Plan for Rio Bosque Wetlands Park*. https://research.utep.edu/Portals/72/Rio%20Bosque/Documents/Rio%20Bosque%20Wetlands%20Park%20Biological%20Management%20Plan.pdf

Córdova, A., & De La Parra, C., eds. 2007. *A Barrier to Our Shared Environment. The Border Fence between the United States and Mexico*. Secretaría de Medio Ambiente y Recursos Naturales. Instituto Nacional de Ecología, El Colegio de la Frontera Norte, Southwest Consortium for Environmental Research and Policy.

De La Parra, C., & Córdova, A. 2012. Chapter 1: Política de muros y los espacios, instumentos y actores olivdados. In A. Córdova & C. De La Parra (Eds.), *El muro fronterizo entre México y Estados Unidos: Espacios, instrumentos y actores para un diálogo constructivo* (pp. 23–46). Tijuana, Baja California: Colegio de la Frontera Norte.

Flores Simental, R., Gutiérrez Roa, E., & Vázquez Reyes, O. 2010. *Paso del Norte en el siglo XXI: breve historia de Ciudad Juárez* (Second). Universidad Autónoma de Ciudad Juárez.

Hammond, G.P., and Rey, Agapito (Eds.). 1927. The Gallegos relation of the Rodriguez expedition to New Mexico. In *Historical Society of N. Mex. Pubs. in History*, Vol. IV, Dec. 1927 (pp. 23). New Mexico: El Palacio Press.
Kaltofen, C., & Acuto, M. 2018 (a). Rebalancing the Encounter between Science Diplomacy and International Relations Theory. *Global Policy*, 9, 15–22. https://doi.org/10.1111/1758-5899.12620
Kaltofen, C., & Acuto, M. 2018 (b). Science Diplomacy: Introduction to a Boundary Problem. *Global Policy* 9: 8–14. https://doi.org/10.1111/1758-5899.12621
Rio Grande—New Mexico, Texas. n.d. *American Rivers*. www.americanrivers.org/river/rio-grande-river/
Ruxton, G. 1973. *Adventures in Mexico and the Rocky Mountains*. Glorieta, NM: Rio Grande Press.
Stotz, N. G. 2000. *Historic Reconstruction of the Ecology of the Rio Grande/Río Bravo Channel and Floodplain in the Chihuahuan Desert* (p. 158). Report prepared for Chihuahuan Desert Program, World Wildlife Fund.
Turekian, V. 2018. "The Evolution of Science Diplomacy." *Global Policy* 9, 5–7. https://doi.org/10.1111/1758-5899.12622
Villagra, G. 1933. *History of New Mexico*. Los Angeles: The Quivira Society.

A MODEL FOR TRANSBOUNDARY WATER COOPERATION: PERSPECTIVES FROM A SCIENCE DIPLOMAT

Katherine E. Himes
James A. and Louise McClure Center for Public Policy Research, University of Idaho

Introduction

Water security is a major challenge facing many regions of the world. Tensions over shared water resources often intersect with national security priorities and connect to a wider set of economic, social, and geopolitical issues. At the international level, such tensions can increase the risk of geopolitical instability. Science diplomacy—water (hydro) diplomacy specifically—enables countries to negotiate agreements on the allocation and management of international watercourses. This dynamic process aims to develop durable solutions to water allocation and management and support or enhance regional cooperation and collaboration, shifting water from a source of tension to a "blue peace."

This piece aims to showcase technical and strategic aspects of science diplomacy through a personal narrative. By sharing these details, I hope the reader will cultivate a deeper understanding of science diplomacy, its formation, and applications.

Connecting Science Policy with Diplomacy

From 2011 to 2015, I served as an American Association for the Advancement of Science (AAAS) Science & Technology Policy Fellow (STPF) at the United States Agency for International Development (USAID). After two years in the Office of Science & Technology, based in Washington, D.C., I accepted an unique opportunity to join the USAID Foreign Service in a term position as an AAAS Overseas STPF.

The AAAS Overseas Fellowship represents an unique approach to international development and diplomacy. Scientists and engineers with advanced degrees are placed in USAID missions or U.S. Department of State embassies, and support all three facets of science diplomacy—science for diplomacy, diplomacy for science, and science in diplomacy (Turekian and Neureiter 2012). Overseas fellows acquire firsthand about policymaking, international development, and diplomacy, and contribute technical knowledge and analytical skills in these arenas. Each mission or embassy utilizes their fellow in a different manner. Portfolios include duties such as managing international research programs, designing and supporting programs, leading monitoring and evaluation activities, and writing talking points, speeches, press releases, and reports.

My expertise in science, government, international development, and diplomacy filled a critical need at the USAID Mission to Central Asia, a regional mission that works in five former republics of the Soviet Union—Kazakhstan, the Kyrgyz Republic, Tajikistan, Turkmenistan, and Uzbekistan. In November 2013, I departed for my post in the former capital city, Almaty, Kazakhstan.

Starting Work in Almaty, Kazakhstan

After several weeks of immersing myself in the mission's work, learning more about the history of the region; meeting with colleagues from USAID, The World Bank, other missions, embassies, and local universities; and studying Cyrillic, I met with mission leadership to discuss the possible major areas of my portfolio.

Mission leadership had identified the need for design and launch of a comprehensive water program that could facilitate transboundary water cooperation across the five countries of Central Asia and neighboring Afghanistan. As the conversation shifted to scope and program details, I thought about the significance of this work, and was eager to capitalize on water-related science policy and science diplomacy efforts I led while in Washington.

Mission leadership shared salient points: USAID supported a water program in Central Asia previously. That program did not meet expectations, so regional sentiments with respect to USAID and water were not exceptionally favorable. The mission hoped to build on a recently launched, fledgling water program that supported local water management councils. That said, transboundary water cooperation between Central Asia and Afghanistan was a new idea, and would have to be well-socialized as part of program development. In addition, international donor efforts related to water in the region were not well-coordinated. Mission leadership sought to develop a well-received water program, with high impact, a small budget, tight coordination with other donors, and emphasis on local capacity-building in less than two years. I was asked to guide this work.

Not long into my tenure at the mission, I learned that the word "water" was like a lightning rod across the region. For instance, diplomats and scientists attended conferences, workshops, and planning sessions with titles such as "The Central Asia Water-Energy Nexus"; "Energy, Water, Food, and the Environment in Central Asia"; and "The Central Asia Climate Change Summit." Attendance often was beyond capacity. Government officials did not meet to talk solely about water. The term hydro-politics provided an apt description.

The next five weeks were filled with conceptualizing possible elements of a transboundary water cooperation program. I spoke with several locally employed USAID colleagues with a combined 50 years of experience in Central Asia and Afghanistan, and sought more information about regional water challenges and past work.

With a growing body of understanding from technical journal articles, government cables, information gathered from other U.S. Government agencies, details about other donor programs, and a list of Central Asia and Afghanistan water priorities, I proposed to mission leadership that a white paper capturing this major research effort would be of value to the U.S. Government. I combed through myriad databases, asking the USAID Knowledge Center (library) for information on past USAID programs, talked with water experts at local universities, met with government leaders, and reached out to other U.S. Government colleagues for additional details.

Water Security in Central Asia and Afghanistan

Many scholars have studied the great rivers of Central Asia and Afghanistan, such as the Amu Darya, Syr Darya, and Pyanj. It was clear from research—my own and others (in academia, government, and business)—that in the coming decades, water management challenges across Central Asia and Afghanistan will grow (Majority Staff Report 2011; Michel 2016; Oregon State University 2020; Putz 2016; Qobil 2016; Trilling 2016). These include intensifying water shortages, escalating demand for hydropower, decreasing water quality, increasing flood risk, and a changing climate. When coupled with dated, inadequate (or no) water-sharing treaties, poverty, social tensions, environmental degradation, and a limited number of capable scientific and education institutions, the risk of instability increases.

The current dynamic reflects the Soviet system: the upstream countries (Kyrgyz Republic, and Tajikistan) and downstream countries (Kazakhstan, Turkmenistan, and Uzbekistan) participated in water-energy sharing arrangements led by Moscow. Many of these agreements dissolved with the collapse of the Soviet Union.

Numerous stakeholders have interests in water across the region, including government policymakers and ministers, international water organizations, international donors, universities and water researchers, technically skilled water resource personnel, future government and academic leaders, and end-users—farmers, households, women and youth, local water basin councils, and water user associations. Interaction and coordination among these stakeholders, especially in a regional context, had been very limited.

Drafts of the white paper were shared with leaders in various U.S. Government agencies. Similar to the scientific peer review process, I incorporated feedback, recirculated revised versions, and (at last) sent a final version of this internal research report to mission leadership.

Designing the Water Program

With mission leadership understanding the complexities of water challenges facing Central Asia and Afghanistan, I proposed that the water program design team should be multi-faceted, with team members bringing multiple skills, perspectives, and experiences to bear. Rather than design a program with two or three staff, in order to create a truly novel USAID water program, which would build local capacity and have staying power (durability), five staff should be included. I recommended my colleague, the locally employed USAID staff member with nearly two decades' experience on related programs, a USAID global water specialist, an AAAS STPF and water engineer, and a USAID conflict mitigation expert. After careful reflection, mission leadership agreed with this approach. We moved forward with trip planning.

For three weeks in September, the five-member water program design team traveled across Central Asia, gathering information by visiting current USAID project sites; meeting with universities, governments, donor organizations, and others; and analyzing best approaches for addressing water challenges in Central Asia and Afghanistan. We sought to learn and see firsthand which programs and approaches were going well, how current efforts could be expanded, and what new elements could be incorporated.

We knew that less attention had been paid to smaller watercourses that connect to the Amu Darya, Syr Darya, and Pyanj Rivers. When planning our trip, we selected locations that would allow us to meet with local water management leaders, whose boots-on-the-ground translated into a wealth of knowledge.

While driving across Tajikistan to meet with these leaders located in a very remote part of the country, we passed through "The Tunnel of Death." The road from Dushanbe, the capital of Tajikistan, to Khujand, the second largest city in the country, includes two mountain passes over 10,000 feet. Travel to Khujand was limited, as Uzbekistan closed the border with Tajikistan, a consequence of diplomatic disputes. The five kilometer long tunnel affords the fastest way through the mountains, as avalanche danger and Anzob Pass are avoided. Funded by Iranian investment, the tunnel had yet to be finished. Traffic was allowed to pass despite significant internal hazards such as deep standing water, no air circulation, and live electrical wire. The program design team felt strongly that these travel risks should not impede opportunities to strengthen transboundary water cooperation by understanding local knowledge and perspectives held by water management leaders.

In Khujand, we met with a small basin water management council started by USAID's current water program, which facilitates cooperation across country boundaries on water management. This small basin council is comprised of local water association members, farmers, community leaders, and others from the Tajik side of the Isfara River border. This council works with a similar council across the river in the Kyrgyz Republic.

In the Kyrgyz Republic, we toured the current USAID project on the Aspara River, seeing firsthand the impact of water monitoring equipment installation on the Kazakh and Kyrgyz villages spanning the river. This locally produced water monitor made possible the enforcement of a 1948 law that called for shared allocation of water across the border in a 38%/62% manner. We also learned that the basin councils had started to work together on sharing water use data and reducing overall water consumption.

Many Waters: Three Legs of a Stool

The program design team proposed a multi-pronged approach to transboundary water cooperation in Central Asia and Afghanistan (Himes 2017). We envisioned our recommendations to be three legs of a stool, each with equal weight.

The first leg of the stool: Expanding the fledgling USAID water program was a top priority; supporting local water governance through transboundary basin councils on small watersheds provided a platform for water dialogue at the local level. By engaging all sectors, uses, and different players, water could be transformed into a source of cooperation, rather than a major reason for conflict. We recommended that USAID continue to utilize the Regional Environmental Centre for Central Asia (CAREC) to build local governance capacity. Specifically, this local implementing partner—with offices in each of the Central Asian countries—should expand training for small basin councils on planning and strategy setting, leading to joint basin planning and vision forming for the basin. CAREC focused on a bottom-up approach that incorporates local needs and challenges, builds planning skills, emphasizes low cost solutions, and places all options on the table.

The second leg of the stool: Strengthening academic research on myriad aspects of water within the region was essential. We proposed that a special call for water research in Central Asia and Afghanistan be incorporated into the USAID Partnerships for Enhanced Engagement in Research (PEER) program.

This competitive, peer-reviewed award program invites scientists and engineers in developing countries to apply for multi-year grants that foster research and capacity-building activities on topics of importance to USAID. The PEER program works closely with the U.S. science, technical, and engineering community, including many federal agencies such as NASA, NIFA, NIH, NOAA, NSF, USDA, USFS, and USGS. The U.S. National Academies of Sciences, Engineering, and Medicine is an important partner, and manages the peer-reviewed grant proposal process and monitors progress of the multi-year grants, among other activities.

Through the PEER program, USAID leverages scientific and engineering research investments that federal agencies already have committed, and advances development priorities. From the start of the application process, developing country researchers are required to partner with a researcher that has received funding from one of the aforementioned federal agencies. Funds can be used to support laboratory equipment, research materials, graduate students, and more. Both the PEER applicant and the U.S. Government-supported researcher benefit from one another's expertise, and partner on truly collaborative research. Since its launch in 2011, PEER has awarded nearly $100 million to more than 350 projects in 50 countries.

The third leg of the stool: Joining a major multilateral diplomatic effort—think science diplomacy with a capital 'S' and capital 'D'—would amplify USAID's programmatic impact and financial investment. Moreover, this approach would facilitate donor coordination. The design team recommended that the U.S. Government join The World Bank's multi-donor trust fund, the Central Asia Energy-Water Development Program (CAEWDP). This effort, already forming, focused on energy, water, and the energy-water nexus across Central Asia and Afghanistan. The World Bank, European Union, Swiss Development Corporation, European Union, and UK Agency for International Development (DFID) already had joined.

The program's budget reflected the spirit of cooperation—multiple USAID missions funded the local governance capacity-building deliverables; the USAID Mission to Afghanistan supported a regional call for PEER proposals focused on water in Afghanistan, Central Asia, India, and Pakistan (U.S. National Academies 2020a), and the U.S. Department of State contributed to CAEWDP.

Science Diplomacy in Action

As planning for Many Waters concluded, and program implementation began, the focus of my portfolio shifted significantly. I traveled to support the three dimensions of Many Waters, as well as the three aspects of science diplomacy.

The first trip was to Brussels, in November, shortly after program design concluded. The major international donors who had joined CAEWDP convened to learn more about energy, water, and energy-water projects at national and subnational levels across the region, and discuss how CAEWDP donor participation could be expanded. Our conversations emphasized the importance of *science for diplomacy*—how scientific cooperation can improve international relations. During this trip, I also met with USAID staff at the U.S. Mission to the European Union, sharing details about Many Waters and the CAEWDP program in particular.

During the winter and spring, I met regularly with CAREC at their Almaty headquarters to learn about plans for expanding small basin councils. We discussed concepts such as multi-council once-yearly convenings, further training for CAREC staff (training of trainers), and additional meetings with federal water ministers and agencies. My USAID colleague (member of the water program design team) and I emphasized how local scientific and technical knowledge could advance each country's foreign policy objectives, supporting *science in diplomacy*.

The following August, I traveled to Kabul, Afghanistan and organized a PEER awardee reception at Embassy Kabul. USAID leadership honored the two Afghanistan PEER awardees—both female—and described expectations for participation in the program. These courageous women presented their proposed work, and impressed the Afghanistan water ministry representative with their project goals. During that same trip, I held a second event—a PEER proposal application writing workshop—for interested scientists and engineers. Over 30 scientists and engineers from across Afghanistan packed the embassy room to learn important pieces such as how to find a U.S. partner, essential elements in the research plan, and desired study topics. I shared with mission and embassy leadership the ways in which science brings a new lens to diplomacy. Through the PEER program, I highlighted the opportunities afforded by science in diplomacy and science for diplomacy, and showcased how diplomacy can facilitate international scientific cooperation—*diplomacy for science*. Upon returning to Almaty, my USAID colleagues and I began planning for the first regional PEER water conference, an important opportunity for PEER awardees to build collaborative relationships and present research findings (U.S. National Academies 2020b).

Toward a Model for Blue Peace

In Central Asia and Afghanistan, USAID advanced “Smart Waters”, a three-pronged regional program, which serves as a model for fostering science diplomacy—water diplomacy in particular. This approach focused on three key levels of engagement: the local, national, and transboundary (regional).

Building local governance capacity through the creation and training of transboundary water councils creates avenues for collaboration. Supporting regional and international water research through the PEER program strengthens connections among universities, governments, and communities. Participating in multi-institutional diplomatic efforts that span the entire region (and are coordinated across international donor efforts) encourages transboundary, basin-level engagement.

This model for blue peace can be applied to other regions of the world, and can be strengthened by viewing water as part of the broader energy-environment-food-water nexus and coordinating with host government strategies. These approaches have the potential to transform water from a source of tension and security risk to water-based solutions.

REFERENCES

Himes, Katherine E. 2017. “Promoting Water Security in Central Asia through International Research Partnerships.” *Seton Hall Journal of Diplomacy and International Relations* 18, no. 1 (Spring 2017). http://blogs.shu.edu/journalofdiplomacy/files/2018/04/Promoting-Water-Security-in-Central-Asia.pdf.

Majority Staff Report. 2011. “Avoiding Water Wars: Water Scarcity and Central Asia’s Growing Importance for Stability in Afghanistan and Pakistan.” United States Senate Foreign Relations Committee. February 2011. www.foreign.senate.gov/publications/download/s-prt-112-10-avoiding-water-wars-water-scarcity-and-centrals-asias-growing-importance-for-stability-in-afghanistan-and-pakistan.

Michel, Casey. 2016. “Tajikistan’s Rogun Dam Rankles Uzbekistan.” *The Diplomat*. July 2, 2016. http://thediplomat.com/2016/07/tajikistans-rogun-dam-rankles-uzbekistan/.

Oregon State University. 2020. “International Freshwater Treaties Database.” Program in Water Conflict Management and Transformation. Last modified November 2020. https://transboundarywaters.science.oregonstate.edu/content/international-fresh water-treaties-database.

Putz, Catherine. 2016. “What Can Be Done about Central Asia’s Water and Electricity Woes.” *The Diplomat*. October 27, 2016. http://thediplomat.com/2016/10/what-can-be-done-about-central-asias-water-and-electricitywoes/.

Qobil, Rustam. 2016. “Will Central Asia Fight over Water?” *BBC News*. October 25, 2016. www.bbc.com/news/magazine-37755985.

Trilling, David. 2016. “Water Wars in Central Asia.” *Foreign Affairs*. August 24, 2016. www.foreignaffairs.com/gallerys/2016-08-24/water-wars-central-asia.

Turekian, Vaughan C., and Neureiter, Norman P. 2012. “Science and Diplomacy: The Past as Prologue.” *Science & Diplomacy* 1, no. 1 (March 2012): www.sciencediplomacy.org/sites/default/files/science_and_diplomacy.pdf.

U.S. National Academies of Sciences, Engineering, and Medicine. 2020(a). "PEER Call for Proposals: Central and South Central Asia/Transboundary Water Research." Last modified November 2020. https://sites.nationalacademies.org/pga/peer/pga_152142.

U.S. National Academies of Sciences, Engineering, and Medicine. 2020(b). "PEER Central Asia Forum." Last modified November 2020. https://sites.nationalacademies.org/PGA/PEER/PGA_190292.

7 Science Diplomacy in Education

DOI: 10.1201/9781003220503-7

SUMMARY

This chapter is a summary of the Science Diplomacy Education Panel Discussion from the conference, chaired by Andrea K. Gerlak, with panelists Hassan A. Vafai, Daniel Ferguson, and Léna Salamé. They discuss the process that The University of Arizona has undertaken to build a science diplomacy program on campus, developing a core course examining science diplomacy, and the development of other related transdisciplinary efforts on campus. Salamé explains the efforts of the Universities' Partnership for Water Cooperation and Diplomacy (UPWCD) as a growing network for strengthening educational and training programs such as these. This is followed by a Q&A with the audience.

Next are three articles that were presented at the conference and subsequently reviewed by a scientific committee to be included in the proceedings: "The Importance of Marginalized Knowledges" by Alberto Arenas, "Expanding Environmental Science Literacy to Include Cultural and Community Knowledge" by Kristin L. Gunckel, and "Environmental Justice Promotion through Experiential Outdoor Education on the River—Part 2: HEAL (Human Rights, Environmental Education, Activism, and Learning) (Abbreviated Version for Publication)" by Dakota Goodman.

SCIENCE DIPLOMACY EDUCATION PANEL DISCUSSION

Chair
Andrea K. Gerlak
Associate Research Professor, Udall Center for Studies in Public Policy, and the School of Geography, Development and Environment, The University of Arizona

Panelists
Hassan A. Vafai
Research Professor, Civil and Architectural Engineering and Mechanics, The University of Arizona

Daniel Ferguson
Director, Climate Assessment for the Southwest, The University of Arizona

Léna Salamé
Lawyer and Conflict Management Specialist; Secretariat of the Global High-Level Panel on Water and Peace, Geneva Water Hub

Building Science Diplomacy at The University of Arizona

Like most university faculty members these days, faculty at The University of Arizona (UArizona) have to wear a lot of hats. As Professor Andrea Gerlak put it, some hats are more fun to wear than others, but as a general rule, faculty need to work in the areas of teaching, research, and administration. She says that her work on science diplomacy hits a sweet spot right in the middle of those aspects: she teaches undergraduate and graduate courses in environmental policy; she helps administer a unique science communication training scholarship program called the Rachel Carson Scholar Program, which is named after Rachel Carson, the "mother of the environmental movement"; and, for two decades now, she has been researching water governance, especially transboundary water issues. Most recently, for the last two years, Dr. Gerlak has been involved with other colleagues on campus to build a program around science diplomacy, and for the last few months, she has been looking specifically at the curriculum—What can we do with science diplomacy curriculum here at The University of Arizona?

The efforts in science diplomacy (SD) really started on the UArizona campus with experiences of hosting groups of Iranian scientists with the U.S. National Academies of Sciences on issues like air pollution in megacities, resilience, wetlands, and climate change and mitigation. A lot of excitement was generated from those interactions and meetings for creating something more on campus with SD. Early on, different organizations from across the campus came together under this umbrella of science, health, and engineering policy and diplomacy, what has now become known as "SPDI" after many variations of acronyms. Those involved in the initiative have been working under this umbrella term, trying to describe and delineate how scientists and policymakers can work collaboratively to enable, advise, and build confidence in science and technology issues and international affairs.

One of the early efforts was the first conference in 2017, which brought leading people in SD onto campus to talk about issues of SD in the Americas. There were a lot of folks from Latin America that joined that conference, including past science advisors to secretaries of state, the ambassador of Costa Rica to the United States, and a lot of leaders in the field; it was deemed a success. Excitement continued to grow along with these efforts, so a speaker series was initiated to invite science diplomats from all industries to campus. The speakers who have come to campus have talked about a vast range of topics, including environmental challenges, water issues, health policy, foreign policy, etc.

The approach to SD at UArizona is based on the National Research Council (NRC) definition of SD, and it builds on and draws from efforts to define SD

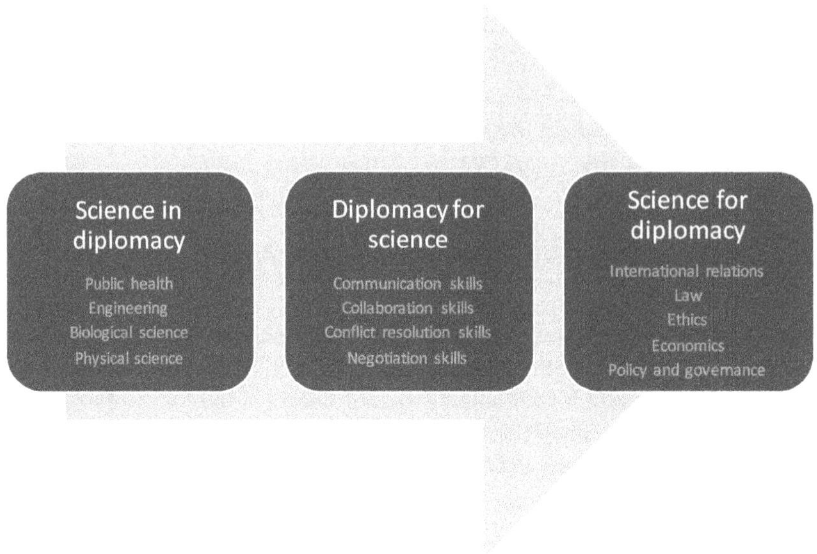

FIGURE 7.1 Science diplomacy activities categories.

from the Royal Society and AAAS (National Research Council 2012). That is, it is focused on three main areas: science in diplomacy, diplomacy for science, and science for diplomacy. Figure 7.1 shows that efforts around SD have been multifaceted on this campus.

So how do you go about developing a significant curriculum for such an interdisciplinary topic? At the core of the curriculum development is a core course, developed by Dr. Hassan Vafai. This course for undergraduate and graduate students acts as an introduction to SD, and this is seen as the heart or the center of the UArizona's SD education (see the next section for further details on this course).

The first order of business was to just examine what already existed. UArizona has a lot of knowledge and expertise in house, but she wanted to look outside the university to see what others are doing around SD and what can be learned from them. Professor Gerlak started an investigation to learn how other universities are being active in SD. The first thing she found is that there are a whole bunch of one-off courses; lots of universities offer a single course that can be viewed as relevant to SD, but it was really just a course here and there. UMass Amherst, University of Massachusetts Boston, Brown, Dartmouth—lots of universities had something, such as "The Role of Science in International Relations and Global Development" at NYU, and "Science Diplomacy: Thinking Globally about Biological and Medical Fields" at Rockefeller University.

In addition to courses, they found some Minor degree programs at some universities, such as Johns Hopkins' Engineering for Sustainable Development minor. They also saw some certificates available. ASU has a certificate that focuses on innovation; a lot of the more traditional diplomacy universities in the United States, like Georgetown, USC, and University of Syracuse, have certificate programs or graduate degrees in what could be seen as traditional public diplomacy. They did find a few graduate programs related to SD: Virginia Tech has a degree at the Master's level in Science, Technology and Politics, and Tuffs offers a Ph.D. in Water Diplomacy.

Looking more globally, some of the big leaders in educational programs related to SD come out of the U.K. University College London and the University of Sussex, two universities that are major players in science, technology, and innovation development. Professor Robert Varady, who is at the Udall Center for Studies in Public Policy at UArizona, recently attended the Budapest Water Forum and the university there is launching a new SD initiative as well. So this initiative is not unique to the United States, but SPDI at UArizona is really trying to observe what the field looks like, and we can see this one-off approach to SD, where they offer a course here and there, all the way to a full degree that speaks to some elements of SD.

From her observations, Gerlak could see that there is a mix of graduate and undergraduate courses at a lot of universities. A big theme is interdisciplinarity—none of this was housed in just one college or school, or focused on just one topic. A lot of the focus they saw was on impact—people were teaching these kinds of courses or introducing them because they were hoping to have some kind of impact, or the students were looking to make an impact. Lastly, they saw that there was an emphasis on leadership, communication, and how these courses could really make someone a leader in the field.

The intention of SPDI at UArizona is to learn from others, but then to also look internally and say, "Well, how can we harness the strengths that we have? We are a Research I University, we have a lot of scientific research happening. How can we harness what's already here in the natural sciences and engineering fields, but also the social science fields, and take advantage of what we have in place already?"

The latest phase has been looking internally at UArizona's set of courses and faculty to think about how they could best build the architecture of SD at UArizona. The goal is to develop a cohort of early career scientists and policy scholars who can apply their expertise to support policymaking and engage in science and society applications of international and domestic diplomatic efforts. They are developing the idea of training a new young cohort of scholars on the physical/biological/engineering side as well as on the social science and policy side.

After lengthy discussions with faculty and reviewing course descriptions and course catalogues, Gerlak compiled a list of courses that could potentially be cross-listed and count toward an SD degree or certificate (Gerlak 2020). UArizona has a really strong program in political science in the School of Government and Public Policy, with a lot of expertise in international relations, conflict resolution, and negotiations at both the graduate and undergraduate levels on which we can build. The College of Public Health has a Health Policy Program with a very active state-level advocacy program in place, where they work with state legislators and regulatory folks; they have a whole range of field-based courses around health policy and analysis; development work in Africa; and a whole lot of activity on the binational level working with the U.S.-Mexico health space. UArizona also has a great many courses around communications, conflict management, and interpersonal communications that can be drawn from.

There is a noticeable gap in terms of addressing SD specifically for the Americas, especially when compared with European efforts, so this is a space that could be filled by UArizona, especially with their expertise as a designated Hispanic Serving Institution. In Latin American Studies, there is a whole slew of courses around health and toxicity; science, technology, and medicine in Latin America; and there have even been recent efforts at UArizona's law school to train Mexican diplomats on issues of U.S. migration, immigration issues, and law in the United States, so that diplomacy is already happening. The School of Geography and Development offers courses on international policy and governance, water-food nexus classes, classes that are more region-based and deal with the border, and also much more local. Finally, the School of Philosophy has a whole range of classes in ethics—Neuro-ethics, medical ethics, environmental ethics, and philosophy of science classes, including Philosophy of Biological Sciences, Philosophy of Physical Science, etc. So even without a formal program yet in place, UArizona really already has a great set of courses than can be drawn from in terms of thinking about what a more robust curriculum could look like.

They are still not sure about the final shape of that curriculum, as there are lots of possibilities here. Everything from a graduate minor to an interdisciplinary minor that's a little more intensive, to a graduate certificate where a graduate student in any field could take four courses from a buffet of courses in addition to the core course to earn a certificate. A minor could also be created for undergraduates. So it is really about finding the right mix, not overburdening students or faculty, and cross-listing courses as much as possible.

SPDI is very much at the beginning of this process. Dr. Gerlak emphasizes that this is a reciprocal relationship between natural and social sciences, and it is not based on just one field. It is important for physical or biological scientists and engineers to take courses in the social science areas, but it is equally important to also train students on the social science side about the big issues in science, whether it be artificial intelligence, the latest water quality assessment, or something about immigration or climate mitigation—whatever the issue may be. It is evident that there is indeed a demand from students and faculty on both sides to have courses that integrate the science to inform and educate more policy-related folks who want to go into SD and develop some of these skills.

At this point, the main focus is to be very open, transparent, and collaborative as the curriculum takes form, but this is really just at the early stages of what can be pulled together at the university to offer something unique and impactful for our students.

Teaching Science, Health, and Engineering Policy and Diplomacy

Research Professor Hassan A. Vafai got involved in SD when he was young. The destructiveness of the War in Vietnam and the accompanying loss of so many lives, including some close friends, probably served as the most important stimulus for him to question war and think about peace. He admired the fact that WWI produced the League of Nations and WWII produced the United Nations. As a young scientist, and with the heart of these treaties being the beacon of hope, he thought that promotion of science and technology would make a path for us to look at human nature in a different way and promote global peace. One of the turning points in his life was when he was invited to the National Academies of Sciences. He met several "giants of science diplomacy" that he remained close friends with, such as Bruce Alberts, Bill Colglazier, Glenn Schweitzer, Norman Neureiter, and others. In this environment, his interest in SD continued to flourish. After joining UArizona, he met Drs. Jeff Goldberg and Kevin Lansey who were also supportive of promoting SD. They started the Science, Health, and Engineering Policy and Diplomacy Initiative (SPDI) through a seed grant from the Lounsbery Foundation and a coalition of faculty from various disciplines across campus, including engineering, environmental sciences, law and public policy, lunar and planetary sciences, health sciences, and more.

As Professor Gerlak mentioned, the SPDI committee decided that a good way to introduce SD would be through holding a lecture series and hosting conferences, focusing on educating and training the next generation of science diplomats. In 2017, SPDI hosted its first international conference. They gave 14 student rapporteurs the opportunity of reporting on that conference and assisting in the publication of the proceedings, as they did for the 2019 conference as they continue to focus on education.

As for the core SD course, Dr. Vafai's years of experience teaching and mentoring undergraduate and graduate students taught him that one of the most valuable ways for students to learn is to engage with each other in discussion about critical topics. He designed the course around this idea, using a model of providing some expert material to the students (such as assigned literature, research articles, treaties or policies, or a lecture by a professor, diplomat, scientist, or industry leader), and then the students were expected to come to class ready to participate in a scholarly discussion with the professor, the guest lecturer, and other students (Lansey and Vafai 2019).

Dr. Vafai utilized expertise at UArizona as well as invited experts from outside the university to contribute to the course. The students heard from a different speaker presenting different perspectives on different issues every week, from a distinguished member of US National Academy of Sciences speaking about "*Science Academies: Origins and Roles in Science Policy and Diplomacy Bridging Science* and *Society*" to a public health professor speaking about Indigenous data sovereignty to a systems engineer discussing sports diplomacy. They heard from professors and professionals at various levels, from national to local, as well as from other countries, particularly emphasizing Mexico and Latin America. Then, the students would discuss in groups, and ultimately submit a term paper about an issue relating to science policy or diplomacy on the topic of their choice. Professor Vafai was pleasantly surprised by how well-thought-out the final papers were. For example, one group of students wrote about North Korea in their term paper. They didn't study North Korea during the course, but the students drew conclusions on how we might improve relations based on the topics that they did discuss. So, using case studies, the students are able to apply those same principles and lessons on solving real-world global challenges.

Throughout the course, Professors Vafai and Lansey sought honest feedback from the students through anonymous surveys in order to improve the next offering of the course. From those surveys, it was clear that hearing "real-world" scenarios or practical applications was an important aspect for the students and gave them an edge that no other courses really had in preparing them for certain careers. In particular, professionals talking about cross-border issues and conducting business with Latin America in practical terms helped delineate how science and policy work together. Students also appreciated having unique or underrepresented perspectives and hearing the stories of how they became involved in SD was inspirational for many.

Some of the lessons learned after the preliminary semester:

- Cross-listing courses for multiple majors is a crucial step to getting the enrollment necessary for truly generating interest in SD. The course was listed as an engineering course (since Professors Vafai and Lansey are both engineering professors), but it severely limited participation from across campus, so the SPDI team has determined the need to rectify this for future offerings.
- Another aspect that restricts participation is the limited access to in-person offerings. As society embraces technology more and more, online courses become increasingly attractive, and they are certainly not unattainable. Accordingly, Dr. Vafai has been developing an online version of this core course so that access is no longer a limiting factor.
- The breadth (and depth) of topics relating to SD to be discussed is quite vast, and a single survey course on SD will not suffice to effectively educate the next generation of science diplomats, though it is an excellent start. Thus, the SPDI committee continues to develop a pathway for a certificate, minor, or graduate degree program to further strengthen the educational curriculum.

The latest effort has been focused on developing this core course into an online course, recognizing a necessity to maximize accessibility to places beyond just Tucson, Arizona. The online course attempts to examine innovative and transdisciplinary topics relating to SD, such as cybersecurity and the outcome of international collaboration to establish norms and laws for cyberspace, artificial intelligence, and other emerging technologies.

The Transdisciplinary Environmental Science for Society (TESS) Program

Professor Daniel Ferguson is a research professor and research scientist at the Institute for the Environment at UArizona, and also directs a regional climate research program called the Climate Assessment for the Southwest. His work is primarily community-based, and his expertise lies in trying to do science that is necessary to solve problems at the scale of decision-making; in short, he tries to fit the science to decisions that are being made.

The Transdisciplinary Environmental Science for Society, or TESS program, is intended to fill a gap that he and his colleagues saw in the undergraduate students coming through, as well as for professionals who work outside of academia. It was developed by four interdisciplinary faculty at UArizona, each with their own experience with transdisciplinary research projects: Connie Woodhouse is currently the leader and is a professor in the School of Geography and Development. She is a paleoclimate researcher and does tree-ring science. Her work over the years has included lots of interesting basic science, but she has also done a lot of work with water management communities in the western United States and internationally, trying to help water managers understand the climate over long periods of time—especially when they don't necessarily have records to look at—and she does that with tree-rings. Monica Ramirez-Andreotta is a professor of environmental science and has done some really interesting work with communities who have been exposed to mining and how communities deal with the impacts of mines. Finally, Gregg Garfin is a professor and climatologist at the School of Natural Resources and Environment. The bulk of his work has to do with science in the community and decision-making. They, along with Professor Ferguson, created the program drawing from their unique backgrounds with transdisciplinary research (The University of Arizona 2020b).

Transdisciplinary research brings together experts from academia, so it's interdisciplinary science, but it's also done in context and with experts from outside of academia, and specifically it is meant to address a socially relevant problem. Each of these professors has this type of work in their portfolio, where they are working with experts across the campus, experts outside of academia, and they are focusing on a problem that is socially relevant. For example, since Dr. Ferguson's work is in climate, his particular research typically has to do with climate issues, particularly in the Southwest.

The goal of TESS is to train researchers, practitioners, political leaders, and educators to be well-equipped to deal with complex environmental challenges. The starting point here was that most of the projects these researchers have worked on are not "solvable" in the traditional or technical sense—for example, climate change is not solvable in the sense that you can just fix it and it goes away. So, it is with these problems that they typically work, and that's really the aim for TESS.

The rationale for creating the TESS program was actually just from the researchers' own experience working at the university, or universities in general, because they found that traditional higher education is not great at a few things. One is integrated ways of dealing with a complex problem. Higher education institutions tend to be a little stovepipe in this way, where different disciplines do great disciplinary things, but they are not necessarily great at crossing those lines. They have historically not been very good at valuing other kinds of knowledge other than academic knowledge, not utilizing traditional or experiential knowledge. There is a movement around this idea of boundary spanning, where people's responsibilities or expertise are fitting in between science and policy or science and decision-making—so "boundary spanners"—and it's difficult to fit that into a typical research project or research environment. Lastly, they found over the years that science graduates, or people who are getting science Ph.Ds. specifically, often don't really understand how science policy actually works, at least in the United States. They don't know where the money is coming from, just that the money is there. So how did that happen? How did those decisions get made? They found that young professionals often are not informed of that, thus the rationale for this program.

There were several machinations of what the TESS program should be because as Dr. Gerlak mentioned, there are different "flavors" of SD programs. The team went through various versions and did a lot of homework to figure out who else in the United States and internationally was doing this kind of training, whether it should be a degree, a degree certificate, a minor, etc. What they eventually landed on was that their initial aim should be at professional development, in part because there are students earning their master's and Ph.D. degrees who aren't continuing on, or are not necessarily getting this training. So, they initially developed an online professional development course. It is intended for a broad audience, including people involved in environmental science, natural resource professionals, early career academics who aren't getting this education in their regular training, and to some extent graduate students—although this is not offered as a for-credit course, but rather as a continuing education credit. The reason for this was that at the time, in discussions with the Provost, this was the most straightforward way to get the project started, but they are now revisiting this aspect to look at other credit options.

The team now has a set of theoretical courses (in development) and one fully developed course which are each four-week self-paced online education courses. In order to get people fully trained on transdisciplinary environmental science in general, they will need some sort of fundamentals course, a practical course, and something about communication, all as the core education curriculum. Then, there are all sorts of quasi-electives—things like ethics, environmental justice (EJ), and so forth—and then obviously this kind of work is experiential anyway, so they will have a capstone experience course to complete the required hours. So, as I said, that is all fantasy except for this fundamentals course right now.

They have had some great support from the university on the technical side of things, but as of now, it doesn't really fit anywhere in the university. They do not have many resources yet, as this effort has been mainly a labor of love and is not yet financially supported by any particular departments. Through just their efforts, they were able to create the Fundamentals in Transdisciplinary Environmental Research course (the fully developed course mentioned previously), which just completed its first four-week term. The first week covers science policy, the next week is about different approaches to doing this work (so how do academics approach this, and how is it approached in the rest of the world), week three is focused specifically on how to do community-engaged work (or how this works in practice), and the last week is a synthesis of what was learned and how to put it all together.

Though the future of this effort is not certain, the team is moving forward and adjusting as they go. They have had to reevaluate some of the funding models and various aspects based on this first run they had. The questions they are now looking to answer are as follows: "how do we develop new courses?" "what partnerships do we need to make on campus?" "where is the market for this or who is interested in this?" and "how often should these courses be offered?" They will continue to ask questions and make improvements as we move forward.

A Role for the Universities' Partnership for Water Cooperation and Diplomacy

Léna Salamé joined the United Nations Educational, Scientific and Cultural Organization (UNESCO) in January 2000, and at that time, the media—with statements from renowned scientists—insisted that the next wars, or the wars of the 21st century, were going to be fought over water. At UNESCO, they felt that they should investigate these claims and see if it was true, and if not, then how could they change this tendency and prove that it was wrong, and if so, then the water issue needs to be addressed. UNESCO launched a project called "From Potential Conflict to Cooperation Potential," or PCCP for short. The objective of PCCP was to use education, research, and science as a way to intervene in places where there are potential conflicts, and to tip the balance away from conflict toward cooperation (IGRAC n.d.). They brought together experts from different disciplines and consulted with them, and the result of this investigation was a series of publications—interdisciplinary publications, case studies, and educational materials, totaling about 3,000 pages—that proves throughout history—through the legal, technical, and diplomatic tools available—that actually there *are* conflicts over water, but so far there are no wars over water. They continued to use the research materials to build educational materials and then to—under the umbrella of education and research—bring people together who otherwise wouldn't talk to each other or wouldn't even meet with each other, but under the umbrella of education and research, or of joint research, they would accept to meet.

One good example is PCCP's work with the Nile River Basin. At some point, they brought together experts from all the countries around the Nile River Basin, because they had a problem with data; they didn't have the same data and they didn't have equal capacity to collect data (Demin 2015). So UNESCO offered for them to come together to collect data, and then to finance a joint database that they could use as a basis for their collective data, for joint management of the river. At that time, the politics of some countries prevented them from actually sharing data—it was against their normal politics. So they had to go back one step and offer capacity-building and training in order to show them the importance of collecting joint databases and to have the same understanding and access to the data in order to then move forward. The most interesting part of this process was when, after having brought together the researchers and the scientists and the experts who collected data together and finally agreed to discuss the data together, they brought the politicians together, but in the room were the experts, as well. The politicians were initially very distant with each other, but when they saw that their countrymen—their own experts—were hugging and greeting each other, this immensely helped the atmosphere. This is one example of how the work with the experts and scientists has really helped to break the ice between the politicians.

Another example is related to other education and training courses. PCCP developed training courses that are region-specific. So, for example, some were for southeast Europe, others were for the Slavic region, the southern African developing communities, others for the Middle East. But for each set of courses, they brought together teachers from the region, and included people from different disciplines, so they had biologists, specialists in conflict resolution, lawyers, engineers, and so forth. They brought them together and asked them to develop region-specific courses.

One of the biggest challenges they encountered in this process was actually the language that people used. Because beyond just speaking different languages, there is also a language barrier between disciplines, and then there is also the problem of strategic language, which is discussed a bit further. In preparation for the course, they examined with great importance the disciplinary language. First, they had to define what (for example) a conflict is, because for the engineer, conflict is something different when compared to the lawyer, and it was yet different when it came to the conflict management specialist. So after 45 minutes of discussion, they figured out how to solve the problem. They came up with these courses that were offered up to various audiences, such as young professionals, or to high-level decision-makers. So this is another example of how the research and education helped bring together people who would not otherwise come together to advance water cooperation and diplomacy.

These courses also led to the creation of an actual embassy specialization at the IHE Delft Institute for Water Education. This is an institute based in the Netherlands that only teaches water matters (IHE Delft 2020). So they created an embassy specialization on Water Conflict Management, which is an 18-month program. People who come to this program have to have some professional experience before they come, so they are not usually young students, and the rate of people going back to their home country is very high. Graduates have then gone on to become ministers, vice ministers, advisors to policymakers, and decision-makers in their countries. This Master's Program is continuing now, and they have actually developed a partnership with Oregon State University and with the University for Peace in Costa Rica (Oregon State University n.d.; University for Peace 2020). In Budapest, the National University of Public Service is also putting together a specialization on water cooperation and diplomacy (UPWCD 2020). This is the university that trains the military and diplomats of Hungary, and they are going to study water issues as well, and have briefings on water matters, and be trained in negotiations and conflict management. At the University of Geneva, there is a new program of water governance and a summer school on water cooperation and diplomacy. So what they started in 2000 has really brought its fruits in terms of developing a method of teaching water cooperation and diplomacy in courses—not only those courses of only water, but vice versa; we also introduced water issues in courses related to peace studies. For example, Salamé was able to visit several universities in Europe as a guest lecturer to introduce water in master programs where only peace studies were given. So there is an opening where water and peace studies are being understood together and offered together.

With all this experience and the copious amounts of information that have been brought together to develop all these programs and these educational materials, Salamé realized that this is not enough. There is a lot yet to be done, because the next generation will have to be water champions. Water champions are water managers who are specialized in their fields, but who can understand other fields as well. For example, an engineer should be able to also understand how the cooperation process goes, how negotiations work, and how the legal treaties can be interpreted, because in a delegation, you can have various experts of various disciplinary backgrounds, but you need to have people who understand each other. As Salamé put it, they will have to work together in their professional life, so they might as well start in the classroom. So those courses that they are preparing are addressed to people of different disciplinary backgrounds, so they have in the same room people from all different backgrounds—the lawyers, engineers, biologists, etc.—because they have to be able to understand each other when they talk and are on a negotiation team with one another. It seems that many negotiation processes have failed because the chief negotiator who is a lawyer doesn't understand the engineer or vice versa.

Now, Salamé and her team are trying to put together this network of universities called the Universities Partnership for Water Cooperation and Diplomacy (UPWCD). She is affiliated with the University of Geneva, who is offering the coordination unit for this network of universities. It is a network of like-minded institutions who want to actually make an impact on the global discourse of water cooperation and diplomacy and who want to advance this discourse and introduce rigorous science into the dialogues that have been created around water cooperation. The aim is really to help policymakers in really making a difference and making the right decisions based on science. They are bringing universities and capacity-building and knowledge development institutions together through this network that will have as its main communication tool an online platform where people who are interested in anything related to water cooperation and diplomacy could go and find the information. So for a student from Latin America who is interested in taking a course on water cooperation and diplomacy, they could go on this website and would find the information. Or a professor from UArizona can go to this website to see if there is an exchange opportunity with IHE Delft. Or a researcher can find out if there is a conference on water diplomacy and cooperation in Thailand on this website. The plan is for it to become the one-stop shop for everything related to water cooperation and diplomacy. Their main project now is the development of a MOOC (Massive Open Online Course) on water cooperation and diplomacy, and the various UPWCD partners are being invited to make contributions to this (Geneva Water Hub n.d.).

Questions and Answers

Question

Glenn Schweitzer: Let me respond to the opening comments on how your program can be strengthened. I'm going to talk from the perspective of someone looking for money. We know funding is important, and it helps if you can really understand what the Foundations want, what they are interested in, and what they might respond to. Hopefully they know the people, but they may not for a program like this. So, the question they're going to ask is: What does The University of Arizona really have to offer that we need? And so that's the question you have to answer. Now we are talking about science policy. My interest now is in raising money for science diplomacy. If I wanted to do something in water, I'd call Oregon State University, no contest (in the United States, anyway). But if I wanted to know about something related with science in one of these countries, who would I call? Say you want to work with Iran. It took me three months to find out that there is one person in the United States who has written a book in English on science technology infrastructure within Iran—there is only one person, and she is a the University of Wisconsin at the Green Bay campus. So you need some kind of capabilities here. I suspect you don't have much of a science diplomacy library, because in reality there really aren't that many publications, let alone in English. So those individuals with the money already know that, so they need to know what sets you apart from other institutions. And obviously if you have students or professors who are interested in it, they need resources and capacity. I built my own library from scratch, which I don't recommend, but there needs to be some capabilities built here in order to attract the funding and the participation.

Answer

Léna Salamé: Honestly, I think that emphasizes my point about the importance of the Universities Partnership. Your example of OSU's excellence in water sciences—well they will point you to the UPWCD website; actually we supported their work when I was at UNESCO and they continue to support it, but they have Aaron Wolf there, who actually had the idea of the UPWCD in the first place. So why am I saying this? What I'm saying is that the *partnerships* and the *networking* of these universities is so important, because alone, they can only do so much. They feel that bringing other universities on board and joining forces with other universities is only going to strengthen their capacities and in turn better serve their own interests. So together we stand and alone we fall.

Gail Burd: I'm Gail Burd, I'm the Senior Vice Provost for Academic Affairs, Teaching, and Learning at the University of Arizona. I'm a biology professor in Molecular and Cellular Biology, and I am also leading a charge right now to redo part of our general education, and one of the main goals that I have and that remains to be seen, but is coming to fruition, is having themes of courses, so courses focused on a centralized theme. We have one going forward this spring on sustainability. It is team-taught with a natural science professor from Planetary Sciences on the Evolution of Inhabitable Worlds, thinking about planets, and the other one is from Geography, in environmental science, basically on environment and society, and those two professors are going to teach in the theme of sustainability. We need more courses within themes that we can pull from. I'd like to make a plug for your course to be part of this theme in sustainability. We want real courses for real credit, so not for the lay person who might be my neighbor down the street. We also hope to focus something on health. It could be global and environmental health, it could be biomedical ethics and health policy, and health policy brings in a lot of the science diplomacy in the sense that we have a really crappy healthcare system right now, and there is a lot of stuff out there in other countries that would be really helpful for our legislators to know about, so if anyone is interested in these initiatives, please let me know.

Question

Christopher Brown: I'm Christopher Brown and I'm a faculty member at New Mexico State. I would strongly encourage you to explore the idea of introducing an interdisciplinary Master's and Ph.D. track. Seven years ago, some colleagues of mine and I set up an interdisciplinary program on Water Science and Management with civil engineering, environmental sciences, agricultural economics, geography, and animal sciences, and it has been wildly successful (NMSU 2020). It is very interdisciplinary, and it sounds like you have the talent on campus to do this with the departments you showed us. I sincerely encourage you to explore it because it has been a wonderful experience for us, and I think it fits with what you want to do.

Question

Itzchak Kornfeld: I am Itzchak Kornfeld from the Hebrew University of Jerusalem. I wanted to speak to the provost and those of you developing curriculum at The University of Arizona. We teach a course which takes lawyers, geographers, geologists, and biologists and we take them out into the field. We have speakers, for example we would invite a hydrologist from Palestine, and then the students are joined into teams to complete a final project. It really opens up the students' eyes, and our eyes as instructors, too.

Question

Sharon Megdal: I invite you all to come to my session on the Middle East. But I wanted to mention, and that is that you only have a small number of programs there. We know that we have so much going on and so much talent. And to the point about what do we have that's unique, well, we are in an arid environment, and we're in a living laboratory, and we have some real similarities to the Middle East that Oregon State will never have. But we also have programs. We have the Arid Lands Resource Sciences program that's in need of a boost, so I think there could be a way to connect some of this to that (The University of Arizona 2020a). We also have a Master's in Water Society Policy program, which I helped create. That is also interdisciplinary and in fact transdisciplinary because they have to do a project on society. So I just want to say that the TESS program is really wonderful and you are to be congratulated on your vision. I think there is tremendous opportunity to have synergies here and build this.

Question

Neha Gupta: I'm a Ph.D. student in the Hydrology department. Growing up, my family traveled a lot internationally, and growing up as a child of immigrants, I became this sort of child diplomat between my home and then the world that I participated in, and I didn't know—up until a year ago—that these worlds could coexist. And so I really thank everyone for their effort, for putting this conference together, the courses, just to show us that there's a different way to orient our careers and bring us all together.

Question

Bill Cragin: I'm Bill Cragin, and I'm here to say that I think you are doing wonderful work, but I think you are starting too late. I have been a lifelong teacher and I can tell you that your students, your graduate students and undergraduate students, should be at the elementary schools once a month, and they should follow these kids through middle school and high school. They should bring fun into the science. We need creative writers who can put your words into words the rest of us can understand. You can start this and build a cadre of people who are really excellent at this, but starting at the undergraduate level is a little late, so I think you should start earlier.

REFERENCES

Demin, A.P. 2015. "Distribution of water resources: A case study of the transboundary Nile river." *Geography and Natural Resources* 36: 198–205.

IHE Delft. 2020. "About IHE Delft." IHE Delft Institute for Water Education Website. www.un-ihe.org/about-ihe-delft

International Groundwater Resources Assessment Centre (IGRAC). n.d. "UNESCO-IHP Groundwater Portal." *PCCP | Unesco Groundwater Portal*. Accessed November 12, 2020. https://groundwaterportal.net/project/pccp

Geneva Water Hub. n.d. MOOC n°3 – International Water Law. *Hydropolitics Towards Peace and Security*. www.genevawaterhub.org/resource/mooc-ndeg3-international-water-law

Gerlak, Andrea. 2020. *Examining Science Diplomacy Educational Opportunities at The University of Arizona*. Report funded by the Lounsbery Foundation. April 2020.

Lansey, Kevin, and Hassan Vafai. 2019. ENGR 495A/595A: Science, Health and Engineering Diplomacy Course Description. Spring 2019. https://shepdiplomacy.arizona.edu/sites/default/files/ENGR_495A-595A.pdf

National Research Council. 2012. *U.S. and International Perspectives on Global Science Policy and Science Diplomacy: Report of a Workshop*. Washington, DC: The National Academies Press. https://doi.org/10.17226/13300.

NMSU. 2020. "Graduate Program Information." Water Science and Management. Accessed November 30, 2020. https://catalogs.nmsu.edu/nmsu/agricultural-consumer-environmental-sciences/water-science-mgmt/

Oregon State University. n.d. Water Cooperation and Diplomacy Joint Education Programme. Accessed November 20, 2020. https://gradschool.oregonstate.edu/water-cooperation-and-diplomacy

The University of Arizona. 2020a. ALRS Home. "WELCOME to the Graduate Interdisciplinary Program in Arid Lands Resources Sciences (ALRS)." Accessed November 12, 2020. https://alrs.arizona.edu/

The University of Arizona. 2020b. "TESS Transdisciplinary Environmental Science for Society." TESS Home Page. Accessed November 12, 2020. https://science4society.sites.arizona.edu/

University for Peace. 2020. Water Cooperation and Diplomacy Scholarship Programme. University for Peace (UPEACE), San José, Costa Rica. Accessed November 20, 2020. www.upeace.org/departments/water-cooperation-and-diplomacy

UPWCD. 2020. "MA in International Water Governance and Water Diplomacy." Universities Partnership for Water Cooperation and Diplomacy. Accessed November 20, 2020. https://upwcd.org/node/335

BRINGING NEW PERSPECTIVES AND KNOWLEDGE INTO EDUCATION

The Importance of Marginalized Knowledges

Alberto Arenas
Professor of Environmental and Sustainability Education, Department of Teaching, Learning and Sociocultural Studies, College of Education, The University of Arizona

The following three anecdotes describe different characteristics of marginalized knowledges:

(1) A community near Lake Titicaca was approached by Inter-American Development Bank officials in the early 1960s. The officials were led by Chilean economist Felipe Herrera, IDB's first president. IDB had loaned funds to Bolivia to build large development projects, and once all the money was spent on the projects, some funds were left unspent. Given the dire needs of many poor communities in Bolivia, bank officials felt that the correct course of action was to ask community leaders how best to use the money. The instruction from bank leaders to community leaders was "You can do with this money whatever you want; you let us know and we'll provide the funds." The community elders discussed the situation among themselves, and eventually came to the following conclusion: "We know what we want: We want musical instruments for our community," the elders replied. IDB officials were puzzled by this request. "No, maybe you didn't understand us; your community lacks many basic services; you don't have paved streets, electricity, sewers. With this money, you can do any of those projects." The community leaders said in response: "No, we understood you very well. But it so happens that in our community everyone plays a musical instrument. On Sundays, after mass, we gather at the local plaza, and we play our music for a long time. Afterwards, we sit down and discuss our problems and how to solve them. Right now, our instruments are old and broken, so we cannot play our music. Without music, we also feel old and broken" (inspired by Weisman 2008).

(2) In Mesoamerican cuisine and culture, amaranth played a primordial role in pre-Hispanic times. Not only it is a high-protein edible plant that contains essential amino acids needed by the human body, but in terms of its cultural value, amaranth was highly revered by the Aztecs that occupied a vital role in religious ceremonies. The Aztec calendar indicated the dates when women had to toast the seed in a tortilla griddle; once toasted, the seed was mixed with honey and human blood of sacrificed children or adults, and they used the paste to form figures of gods, birds, and snakes to be consumed in the ceremonies of the great temples. When Hernán Cortés invaded Mexico in 1519 and began the conquest of the Aztec empire, these ceremonies were considered a perversion of the newly imposed Catholic doctrine and were banned from the Aztec culture. In order to destroy the religion and rituals of the Aztecs, Cortés prohibited all cultivation of amaranth, ordered that the crops

be burned or crushed, and ordered that the hands be cut off from people who were caught in possession or cultivation of this grain. Thus, one of continent's most important crops was suddenly forced into oblivion, and it remained so for centuries (Barba de la Rosa, Silva-Sánchez, and González de Mejia 2006). To be precise, amaranth really never disappeared from Mesoamerican cuisine. For instance, throughout the centuries, it could be found in the form of a candy in the streets of Mexico City called *alegría* ("happiness" in Spanish), a mix of amaranth seed with honey or sugar, very popular among lower classes in Mexico (Cardona 2007). Nowadays, the amaranth has experienced a resurgence of sorts as a super-food by health-conscious consumers in North America and Europe, but it has never recuperated the position of excellence it once occupied in Mesoamerican cultures.

(3) The U'wa Indigenous people from Colombia inhabit the cloud forests of northeastern Colombia. They gained international visibility in 1995 when they threatened to commit mass suicide if Royal Dutch Shell and Occidental Petroleum continued with their plans to drill oil in their ancestral territory. In an Open Letter to the Colombian people and the international community, the U'wa explained how the drilling project would damage the delicate cloud forest ecosystem sustaining them and would destroy the cultural and physical integrity of the community. The letter in part read (U'wa People 2010):

> We are born as children of the Earth. More than a thousand times and in a thousand different ways we have told the white man (Riowa) that the land is our mother, that we cannot and do not want to sell it, but the white man seems to not have understood, insists that we yield, sell or mistreat our land. We know that the Riowa has put a price on everything alive and even on the same stone, it trades with its own blood and wants us to do the same in our sacred territory. Some White Chiefs have expressed horror at our decision of collective suicide as a last resort to defend our Mother Earth. Once again they present us as savages. To this we say: U'wa commits suicide for life, white men commit suicide for coins. Who is the savage one?

Fortunately, the U'wa never arrived at the extreme of mass suicide because the corporations in question abandoned their ancestral territory. It should be noted, however, that the U'wa were not just issuing an idle threat: U'wa oral history tells how 400 years ago, a number of U'was committed mass suicide rather than give themselves up to the Conquistadores—and centuries later, they were willing to execute such a heart-wrenching action again.

What do these three anecdotes have in common: in each case, groups of people rooted to the land throughout millennia have used alternative ways of relating to one another and to nature (i.e. arts; food production; and love for the earth) to sustain and protect the collectivity and the larger biosphere. These anecdotes offer examples of marginalized knowledges, which are the collection of intergenerational knowledges and skills from communities worldwide that hegemonic forces have pushed to the margins of society. I am using the term "knowledge" in the traditional sense, that is, knowledge refers to information, facts, skills, and behaviors, but also in a wide encompassing way, that is, knowledge also refers to cultural artifacts, ways of relating to oneself, to one another, to nature, and to the cosmos in general. All communities worldwide produce these knowledges and attempt to preserve and pass them on to future generations. It is the human capital that materially poor rural and urban peoples have developed over time—both Indigenous and non-Indigenous communities—while being part of a given ecosystem and locality. Marginalized knowledges serve to fulfill economic, social, environmental, spiritual, or cultural needs. Although many consider poor people to be lacking adequate amounts of money or enough material possessions, they in fact may be wealthy, thanks to the knowledge and skills they may possess to fulfill basic subsistence needs along with a unique sense of identity. Some examples include knowledge that helps its users grow food, provide shelter, promote health care, make house wares, take care of plants and animals, fulfill cultural and spiritual needs, protect the land, and in general, improve the community's livelihood and environmental well-being. Because these knowledges sometimes lack a real or perceived value within the market economy, they have often been prevented from being viewed as a vital asset by society at large.

Today, many marginalized, local, parochial knowledges are at risk of becoming extinct. The disappearance of these knowledges stems from various reasons: it is partly due to an educational system that sees little use for these forms of knowledge (see, for instance, Bowers 2006). Partly as a result of their lack of real or perceived monetary value, they have no role in mainstream commerce. And also due to fast-paced economic, political, social, and cultural changes that make it increasingly difficult for local knowledges to be incorporated into the daily lives of new and even old generations (see Apffel-Marglin and PRATEC 1998). In some cases, the disruption of more sustainable ways of living occurred as the result of the imposition of more powerful forces at play, as in the case of amaranth that was prohibited by the Spanish Crown and the Catholic Church. In other cases, it has occurred as a result of the voluntary adoption of modern technologies that have promised easier and faster solutions to problems—real or perceived—or the adoption of Western lifestyles that are viewed as more glamorous and alluring, strategies that have pushed many ancestral practices to the margins of economic, social, and cultural lives. The tragedy of the eventual disappearance of marginalized knowledges is most obvious to those who have developed them, made their living from them, or who belong to a culture where the knowledge occupies a primordial place. But the negative consequences are felt by society as a whole because valuable skills, technologies, artifacts, problem-solving strategies, and expertise become lost. Some examples of local knowledges that are indispensable for a good living include:

- Non-commodified art forms that bring the community together in ways that allow the community to support one another and live more lightly on the planet.
- Sustainable agriculture, knowledge related to marginal crops, crop diversity, intercropping, erosion prevention, and safeguarding of the biological richness of the soil.
- Sustainable animal husbandry, knowledge of breeding strategies, livestock characteristics and requirements, humane forms of sacrificing animals, and ethnobotanical knowledge for treating common illnesses.

- Responsible management of natural resources, protection of rivers and forests, sustainable management of wild species, and preservation of native species.
- Promotion of preventable health care and knowledge and use of plant properties for medicinal purposes.
- Ecological construction and design of housing and other dwellings that use low-impact materials.
- Community well-being, common or shared knowledge that provides links between community members and generations (e.g. festivals, music, arts, clothing).
- Poverty alleviation, knowledge of survival strategies based on local resources.
- Conservation and revitalization of Indigenous languages.
- Treating nature with respect and reverence, perhaps even with veneration, as many cultures have done throughout history.

Neglected and Underutilized Species

Neglected and Underutilized Species (NUS) serve as a useful platform for exploring the topic of marginalized knowledges, given that they tend to include many of the advantages identified above. The importance of NUS cannot be understated. Fewer crops are feeding more people worldwide, and the dangers to biological and cultural diversities are staggering (Diario de Gastronomía 2012; Zimmerer 2017). Human groups have used about 7,000 different crops for consumption throughout history; today, that number of plants has been reduced to about 150, and of these, only three crops (rice, wheat, and corn) provide about 50% of the world's food energy intake. Agricultural diversity has collapsed to such a degree that only 12 crops and five species of animals provide 75% of all the food consumed worldwide (Bioversity International 2014).

Agrobiodiversity, a subset of biodiversity, refers to the rich genetic legacy that results from the interaction between natural selection and cultural practices of inventive farmers, herders, and fishermen throughout millennia. This agrobiodiversity is declining rapidly due to the overwhelming presence of industrial forms of agriculture, husbandry, and fishing that pushes to the margins the knowledge and skills of countless peasant and Indigenous populations who have an intimate knowledge of the land. It is estimated that Mexican farmers are cultivating only 20% of the corn types that were grown there in 1930. Chinese farmers are producing only 10% of 10,000 varieties of wheat that were recorded there in 1949. And farmers in the United States are producing only 5% of the apple varieties that existed in the country in the 1800s. The dangers to the reduction of agrobiodiversity are numerous. These include:

- The diminished arrays of nutrients, causing the development of a poor diet for people worldwide.
- The production of fewer crops and fewer varieties of the same crop (all for export) as tastes become homogenized, making local communities dangerously dependent on global markets.
- The biological impoverishment of soils.
- The loss of cultivation techniques, and local know-how on how to maintain the richness of the local agrobiodiversity.
- The loss of genetic diversity in staple food crops.

Quinoa: A Marginal Crop Moving into the Mainstream

One of these NUS crops is quinoa (amaranth, mentioned earlier, is another NUS), and it is worth highlighting, given that the United Nations declared 2013 as the Year of Quinoa (FAO and CIRAD 2015). Quinoa, from the highlands of Peru and Bolivia, has been dubbed the mother of all grains, given its dietary and cultural importance in Inca culture. The German geographer and explorer Alexander von Humboldt wrote that quinoa was to ancient Andean societies what "wine was to the Greeks, wheat to the Romans, cotton to the Arabs" (Erdos 1999). Quinoa has played a vital role in the nutrition of Andean communities for centuries. This food contains all nine essential amino acids, rare in plants but common in meat and its derivatives. Quinoa also exhibits a very high content of carbohydrates, minerals, and vitamins, and it does not produce all of the negative environmental externalities associated with meat production.

In terms of its history, scholars have attributed the success of the Incan empire, in part, to the environmentally sustainable cultivation, storage, and distribution of Indigenous plants, including quinoa, which allowed it to feed both Incas and conquered groups (Mujica et al. 2001). With the arrival of the Spanish conquest and eventual colonization, quinoa suffered a similar fate to that of amaranth. Given the importance of quinoa in Inca religious ceremonies and cultural life, Francisco Pizarro as the Spanish Crown's representative and the Catholic Church decided to outlaw the cultivation of quinoa in the 1530s and burned all the quinoa fields. This was one of the strategies used by the Spaniards to subvert Inca culture. The Inca people were then forced to subsist on maize and potatoes; given that these foods did not provide all the nutrients needed for a balanced diet, malnutrition and infant mortality rates became rampant. Over time, the Spaniards imported barley and wheat, crops that the conquerors were more familiar with, but these were not fully embraced by the Inca. Despite efforts from the Spanish Crown to eradicate quinoa, it continued to be cultivated illegally high in the Andean mountains and eventually it fell into obscurity.

The marginal status of quinoa started to change in the 1970s, thanks to an influx of development funds from national and international organizations that recognized the cultural and nutritional importance of quinoa. At the dawn of the 21st century, health-conscious American and European consumers have been paramount in the resurgence of quinoa production, and the grain (or seed, to be more precise) has now moved from being an inconsequential and little-known food in international markets to being a sought-after grain by consumers in the West and also, to a certain degree, by consumers in its place of origin, Bolivia and Peru.

It should be noted that when a knowledge moves from a marginal status to the mainstream, it carries with it important benefits and also potential dangers. In terms of important benefits related to increased quinoa production, there have been both nutritional benefits to consumers and also economic benefits to producers of quinoa (Cherfas 2016). For instance, in one study in Puno, Peru (where 80% of Peruvian quinoa is produced), despite an increase in the market price of quinoa, Puno residents are actually eating more of the crop. And in terms of economic benefits, the welfare of Peruvian quinoa growers as a whole has increased more rapidly than that of the rest of the Peruvian population. In terms of potential dangers, there are several (Cherfas 2016; FAO and CIRAD 2015):

(1) There is a danger that the soaring price of quinoa leads farmers to abandon other agricultural commodities to focus on the mono-cropping of quinoa and to increase the use of industrial fertilizers and pesticides; these chemical additives would make their way into groundwater or become airborne, leading to a toxic environment that affects fauna and flora.
(2) A related problem is that in the past, tracts of land remained fallow to allow the soil to recuperate its biological richness. Now, farmers want to take advantage of quinoa's high prices and have adopted intensive farming that is depleting the richness of the soil and causing erosion.
(3) While the evidence in a place like Puno so far does not indicate a decrease in quinoa consumption as a result of price increases, it is possible that eventually poor non-quinoa-producing families may not be able to afford it anymore.
(4) There are about 3,000 different varieties of quinoa, and it is this genetic diversity that allows for the grain to withstand pathogens that may eventually decimate the plant, a threat made even greater with climate change. However, the emphasis on food exports has made it so that farmers are concentrating on just a few varieties of quinoa, abandoning most other varieties.
(5) Quinoa is now being grown outside of Peru and Bolivia. Countries such as the United States, China, India, and Nepal are all growing quinoa, and consequently prices are going down—according to Cherfas (2016), after a continued rise in prices, the prices of 2015 are similar to those of 2012. A decrease in prices may have the salutary effect of having farmers going back to more traditional and healthier methods of production, but a volatility in prices may present a less clear picture for ensuring the long-term sustainability of the crop.

Concluding Thoughts

Marginalized knowledges are an integral part of the cultural wealth of poor communities worldwide, and they have enabled members of these communities to live dignified lives with a sense of identity and belongingness. These knowledges, found at the edges of society, help its users satisfy basic material, social, cultural, economic, and spiritual needs. However, because they have been historically subjugated by larger hegemonic forces (either by force or by mechanisms of the capitalist market), these marginalized knowledges have been ignored, banned, vilified, or considered trite and useless by society at large. This chapter has concentrated on NUS as examples of cultural artifacts that, if given their proper attention, can greatly aid in addressing intractable social and environmental problems, such as poverty, climate change, and biocultural extinction.

The examples used in this chapter all come from Latin America, and it is only appropriate to end with an excerpt of a song co-written and sang by René Pérez (Residente), co-founder of the Puerto Rican band Calle 13 entitled "Latin America" (Cabra, Pérez, and Arcaute 2011). In this song/poem, Calle 13 rejects the capitalist approach of commodifying knowledge and commodifying life itself, which is the complete opposite of the vision represented by marginalized knowledges. The lyrics, written in the first person as if Latin America was a single voice speaking to the listener, defends the land as belonging to everyone and to no one at the same time. Just as the land is not for sale, so are the cultural and biological commons which belong to everyone in the continent, and these should not be privatized or exploited in ways that hurt the commons. The song defends the roots and traditions that have persisted throughout time, despite centuries of exploitation, subjugation, and abandonment. As the chorus read:

Tú no puedes comprar el viento Tú no puedes comprar el sol Tú no puedes comprar la lluvia Tú no puedes comprar el calor Tú no puedes comprar las nubes Tú no puedes comprar los colores Tú no puedes comprar mi alegría Tú no puedes comprar mis dolores	You cannot buy the wind You cannot buy the sun You cannot buy the rain You cannot buy the warmth You cannot buy the clouds You cannot buy the colors You cannot buy my happiness You cannot buy my pain

REFERENCES

Apffel-Marglin, F. and PRATEC, eds. 1998. The Spirit of Regeneration: Andean Culture Confronting Western Notions of Development. London: Zed Books.

Barba de la Rosa, A.P., Silva-Sánchez, C., and González de Mejia, E. 2006. "Amaranth: An ancient crop for modern technology." In Hispanic Foods: Chemistry and Flavor, edited by Tunick, M. H and González de Mejia, E., 103–116. Washington, DC: ACS Publications.

Bioversity International 2014. Bioversity International's 10-Year Strategy 2014–2024. Rome: Bioversity International.

Bowers, C.A. 2006. Revitalizing the Commons: Cultural and Educational Sites of Resistance and Affirmation. Lanham, MD: Lexington Books.

Cabra, E., Pérez, R., and Arcaute, R. 2011. Latinoamérica. On Entren los que quieran [CD]. Sony Latin.

Cardona, G. 2007. Delicias Vegetarianas de México (Vegetarian Delicacies from Mexico). Mexico City: Editorial Pax.

Cherfas, J. 2016. "Your Quinoa Habit Really Did Help Peru's Poor. But There's Trouble." National Public Radio. March 31, 2016. Online: www.npr.org/sections/thesalt/2016/03/31/472453674/your-quinoa-habit-really-did-help-perus-poor-but-theres-trouble-ahead

Diario de Gastronomía. 2012. "Los cultivos olvidados podrían ser la despensa del future (Forgotten crops could be the food of the future)." December 11, 2012. http://diariodegastronomia.com/los-cultivos-olvidados-podrian-ser-la-despensa-del-futuro/

Erdos, J. 1999. "Atawallpap Mikhunan: Quinoa, Mother Grain of the Incas." Planeta.com. http://old.planeta.com/planeta/99/1199quinoa.html

FAO and CIRAD. 2015. State of the Art Report of Quinoa in the World in 2013. Rome: FAO.

Mujica S., A., Jacobsen, S-E., Izquierdo, J., and Marathee, J.P. 2001. Quinua: Chenopodium quinoa Willd: Ancestral Cultivo Andino: Alimento del Presente y Futuro (Quinoa: Andean Ancestral Crop: Food for the Present and the Future). FAO: Santiago de Chile.

U'wa People. 2010. "Carta de los U'wa que Crea Consciencia sobre la Destrucción de los Blancos a la Tierra (Letter from the U'wa to Raise Consciousness about the Destruction of the Earth by the White Man)." Oil, the Blood of the Earth. www.bibliotecapleyades.net/gaia/esp_gaia32.htm

Weisman, A. 2008. Gaviotas: A Village to Re-Invent the World, 2nd edition. White River Junction, VT: Chelsea Green Publishing.
Zimmerer, K. 2017. "Fewer crops are feeding more people worldwide – and that's not good." *The Conversation*. November 28, 2017. http://theconversation.com/fewer-crops-are-feeding-more-people-worldwide-and-thats-not-good-86105

EXPANDING ENVIRONMENTAL SCIENCE LITERACY TO INCLUDE CULTURAL AND COMMUNITY KNOWLEDGE

Kristin L. Gunckel
Professor, Department of Teaching, Learning and Sociocultural Studies, College of Education, The University of Arizona

Public participation in decision-making about environmental issues at the local level is a foundational tenant of sustainability science (Kasemir et al. 2003). Such participation requires environmental science literacy, defined as the capacity to use model-based reasoning to make decisions about socio-ecological issues (Arvai et al. 2004; Gunckel et al. 2012; Mohan, Chen, and Anderson 2009; Sabel et al. 2017). Major science education policy and standards documents, such as the *Framework for K-12 Science Education* (National Research Council 2012) and the Next Generation Science Standards (NGSS Lead States 2013), promote this vision of environmental science literacy. They argue that science education must prepare students to engage in scientific practices such as modeling and making arguments from evidence in order to make decisions that address the world's most pressing problems, including addressing global climate change and ensuring adequate supplies of healthy food, fresh water, and clean air.

Yet, communities most impacted by environmental issues often feel marginalized from and by science. They may feel that science does not acknowledge the important role that their cultural knowledge plays both in doing rigorous science and in being responsive to the concerns and perspectives of the communities most impacted by environmental issues. A more inclusive definition of environmental science literacy grounded in a broader perspective of the role of cultural and community knowledge in conducting science and using science to inform public decision-making is needed.

In this paper, I point out the consequences of a nature-culture divide in science and provide two illustrative examples. I then offer an expanded framework for environmental science literacy that could possibly break down this divide and prepare students to use scientific model-based reasoning more justly when participating in democratic conversations about environmental issues.

Model-Based Reasoning in Environmental Science and the Consequences of the Nature-Culture Divide

A core practice of science is developing explanations based on evidence for how the world works (Giere 2010). Central to this practice is the construction of models, abstractions of complex systems that can be used to explain and predict phenomena. Constructing models involves iterative rounds of testing and revising models to improve their alignment with observations. Model-based reasoning is scientific thinking grounded in an understanding of the relationships represented in the model (Windschitl, Thompson, and Braaten 2008). It relies on identifying the mechanisms that explain how and why elements of a model work together. Such thinking is essential for making sense of complex environmental problems, making predictions about the impacts of environmental issues on people and ecosystems, and developing strategies for preventing and mitigating problems (Arvai et al. 2004; Gunckel et al. 2012; Mohan et al. 2009; Sabel et al. 2017). Therefore, a goal of science education is the development of model-based reasoning so that the public can appreciate and evaluate conclusions and advice from the scientific community.

Bang et al. (2012), however, argue that within the normative practice of science and the descriptions of science practice prevalent in science education, a nature-culture divide restricts the types and sources of knowledge valued in modeling and model-based reasoning. As such, cultural and community knowledge about the environment and people's relationships to the environment is devalued and excluded from science practice and science education. A consequence of this divide is the alienation of marginalized communities from scientific knowledge, resulting in skepticism within these communities of scientific findings and limits on their meaningful participation in preventing and mitigating environmental problems. The divide also segregates marginalized communities from potential allies because dominant, non-scientists who do not recognize and appreciate cultural knowledge of marginalized communities may ignore, or worse, work against local communities when participating in decision-making addressing environmental issues. Below, I offer two illustrative examples.

The Gold King Mine Spill

A mine accident in Colorado provides an instructive example of the consequences of the nature-culture divide and the possibilities that result from bridging scientific and cultural knowledge. In 2015, a cleanup operation at the Gold King Mine in Colorado resulted in an accidental spill of three million gallons of acidic, metal-laden water into the Animas River, a tributary to the San Juan River. The U.S. Environmental Protection Agency (EPA) conducted risk assessment studies to determine the impact of the spill on the downstream communities. They concluded that the spill did not pose a significant short-term health risk (U.S. EPA 2015). This finding was rejected by farmers on the Navajo Nation downstream from the spill because it did not reflect their cultural and spiritual realities (Beamer et al. 2016).

In conducting the risk assessment, the EPA used a model for exposure that assumed the main source of public contact with the water was from recreation (U.S. EPA 2015). The EPA justified this assumption as reasonable because the first communities impacted by the spill were towns with recreation-based economies from fishing, kayaking and rafting, and hiking along the river. However, the EPA did not take into account all of the ways that the Indigenous communities along the river used the water, including for farming, drinking, and spiritual practices (Chief 2016). From the perspective of the Navajo farmers affected by the spill, the limited assumptions about who used the water and for what purposes undermined the validity of the scientific model for making decisions about whether to use the river water to irrigate their corn.

In contrast, a more culturally inclusive study was conducted by Chief et al. (2017) that incorporated a Diné perspective of the San Juan River as a life-giving force.

Based on this cultural knowledge about the river, the researchers identified over 40 ways in which the Navajo community members used the water from the San Juan River. The researchers then investigated how lead and arsenic from the spill moved along the pathways that had the greatest potential to impact community members. This example shows how cultural knowledge and scientific models are not necessarily binary (Bang et al. 2012) and can be coordinated to produce rigorous science that is relevant and responsive to community concerns.

Chief et al. (2017) also recognized the role that cultural and historical contexts played in community acceptance of scientific research. Lay person acceptance or rejection of scientific models involves issues of trust and credibility of social institutions based on historic interactions (Wynne 1992). For the Navajo farmers, the EPA was suspect because of the centuries-long legacy of harm inflicted on Indigenous communities by the U.S. Government. The failure of the EPA risk assessment models to incorporate cultural knowledge was another injury. Sensitive to this context, Chief and her colleagues held regular community meetings, called teach-ins, which incorporated feedback mechanisms from the community. This dialogic structure functioned to create an opportunity for a historically marginalized community to participate in science discourse and begin disrupting the nature-culture divide that had contributed to their alienation.

Tucson TCE and 1,4 Dioxane Groundwater Contamination

Another example, from a different community, shows the dangers of ignoring local knowledge identifying and responding to environmental issues. In Tucson, Arizona, scientists from multiple agencies, including U.S. EPA, the Arizona Department of Water Quality, and Tucson Water (the local municipal water utility), and various environmental consultants for the Air National Guard, the U.S. Air Force, and companies such as Raytheon, meet regularly with community members to present data monitoring the cleanup of TCE and 1,4 dioxane from groundwater. The contamination is the legacy of dumping on an open desert of chemicals used to clean airplanes in the 1950s. In a story that has echoes in large and small communities around the world, the chemicals infiltrated into the groundwater and contaminated the drinking water of the nearby residents. In this case, it took 30 years for the problem to be publicly acknowledged and another 30 for remediation efforts to come on line. In that time, several generations of people were adversely affected with cancer, autoimmune diseases, and dental problems.

At these public meetings, the scientists give presentations filled with tables of numbers to show that the wells for which they are responsible are pumping out the contamination and the aquifer is being cleaned. But the local residents who attend the meetings question the numbers and the trends that are shown. Furthermore, they do not feel that the scientists understand their plight. For example, in the Tucson community affected, a family may live in the same house for multiple generations. People are born, grow up, and die in the same house, often living there for over 70 years. The consultants claim that the water is safe, but many families in the community believe that the effects from drinking contaminated water for 30 years span generations, and that even though the water now meets drinking water standards, their children are affected by the contamination that was present years ago (Davis 2017). The people express anger because the numbers that the consultants show, the evidence they present, do not match the people's experience and they fear that once again, science has let them down. The level of distrust in the community is high. Nevertheless, the community advisory board, which includes representatives from the agencies as well as local residents, is an attempt to move past the divisions and work together toward solutions.

A More Inclusive Framework for Environmental Science Literacy

Both of these examples show how model-based science failed to acknowledge or incorporate the local and cultural knowledge of the communities whom they were working to protect. Scientific model-based decision-making does not exist outside of the socio-cultural context and failure to take that into account when using scientific models to address local problems leads to increased mistrust and the perpetuation of environmental injustice. One could develop a list of lessons for scientists working in these communities to help them better communicate with the local people, to build trust, and to begin to address the problems that in these cases science helped to create.

However, there are also lessons here for building environmental science literacy among people who are not scientists, particularly related to what we should expect students to understand and be able to do by the time they finish high school. A definition of environmental science literacy that only includes a model-based understanding of science limits the usefulness of environmental science literacy for making decisions about socio-ecological issues because it hides the reality that science is a social and political endeavor, that people are part of systems, that local and community knowledge are valuable, and that sustainability has moral and ethical dimensions. Without these elements, model-based science is insufficient as a tool for promoting sustainability and environmental decision-making.

I propose a framework for environmental science literacy that expands on model-based reasoning as the basis for environmental decision-making to include model-based reasoning, knowledge of the socio-cultural-historical context in which the issue is embedded and science will be used, incorporation of cultural and community knowledge, and recognition that science can both perpetuate and address social injustices and environmental issues.

Scientific Model-based Reasoning

Scientific models and explanations are still foundational to an expanded definition of environmental science literacy. Students are served well by having the ability to interpret and evaluate the arguments presented by scientific experts. The first step in determining the validity of arguments presented at public meetings or by government agencies is to be able to make sense of the models and explanations used in the arguments. Students must still be able to interpret graphs, charts, and other data representations to understand scientific claims, the evidence for those claims, and the reasoning to support the argument. Model-based reasoning is necessary to evaluate claims made by experts and then use those claims to inform personal and collective decisions relative to an environmental issue that affects one's life. In the case of the Gold King Mine spill, understanding the mechanisms of acid mine drainage, the interconnected nature of watershed, the flow of water through aquifers, and principles of solution and dilution are essential for recognizing and responding to the crisis. Similarly, evaluating the potential threat from TCE and 1,4 dioxane in the groundwater requires understanding principles of groundwater flow and contamination concentrations.

Understanding the Socio-cultural-historical Context

Neither the U.S. EPA nor the local municipal water utility in Tucson recognized the socio-cultural-historical context in which the science they were conducting was situated. Nor, would I argue, did most non-residents interested in these cases. In the aftermath of the Gold King Mine spill, the Navajo Nation sued the EPA for damages, not only for the initial spill, but because from their cultural perspective, the water was damaged. However, the EPA's standards for water quality were not questioned by those off the Navajo Nation. They wondered why the Navajo Nation was pursuing the litigation against the U.S. Government, because by the EPA standards, after several weeks, the water was no longer a hazard. Similarly in Tucson, after the municipal water utility installed water treatment infrastructure, the water met government standards for water quality. Yet, the residents continue to complain about intergenerational impacts of drinking contaminated water for so many years.

There is a common perception, perpetuated in science and science education, that science is objectively neutral and can provide absolute criteria for decision-making, such as water quality standards and risk assessment models. On the contrary, science is a social construct and environmental problems are socially, culturally, and historically situated (Lederman, Lederman, and Antink 2013). Recognizing this social, cultural, and historical perspective counters the nature-culture divide and positions environmental science literacy as a community-based asset (Calabrese Baron and Roth 2004). Non-native people questioning the Navajo Nation lawsuits failed to recognize the multiple ways the Diné used the water and did not recognize that the EPA standards did not account for these contamination impacts. They also failed to consider the past water disasters, such as a uranium spill into the same river in 1979, and multiple other violations at the hands of the U.S. Government, against which a legacy of mistrust was built (Black 2018). In Tucson, recognizing the impact of over 50 years of contaminated water on families and communities is important for considering how and why the science failed the residents affected by the contamination. Water quality standards, while based in science and scientific models, will always be subjective criteria and recognizing when and how standards and risk assessments are responsive or not to the social, cultural, and historical concerns of the people affected is a necessary part of being literate in environmental science.

Incorporation of Cultural and Community Knowledge

For environmental science literacy to be useful to the public for participating in the prevention and mitigation of environmental problems, it must include the capacity to recognize and include cultural and community knowledge. This is not a relativistic argument that all knowledge is epistemically equivalent. It is, rather, an acknowledgment that the divide between cultural knowledge and scientific knowledge about the environment is and should be permeable. Just as scientific models can and should incorporate cultural and community knowledge, so should the public when considering the arguments made by experts on the basis of such models. Thus, a public literate in environmental science should recognize and value the ways that the local community's experiences, perspectives, and ways of understanding the world can and should be incorporated into scientific models, and hold scientists accountable when it is not. Scientific communities and lay communities play a role in making environmental science both more rigorous and more responsive to all whom are affected by environmental problems.

Recognition Science Is an Instrument both Perpetuating and Addressing Social and Ecological Justice

Finally, environmental science literacy should include a critical dimension that recognizes the ways that normative definitions of science and scientific models can contribute to the alienation and marginalization of non-dominant groups affected by environmental problems. Both the Gold King Mine spill and the Tucson TCE and 1,4 dioxane contamination stories are illustrations of how model-based science was complicit in perpetuating social injustices at the same time that scientists were purportedly addressing environmental disasters. Dismantling the nature-culture divide can expose how science reflects the perspectives and actions of those who use its tools and how when those who wield its power do not recognize and include the knowledge of those most impacted by environmental issues, the science can do as much or more harm as good. Environmental science literacy must include this critical dimension or it too will fail to achieve its promise to help individuals and communities address their environmental concerns and will instead perpetuate social and ecological injustices.

Conclusion

In the current political context, the legitimacy of science is under attack. There are many who reject its tenants and its findings as irrelevant, immoral, or untrue. Frequently, these attacks are baseless and emanate from the far right for purposes of political gain. Nevertheless, the practice of science is not without problems. Retaining its status as a public good requires that science acknowledges the ways that the institution has been complicit in social injustices. If sustainability science has a hope of productively engaging the public in participating in decision-making around important topics such as how to confront climate change and ensure access to good food, air, and water, then the definition of environmental science literacy must expand. Including in this definition cultural and community knowledge may help environmental and sustainability science transcend the nature-culture divide that has hindered public participation of marginalized communities in using science to address the environmental issues that they face.

REFERENCES

Arvai, J., Campbell, V. E. A., Baird, A., & Rivers, L. 2004. Teaching students to make better decisions about the environment: Lessons from the decision sciences. *The Journal of Environmental Education* 36 (1): 33–44.

Bang, M., Warren, B., Rosebery, A. S., & Medin, D. 2012. Desettling expectations in science education. *Human Development* 55 (5–6): 302–318.

Beamer, P., Chief, K., Borrero, N., & Rivera, B. 2016. "Water is our life": How a mining disaster affected the Navajo Nation, *Truthout.org*. Retrieved from https://truthout.org/articles/water-is-our-life-how-a-mining-disaster-affected-the-navajo-nation/#

Black, E.E. 2018. Case study of the Gold King Mine spill. North Dakota State University.

Calabrese Barton, A., & Roth, W.-M. 2004. *Science Education as and for Citizen Science Rethinking Scientific Literacy* (pp. 157–179). New York: Routledge Falmer.

Chief, K. 2016. Navajo Gold King mine exposure project (Vol. Fall 2016): University of Arizona.

Chief, K., Beamer, P., Ingram, J., Billheimer, D., Torabzadehkhorasani, E., Settimo, A., & Ornelas, Y. 2017. *Incorporating Diné perspectives in assessing temporal and spatial changes of contaminants after the Gold King Mine spill in Navajo agricultural communities*. Paper presented at the AGU Fall Meeting Abstracts.

Davis, T. 2017. "South-side Tucsonans mobilize for another water-pollution struggle." *Tucson Daily Star*. Retrieved from https://tucson.com/news/local/south-side-tucsonans-mobilize-for-another-water-pollution-struggle/article_25aafd92-3a5a-5550-bbd6-9fc0d02cd0dd.html

Giere, R.N. 2010. *Explaining Science: A Cognitive Approach*. Chicago, IL: University of Chicago Press.

Gunckel, K. L., Covitt, B. A., Salinas, I., and Anderson, C.W. 2012. "A learning progression for water in socio-ecological systems." *Journal of Research in Science Teaching* 49 (7): 843–868. doi: 10.1002/tea.21024

Kasemir, B., Jäger, J., Jaeger, C.C., and Gardner, M.T. 2003. *Public Participation in Sustainability Science: A Handbook*. Cambridge: Cambridge University Press.

Lederman, N.G., Lederman, J.S., and Antink, A. 2013. "Nature of science and scientific inquiry as contexts for the learning of science and achievement of scientific literacy." *International Journal of Education in Mathematics Science and Technology* 1 (3): 138–147.

Mohan, L., Chen, J., and Anderson, C.W. 2009. "Developing a multi-year learning progression for carbon cycling in socio-ecological systems." *Journal of Research in Science Teaching* 46 (6): 675–698.

National Research Council. 2012. *A Framework for K-12 Science Education: Practices, Crosscutting Concepts, and Core Ideas*. Washington, DC: National Academies Press.

NGSS Lead States. 2013. *Next Generation Science Standards: For States, By States*. Washington, DC: National Academies Press.

Sabel, J.L., Vo, T., Alred, A., Dauer, J.M., and Forbes, C.T. 2017. "Undergraduate Students' Scientifically Informed Decision Making About Socio-Hydrological Issues." *Journal of College Science Teaching* 46 (6): 64–72.

U.S. EPA. 2015. "Frequent questions related to Gold King Mine response." United States Environmental Protection Agency. Retrieved from www.epa.gov/goldkingmine/frequent-questions-related-gold-king-mine-response

Windschitl, M., Thompson, J., and Braaten, M. (2008). "Beyond the scientific method: Model-based inquiry as a new paradigm of preference for school science investigations." *Science Education* 92 (5): 941–967.

Wynne, B. 1992. "Misunderstood misunderstanding: Social identities and public uptake of science." *Public Understanding of Science* 1 (3): 281–304.

ENVIRONMENTAL JUSTICE PROMOTION THROUGH EXPERIENTIAL OUTDOOR EDUCATION ON THE RIVER—PART 2: HEAL (HUMAN RIGHTS, ENVIRONMENTAL EDUCATION, ACTIVISM, AND LEARNING)

Abbreviated Version for Publication

Dakota Goodman
College of Natural Resources, University of California, Berkeley

Senior Thesis, Spring 2018

Advisor: Kurt Spreyer

AUTHOR'S NOTE

Environmental psychology research has shown that exposure to nature changes the individual's psyche. The following text contains excerpts from HEAL, which was part 2 of what is intended as a three-part project. Part 1 was my UC Berkeley, Human Right Interdisciplinary minor thesis and nonprofit proposal (Goodman 2016). Part 2 was my UC Berkeley, Society and Environment senior thesis which presented observational and statistical data, testing my theory with an outline for the next step of field research. Part 3 will include reports on the testing of this theory in the field, on river trips.

The important findings to note include the concept of awe, the three-day effect, the importance of the river context, the history of outdoor education on river trips and gaps that need to be filled. This is a social justice, environmental justice, and public interest-oriented project. The results in this thesis present data from a pilot study.

INTRODUCTION

Modern citizens of the world have lost connection with each other and with the environment. With the 21st-century age of technology comes alienation. We have become disillusioned. Long-lasting connection with other people and places is much harder than it used to be. The purpose of this research is to establish a program that supports compromise and education that will give people of all ages, genders, ethnicities, and backgrounds the tools to build connections to the people around them and to the environment, both local and farther away. The classroom context is a river canyon and that is where initial connection to environment will happen, if not already established. Once an individual has gone through the program, they will be able to connect to any environment they situate themselves in.[1]

Overall, there are plenty of amazing programs that seek to help at-risk individuals or have targeted educational and political goals. However, to combine the idea of EJ

1 Note that environment in this context is referring to the natural world but the techniques will stimulate connection to natural and urban environments. The psychology research that will be discussed is referenced as environmental psychology. This is also sometimes understood as environmental ecopsychology—an emphasis on the natural world's impact on human consciousness.

promotion through experiential outdoor education (EOE) on the river, to teach human rights education (HRE) and environmental awareness to all sectors of society, specifically utilizing the power of river canyons on the psyche, has yet to be done. There is a gap in current theories and practices.

Gap Defined

There have been/are programs to get underprivileged, at-risk youth, and veterans on river trips, organizations that have brought politicians on river trips, promotion of social justice while rafting, seminars for guides to learn to teach environmental science to promote and protect environmental rights. This suggests that there is a possibility to teach human rights, with a focus on EJ, to all sectors of society, on multi-day river trips (Goodman 2016). The educational techniques that can be used include Environmental Education (EE), HRE, and EOE. There is a combination of education and psychological research findings that point to the transformative experience observed by those who have spent time on rivers (APHA 2013; Aasen, Grindheim, and Waters 2009; Bricker and Kerstetter 2000). These findings provide additional support to the hypothesis on *environmental justice promotion through experiential outdoor education on the river.*

The stressors of the modern world are causing a mental health epidemic. Certain sectors of society are at greater risk than others, and those groups of people when tested will reveal more substantial changes in their psyches (Anderson 2018). However, all people benefit from exposure to nature, urban and rural. The rural exposure, in a wilderness context, deep within a river canyon, is the most effective. This is especially true if the individuals are exposed to the area for three days or longer (Great Outdoors Lab Conference 2016). For the urban context, groups associated with the Great Outdoors Lab, over the past couple of years (2015–2017), have proposed prescribing time in parks instead of medications as a treatment for anxiety and trauma (Great Outdoors Lab Conference 2017).

The human rights, EJ component, as defined in this hypothesis, look at EJ as a way to define the human rights to equal access to nature and the environment.[2] Time in nature teaches an individual what it means to fail and succeed. The health and well-being of the individual correlate with the health and well-being of the earth. By promoting the use of nature for the health of human beings, there is a connection being established, which should lead to the further desire to protect the environment.

HEAL Hypothesis

"Environmental justice promotion through experiential outdoor education on the river." EJ, as a human right for all individuals to have equal access to nature and environmental benefits, can be promoted through EOE on multi-day river trips. How do we incorporate a way to break down social discrimination behaviors that are part of most individuals subconscious, while educating and reconnecting individuals to their public lands and waterways?

2 In 2016, access to nature, food, water, and shelter were at the top of the list of basic needs for patients that circulated through the Oakland Children's Hospital. From the data collected by the Great Outdoors Lab and my previous research, I believe that nature can be used as a medicine.

HEAL Intervention

The intervention approach accesses people through an educational program that targets EE and team building education (HRE) to promote connections between individuals and from individuals to their environment.

HEAL Central Research Question

How effective is the curriculum in achieving emotional, behavioral, and value changes? How will the observational and survey data help refine the curriculum?

HEAL Sub-Questions

Can changes be observed through testing the proposed hypothesis? What kind of responses resulted from the hypothesis' approach? Can the intervention evoke behavioral and/or emotional change? Does this change reflect a transformation in the individual's values? Differences in different groups/individuals responses?

The indicators measured and assessed were for the two main categories of behavioral and emotional transformation. Behavioral change is seen as an alteration in individual actions and interactions with other people or places. Emotional change is seen through new emotional openness to people and places; "reactions to threats and opportunities in our environment" (Anderson 2018). First, the question of if there were changes and how to account for those changes was asked. Then, if there were changes, the question of, what were the drivers and manifestations of these changes, was presented. "Is awe an effective driver of the psychological changes observed?" The level of effectiveness and lessons for the future is the concluding question for this phase of the research and the question that will be tested on river trips. To compare the observational change data with harder evidence, a survey was sent out to guides and customers to better interpret profiles of knowledge, attitudes and values, and behavior and practices before and after the river trips.

Methodology

As a commercial guide, I wanted this dual-purpose methodology for personal growth within the professional field and for observational data collection to help answer my research question. Observing while guiding allowed me to observe many different groups of people over a 3.5-month period, for several summers in a row. My end goal with this research is to offer the program as a specialty trip that commercial companies can run. The program I am proposing would train guides to go into the field prepared to integrate the educational curriculum into the trip (Goodman 2016, "Literature Review").

HEAL Research

(Author's Note: This section contains new research that was not included in part 1 of this study.)

Exposure to nature has been scientifically proven to benefit human health. Gregory Bratman et al. published a paper in 2015 on how "Nature Experience Reduces Rumination and Subgenual Prefrontal Cortex Activation." This study looked at the impacts of urbanization on mental health. With increasing urbanization, increasing mental health illness levels, such as depression and anxiety, have also increased. Additionally, Steller et al. in their article "Positive Affect and Markers of Inflammation: Discrete Positive Emotions Predict Lower Levels of Inflammatory Cytokines" discussed why studying cytokines is an important marker of health (Stellar et al. 2015, 129). The identification of awe as a strong indicator of proinflammatory cytokines links the importance of access to nature. In nature, on a river canyon, one can both heal and change emotionally and behaviorally from these natural effects.

In 2017, members of the BSI lab published the chapter "Emotion in the Wilds of Nature: The Coherence and Contagion of Fear During Threatening Group-Based Outdoors Experiences" (Anderson, Monroy, and Keltner 2017). This team took the research on awe and the impact of nature on the psyche specifically to a river context (Anderson, Monroy, and Keltner 2017). The researchers conducted this study to look at emotional expressions. They analyzed two different social properties of emotion, informative and evocative. Informative has to do with the social context and evocative has to do with the systematic responses evoked from emotional response. The specific emotion they decided to target was fear and also how fear related to cortisol levels. Previously, the BSI lab showed that there are higher levels of cortisol in individuals exposed to nature through the impact of awe. "We recorded spontaneous expressions in the face, body, and voice of four emotions relevant to the emotional context of white- water rafting: fear, amusement, pride, and awe" (Anderson, Monroy, and Keltner 2017; Keltner and Bonanno 1997; Piff et al. 2015). Fear was shown through eyes and lip movement. Amusement was shown through cheek movement. Pride was shown through arm raising and posture. Awe was shown through facial expression. Fear was also shown through voice reactions such as screaming. This research took into account different intensities of rapids and had the participants record a rafting diary. To collect the cortisol data, saliva samples were taken. These findings of emotional response from a river trip can be paired with the previous research on the impact of awe on the psyche when exposed to nature, to add further proof for the hypothesis proposed in this paper.

In addition to the paper on emotion, Craig Anderson and the rest of the team looked at how "Awe in Nature Heals: Evidence from Military Veterans, At-Risk Youth and College Students" (Anderson, Monroy, and Keltner 2018). In the first study, they looked at veterans and at-risk youth to see how a river trip evoked awe and how that impacted those two groups. In the second study, they looked at the impact of exposure to nature on the daily and the evocation of awe, and impacted a student population. This research team, taking into account that there are many factors effecting emotional response, showed the importance of evoking awe on the daily and in "extraordinary" events (white water rafting). "These findings suggest that awe may be one active ingredient in the remedy that is time spent outdoors" (Anderson, Monroy, and Keltner 2017).

At the 2017 Great Outdoors Lab Conference, Dr. Dacher Keltner, Professor of Psychology at UC Berkeley and Director of the Greater Good Science Center, discussed the importance of awe for physical and mental health. The studies from the Berkeley lab showed that the cytokine system becomes less inflamed from feeling awe. Keltner stated that "thirty percent of our awe experiences come from interactions with nature" (Berkeley Social Interaction Laboratory 2005). Even though awe can be evoked by everyday nature, and there is a small impact on the psyche, it is not the same as full immersion in nature. Dr. Nooshin Razani, a Pediatrician at UCSF Benioff Children's Hospital, discussed how "social inequalities in access to nature doesn't correlate with desire to be in nature" and that "green space was one of the top three social needs" for patients. Razani and Keltner both suggested that mental illness, anxiety, and trauma could be treated with exposure and time spent in nature. Peter Mayfield, the founder of the Gateway Mountain Center, discussed the importance of full immersion for total impact of awe on the psyche. He also pointed to the opportunities for risk management and social work to be conducted in the full emersion settings.

Methods

(See also "Tables and Site Description" from a full-length version of "HEAL.")

Data collection included observational data collection and survey statistics analysis. The data was collected to understand responses through three categories: emotional, behavioral, and value changes. Demographics were also taken into account.

To further test my hypothesis, after completing my HRI thesis in 2016, I collected observational data throughout two commercial river seasons in Idaho. In addition to the observational data, surveys were released to customers and guides from the 2017 season for statistical data. The observational data was collected on trips I worked on as a guide and trips I lead. The observational data from 2016 is mostly from a guide's perspective and from 2017 there is more data from a trip leader's perspective. It is important to note that on my trip leads, I was able to control the layout of the trip a little bit more than just working as a guide.[3] The educational curriculum I wove into my guiding was meant to appear as just a normal part of how I worked as a river guide or trip leader. I opened up the conversation on collecting observational data with the guides in the beginning of the 2017 season. With this conversation started, I was able to have them share some of their observations in the moment. Additionally, providing a brief introduction to the material I was studying allowed them to already have an idea of why I was asking the questions I did with the survey, which increased the amount of responses for the statistical data. Also to be noted is the difference in types of trips, whether they were family trips, adult-only or specialty trips, and where they were located. The OARS Idaho operation runs trips on the Middle, Main and Lower Salmon River, Hells Canyon, and the Owyhee River when water levels permit. These trips are typically four to six days long with the exception of some three-day Hells Canyon trips. The Idaho operation had begun an outreach program for youth day trips but none of those trips are included in this data set.

The observational data was collected via an educational study with tools of psychology research interwoven. The intervention was not meant to be obvious, or in full form of what the education program will eventually shape up to be. Through the

3 For part 2 of this larger project, the data collected had no true "control variable." The field research that is proposed at the conclusion will have a defined control variable. The purpose of this stage of the research is to prove that initial testing of this hypothesis works.

observational study, the intervention was subtle, a personal twist on guide technique. The monitoring came in the form of notetaking every evening after seeing the interactions of the trip participants throughout the day (see also Appendix 1 from a full-length version of "HEAL").

During the 2016 season, I wrote down everything I observed in terms of behavioral and emotional change. In 2017, I established a more quantifiable method of observing the participants.

Data Collection

(See also "Appendix 1" from a full-length version of "HEAL.")

Observational Data Tools

To supplement the interpretation (interp) talks that OARS Idaho already highly encourages guides to provide for the customers, I made sure to include "team building exercises" as a form of the human rights educational curriculum. Riddles, games, and for my trip leads a closing ceremony.[4] For the trips I led, a typical intern list would include information about the Native Americans, the Chinese miners, the frontiersmen and women, salmon lifecycles, geology, river restoration and preservation/ environmental history, history of OARS and Dories, and land and resource management. I would also try to include information on ecology, botany, hydrology, public policy, and other environmental or earth sciences when the guides could speak on those topics.

On every trip, whether or not I was the trip leader, interp was incorporated into the daily schedule. OARS Idaho promotes the education of the river basin of which people are traveling through. The interp taught on Idaho OARS river trips includes Native American history, ecology, salmon, geology, and environmental history with other specialties highlighted depending on the crew. When I trip lead, I talk to my customers about the story of the dory I am rowing, the name and why it was named after the harmed or forgotten place, and why Idaho OARS still runs dories. As the modern river conservation movement picked up speed in the late 1960s, the commercial guiding industry played a huge role in spreading awareness and education of these places to customers and local residents that the guides lived near during the off-season. Martin Litton founded Grand Canyon Dories with the idea that each boat would be a symbol for a harmed, forgotten, or needing to be restored river or waterscape. Curt Chang, with Litton's blessing, went up to Idaho to start Northwest Dories, and save the Idaho Rivers. Litton worked with David Brower in the initial modern dam fights. George Wendt, Litton's successor was also part of the battle to save Glen Canyon (Goodman 2017). When Wendt took over the company and merged OARS and Dories, the legacy continued, which is why I was able to row the Owyhee, the Malibu Canyon, and the Glen Canyon dories throughout the 2017 season. The impact of Western culture can be traced farther back than dams, in

4 When I was the trip leader, I would always end my trips with water cheers and a poem for them to reflect on. Water cheers was a way to bring everyone together at the end of the trip in another group setting and have everyone reflect on their experience. The question posed was to "cheers" to one of your favorite memories or takeaways from the trip.

these river canyons. History of the frontier, mining, and railroad, and the social implications of those events are all discussed throughout the trips. These canyons where the initial observational and survey data research was conducted are perfect testing locations because of the length of the river trips and the immense natural and human histories that can be taught, in addition to the modern-day environmental science education.

Observational Data Analysis

To analyze this portion of the data, I transcribed all of my handwritten notes from the summer and looked for common themes. Some of the main trends I was looking for included the three-day effect, signs of comfortability with people and the environment and how those signs evolved over the four to six days on the river, interactions out of the norm, openness, reflection of the interp education, and reflection of where the individuals were in time and place.

Survey Data Tools

In addition to the observational data, I created and distributed surveys for statistical data. The survey was split into two questionnaires: one for guides, the other for customers. The survey was sent out to the 2017 guide listserv and to all the trip email strands I participated in during the 2017 season. They were asked a series of demographic questions. The questions, besides the fill in the blank open-ended responses, were based on the Likert scale. (The questions can be seen in Appendix 1 from a full-length version of "HEAL.")

The main research question and sub-questions drove the development of the survey. When developing the survey, the sub-questions were further flushed out to target each interdisciplinary category in this research. For HRE, the question was: Does the behavioral and emotional, psychological change, lead to more openness in an individual's perspective on new values, ideas, and actions? What changes occur? For EE: participants report/recognize the impact of awe from the river canyon? How did awe play a factor in the absorption of information from the EE curriculum? Team building curriculum? Lastly, I composed a question that targeted the issue of a divided society and the hope for the upcoming generation to be a generation of compromise. For alienation: Will this technique promote growth of compromise skills for leaders to utilize while building social, political, and economic policy? More specifically, can this technique work as a call to action for protecting waterways because a better understanding of watersheds has been laid down and an emotional connection to the place itself/the people whom an individual shared the place with has been established?

Survey Data Analysis

The goal of the survey was to collect data on emotional, behavioral, and value changes in customers and guides. The guides' survey had one more category of questions to answer about their observations of how the guests reacted and changed in the four- to six-day river trip. The two different versions of the survey were built on the Qualtrics platform and distributed via email listservs from the trips of the 2017 season that I was a guide for or lead. The survey was open for two and a half weeks and there were 61 responses from the customers and 26 responses from the guides. Once the survey was closed, the data and reports were exported. (The reports can be seen in Appendix 1 from a full-length version of "HEAL.") To analyze the survey data, I looked at the statistics of the responses to questions posed and the open-ended responses wording/meaning. When analyzing wording and meaning of the open-ended responses, some of the things I was looking for were signs of awe, increased openness, and self-reflection. I also compared the statistics from the guides' responses to that of my observational data.

Results

Overall, the data sets showed that there is an impact of multi-day river trips on the individual's psyche. These impacts were analyzed in terms of being able to benefit from emotional, behavioral, and value changes. These changes were a result of the educational context, the river, and the education being presented. The changes were amplified, in comparison to the previous two seasons where the program was being thought of but not yet observationally tested. (See also Appendix 1 from a full-length version of "HEAL.") Additionally, it should be noted that these results came from a limited data pool. In the next steps of this study, I intend to ensure more diversity in the sample sets, which, in turn, will reflect the purpose of this research—to promote all individuals' access to healing on the river.

Observational Data Results

The three-day effect worked. It was clear that the river trip participants' behavior and emotional state began to significantly change after day 3. In the case that a sense of teamwork, comfortability with their surroundings and the people they were sharing the experience with, was clearly developing before three days, it usually had to do with a "significant" event. These significant events include boat flips or other unexpected events. It is noteworthy that even when the trip consisted of a big family group or a big group of friends, people who already knew each other, similar changes could be observed. (See more from a full-length version of "HEAL.")

Overall, even in the trips where the three-day effect was not obvious, there were still changes observed. The connection to place and openness observations were recorded throughout all of the river trips, but it is significant when within the week on the river, an individual explicitly states the change, which is why those two categories above are highlighted.

Survey Data Results

The survey data was consistent with the observational data. Overall, there were 61 customer responses and 26 guide responses. Both groups of people showed that there is in fact a huge change in psyche from a multi-day river trip with different educational techniques woven into the everyday technique of the guides. Survey results showed positive emotional, behavioral, and value changes in customers and guides. The customers showed a more drastic response because river trips are a rarer experience (confirmed with data), whereas for guides the river experience has been normalized.

These levels of change helped identify which areas need to be concentrated on more in following studies.

Customers

The majority of respondents from the customer survey were between ages 50 and 70 or 19 and 30. An almost equal number of responses came from males and females. The majority of respondents identified as white. 37% of respondents were from the Pacific Northwest and 31% were from the Eastern part of the United States. There were also people from the Southwest, Midwest, and outside of the United States—the U.K. The largest percentage of respondents at 34% had earned a master's degree and 25% had earned a bachelor's degree. The largest percentage for bracket of income was 35% at $100,000–149,000 then 28% at over $200,000.

The survey results from the customers showed that 39.2% had been on 2–5 river trips, 27.5% had been on 1 river trip, 13.7% had been on 6–10 river trips, 13.7% had been on 20-plus river trips, and 5.9% had been on 11–20 river trips. The largest group of people had been on either 2–5 river trips or just 1 river trip. The levels of outdoor activity with the last year, five years, and life as an adult were pretty consistent across time span categories and the given activity. Backpacking, various snow sports, and hunting were three categories where it appeared that there was more activity throughout their life as an adult versus the past year or five years. The majority of customers, 85%, were "extremely pleased" when asked how much they enjoyed their trip. There were a few outliers. The knowledge categories showed that a "moderate amount of knowledge" was gained about ecology, land and natural resource management, public policy, history of peoples who inhabited the canyons, and geology. Additionally, "a lot" of knowledge was gained in the categories of history of peoples who inhabited the canyons.

The fill in the blank answers showed huge takeaways with knowledge, personal growth, and relaxation. These comments showed a connection to the environment and other people. The responses about personal growth showed significant psychological change and self-reflection. The responses about their biggest takeaway in terms of relaxation showed significant impacts on the individual from time spent in nature, whether those impacts were reflections on technology, the high pace life of urban environments, or interactions with other people. Lastly, there was a fill-in-the-blank category asking about the biggest takeaway from the trip not listed in the previous three categories. These responses showed the first thing that comes to mind for the individuals who responded to the survey, after being asked a set of questions on educational, personal growth, and relaxation takeaway prompts. The open response takeaways further reinforce the observational conclusions of psychological change on emotional, behavioral, and value levels. These are further explored and proven with the next set of survey response data below.

The next set of survey questions displayed the following results. The majority of customers said that they would go on another multi-day trip or both (a single and a multi-day). There were zero responses recorded for "would not go on another trip." 95.2% of customers who responded answered "yes" to whether or not they had fond memories associated with the experience. Overall, the responses above continued to display a positive impact of the river on an individual's psyche.

Customers: Emotional Change

The next set of questions were aimed at analyzing emotional change. The question, "have you felt more comfortable around new people," got the highest responses for "not at all" and "some"; both responses were 34%. The question, "have you connected with 'strangers' in passing conversation more easily," got the highest response with "some" at 43%. The last question in this section was "have you been able to more easily imagine yourself in another's 'shoes,'" and the highest response rate was "some" at a percentage of 32%.

I then asked if the customers' daily habits had changed after the trip and allowed multiple answers. There were high response rates for all of the categories which included, going outside more, being more active, more involved in community events and other. Either the respondents already were doing the above behaviors, or they increased. One of the interesting responses included: I am much more conscious of the waste I personally produce, and have abandoned completely using foam and paper coffee cups—I use my stainless steel "No Trace" cup every day for my coffee.

Customers: Behavioral Change

The next set of questions were about behavioral change. 46% of customers responded "some" to the question: have you found yourself more willing to listen to opposing opinions or comments? The majority of respondents did not notice "any other behavioral changes" without being prompted. 59% of respondents said that they did not make resolutions after the trip but 25% did. Overall, the respondents did not become more active in community service.

Customers: Value Change

Value change was evaluated next. 31% of respondents said "a little" in terms of the question: do you value the environment more after the trip? This can be interpreted as most people already valued the environment if they voluntarily were coming on the trip. 35% of respondents said that they valued public lands "a good deal" more after the trip. Before the trip, 24% of respondents were inclined to support river conservation "a good deal" and after 44%. Before the trip, 29% of people were inclined to support river conservation "a great deal" and after 38%.

Overall, the customers' responses displayed significant behavioral, emotional, and value changes after their river trip. These responses also identified gaps in the study. Further analysis will include psychological impacts of "unplugging"—time away from technology—and how that impacts the effect of awe. Additionally, in part 3, the field testing will look at continued or maintained positive change post-trip after a month, a year, and five years. Lastly, the field studies will look at how to better target emotional change.

Guides

The guides' survey results were in agreement with the observational analysis data that I collected. When asked different questions about throughout the season observations, "a good deal" was the highest response rate at 46% to the question of whether or not the guides saw customers opening up more after day 3 of the river trip. The example was "showing their personality more." The next question was about customers interacting more after day 3 with other customers and 38% of guides said that this occurred "a good deal." When asked the same question but interaction with guides instead of other customers, 30% said "a good deal." The guides did not observe a high rate of customers asking for the email contact list before it was presented at the end of the trip—an indicator I used to judge how quickly they were connecting and a desire to stay in touch.

The majority of guides who responded were in the 19–30 age range. There was an even gender ratio response and the majority identified as white. Everyone who responded said that it was not their first commercial season and the majority of respondents were "extremely pleased" with their season. Most guides said that they would work another multi-day season again. In terms of knowledge gained, there was a spike after the season with environmental science, land and natural resource management, public policy, history of peoples, and earth science. There was "a little" knowledge gained after the season with EJ. Overall, the personal growth and relaxation takeaways for the guides showed a continued psychological change and impact from the river, even after an extended amount of time and experience outdoors.

Guides' Emotional, Behavioral, and Value Changes

For the emotional, behavioral, and value changes, the guides answered the same questions as the customers, just directed toward the overall season not just one trip. The highest response for the emotional change questions was "some" to the questions: have you felt more comfortable around new people? Have you connected with "strangers" in passing conversation more easily? Have you been able to more easily imagine yourself in another's "shoes"? When asked if their daily habits changed after the season, most guides recorded answers that did not show significant change in their daily habits. There were several responses recorded about the use of less technology and more involvement in environmental activism. In the table of questions about behavioral change, the statistics reflected the same results as the open-ended questions. One recorded answer was the simple word of "happier" after the season. In terms of value change, the guides responded "a great deal" to the questions: Do you value public lands more after the season? How inclined were you to support river conservation before the season and after? The majority of respondents said "a good deal" to the question: do you value the environment more after the season? Overall, the guides' responses about customer emotional, behavioral, and value development aligned with my observational data. Additionally, their responses displayed individual change from working on a river for a commercial season.

Discussion

The changes identified included a significant increase in openness toward new ideas and people, the river trip providing and facilitating important reflection time. The results also reflected decreases in materialism, self-focus, and stress.[5] Out of the three

5 Psychological change indicators listed in HRI Thesis Defense: Restorative benefits of nature include giving one's mind time to reflect, higher levels of concentration, increase in curiosity and intellectual purpose, sacrifice for group well-being, expanded self, generosity, purpose, perspective, and robust health.

categories targeted through the survey, behavioral, emotional, and value changes, both customers and guides' responses showed changes in all three categories. Reported behavioral change was the highest out of the three categories. To better target emotional and value changes in the future, continued monitoring and alterations in the curriculum will need to be implemented.

Hypothesis to Theory

Overall, this is an effective approach. The spreading of awareness to the commercial rafting community of the impact every-day-guide actions can have on our customers is significant. The educational curriculum is not a crazy new phenomenon; it is a combination of techniques already utilized so as to be integrated in a non-obvious way. It is the combination of educational techniques, the place setting, and the awareness of what to observe, that allowed changes to be made and seen. The new thinking around this project is how the research methods and techniques are combined and the target groups at play.

The guides are of great importance because they are a subgroup that reflects a change from long exposure to the river and they share their passion with the customers. The importance of guides in building a movement of activism to protect rivers cannot be understated. The importance of the commercial river guiding community in the modern river conservation fight has been displayed in the survey data and historically (see Stanislaus thesis for further details of historic impact; Goodman 2017). It is the guides who are the teachers in this scenario. They are able to share their knowledge and guiding skills in a context that naturally leads to positive emotional, behavioral, and value changes. Emotionally, customers open up to new ideas and experiences. Behaviorally, customers are able to connect to people who they would not have connected with in an urban setting. Everyone has some level of implicit bias and the subgrouping of modern society and alienation further separates individuals from connecting on a human level. Human rights is almost a taboo word. These rights are believed to be innate yet are violated on a daily basis. Historically, the Western society has been built on a ranking system and overall, society has always been organized in terms of some form of hierarchy. Taking away competition and ranking from society is not the goal of this project. The goal is to show that despite differences, it is possible to connect to strangers and to unknown places, in order to form progressive compromises that will allow us to fix some of the issues the world faces in the 21st century.

The masses' voting support plays an important role and gives them power to convince legislators to support positive water resource management. If this program operates as a way to build compromise skills and promote EJ, then it could target the human rights violations connected to environment, such as health issues and access to resources, and issues of protection and restoration of watersheds.

BIBLIOGRAPHY

Anderson, Craig L., Monroy, Maria, and Keltner, Dacher. 2018. "Awe in nature heals: Evidence from military veterans, at-risk youth, and college students." Greater Good Science Center.

Anderson, Craig L., Monroy, Maria, and Keltner, Dacher. 2017. "Emotion in the wilds of nature: The coherence and contagion of fear during threatening group-based outdoors experiences." *Emotion* 18, 3: 355–368. https://doi.org/10.1037/emo0000378

Anderson, Craig. 2018. Interview. UC Berkeley. February, 2018.

APHA. 2013. "Improving Health and Wellness through Access to Nature". American Public Health Association: 20137. November, 3 2013. www.apha.org/policies-and-advocacy/public-health-policy-statements/policy-database/2014/07/08/09/18/improving-health-and-wellness-through-access-to-nature.

Aasen, Wenche, Liv Torunn Grindheim, and Jane Waters. 2009. "The outdoor environment as a site for children's participation, meaning-making and democratic learning: examples from Norwegian kindergartens." *Education 3–13* 37, 1: 5–13.

Berkeley Social Interaction Laboratory, Keltner Lab, 2005, http://socrates.berkeley.edu/~keltner/publications.htm.

Bratman, Gregory N., J. Paul Hamilton, Kevin S. Hahn, Gretchen C. Daily, and James J. Gross. 2015. "Nature experience reduces rumination and subgenual prefrontal cortex activation." *Proceedings of the National Academy of Sciences* 112, 28: 8567–8572.

Bricker, Kelly S., and Deborah L. Kerstetter. 2000. "Level of specialization and place attachment: An exploratory study of whitewater recreationists." *Leisure Sciences* 22, 4: 233–257.

Buckley, Ralf, Aishath Shakeela, and Daniela Guitart. 2014. "Adventure tourism and local livelihoods." *Annals of Tourism Research* 48, C: 269–272.

Canter, David V., and Kenneth H. Craik. 1981. "Environmental psychology." *Journal of Environmental Psychology* 1, 1: 1–11.

Castillo, Stephanie. 2015. "Awe-Inspiring Moments Lower Inflammation Marker Cytokines, Positively Impact Health." Medical Daily. April, 3 2015. www.medicaldaily.com/awe-inspiring-moments-lower-inflammation-marker-cytokines-positively-impact-health-328092.

Cullet, Philippe. 1995. "Definition of an environmental right in a human rights context." *Netherlands Quarterly of Human Rights* 13: 25.

ETC. n.d. "ETC – About." *Environmental Traveling Companions*. www.etctrips.org/about.

Forsythe, David P. 2009. "Encyclopedia of Human Rights." Volume 5. New York: UC Berkeley Law Library/Oxford University Press. www.oxfordreference.com/view/10.1093/acref/9780195334029.001.0001/acref-9780195334029-e-83?rskey=mBoKwK&result=6.

Grand Canyon Youth. 2019. "Grand Canyon Youth." www.gcyouth.org/.

Frumkin, Howard, Bratman, Gregory N., Breslow, Sara Jo, Cochran, Bobby, Kahn Jr., Peter H., Lawler, Joshua J., Levin, Phillip S., et al. 2017. "Nature contact and human health: A research agenda." *Environmental Health Perspectives* 125 (7).

Goodman, Dakota. 2017. "The Personification of Natural Waterscapes: A Brief History of Friends of the River (1970–1992)." The Stanislaus Online Archive. http://stanislausriver.org/items/show/3419.

Goodman, Dakota. 2016. "Water Watchers-Education Advocacy Adventure! Environmental Justice Promotion through Experiential Outdoor Education on the River." Human Rights Interdisciplinary Minor Thesis, UC Berkeley. May 11, 2016.

Great Outdoors Lab Conference. 2016. *Sierra Club*, San Francisco. April, 2016

Great Outdoors Lab Conference. 2017. *Sierra Club,* Webinar. October, 2017.

Great Outdoors Lab. 2018. Sierra Club. https://content.sierraclub.org/outings/sierra-club-outdoors/great-outdoors-lab.
Hicks, Tom. 2006. "Headwaters Institute 2006 Curriculum".
Hicks, Tom. 2006. "Headwaters Institute 2006 Planning Guidelines".
"High Level Expert Meeting on the future of Human Rights and Environment: Moving the Global Agenda Forward," *United Nations Environmental Programme,* 2009, www.unep.org/environmentalgovernance/Events/HumanRightsandEnvironment/tabid/2046/Default.aspx.
"History of Outward Bound," *Outward Bound Peacekeeping,* 2014, http://outwardboundpeace.org/about/history-of-outward-bound/.
Hopkins, Karen. 2011. "Amnesty International's methods of engaging youth in human rights education: Curriculum in the United States and experiential learning in Burkina Faso." *Journal of Human Rights Practice* 3 (1): 71–92.
"Human Rights and the Environment." 2011. *United Nations Human Rights Office of the High Commissioner.* www.ohchr.org/EN/Issues/Environment/HREnvironment/Pages/HRandEnvironmentIndex.aspx.
"Human Rights Education and Training." *United Nations Human Rights Office of the High Commissioner,* 1996-2016. www.ohchr.org/EN/Issues/Education/Training/Pages/HREducationTrainingIndex.aspx.
Hungerford, Harold R., and Volk, Trudi L. 1990. "Changing learner behavior through environmental education." *The Journal of Environmental Education* 21 (3): 8–21.
Kaplan, Stephen. 1995. "The restorative benefits of nature: Toward an integrative framework." *Journal of Environmental Psychology* 15 (3): 169–182.
Keltner, D., and Bonanno, G. A. 1997. A study of laughter and dissociation: Distinct correlates of laughter and smiling during bereavement. *Journal of Personality and Social Psychology* 73(4): 687–702. https://doi.org/10.1037/0022-3514.73.4.687
Massey, Doreen. 2010. *A global sense of place*. Aughty.org, 2010.
Mcquoid-Mason, David. 2007. "Methods of teaching human rights: Some examples from street law legal literacy programmes." *Journal of Commonwealth Law and Legal Education* 5 (2): 127–143.
Mihaylov, Nikolay L., and Perkins, Douglas D. 2015. "Local environmental grassroots activism: Contributions from environmental psychology, sociology and politics." *Behavioral Sciences* 5 (1): 121–153.
Moye, Jayme. 2016. "Rough water: A river guide tells her side of the story." *Men's Journal.* www.mensjournal.com/adventure/collection/rough-water-a-river-guide-tells-her-side-of-the-story-20160120.
"Outward Bound – Our Mission." 2016. *Outward Bound.* www.outwardbound.org/about-outward-bound/outward-bound-today/.
Phillips, Brian, and Gready, Paul. 2013. "Introduction–Why a special issue on human rights education (HRE) and training?" *Journal of Human Rights Practice* 5 (2): 215–219.
Piff, P.K., Zhang, J.W., and Keltner, D. 2015. "On nature and prosocial behavior." In Bird, W., & van den Bosch, M. (Eds.), *Nature and Public Health: The Role of Nature in Improving the Health of a Population*. Oxford University Press: Oxford, UK.
Pooley, Julie Ann, and Moira O'Connor. 2000. "Environmental education and attitudes emotions and beliefs are what is needed." *Environment and Behavior* 32 (5): 711–723.
Schlosberg, David. 2007. "Defining environmental justice: Theories, movements, and nature."
Simmons, Deborah. 1998. "Using natural settings for environmental education: Perceived benefits and barriers." *The Journal of Environmental Education* 29 (3): 23–31.
Snake River Basin Image. 2017. "American Rivers." www.americanrivers.org/river/snake-river/.

"Special Rapporteur on human rights and the environment (former Independent Expert on human rights and the environment)." 2012. *United Nations Human Rights Office of the High Commissioner.* www.ohchr.org/EN/Issues/Environment/SREnvironment/Pages/SRenvironmentIndex.aspx.

Stellar, Jennifer E., John-Henderson, Neha, Anderson, Craig L., Gordon, Amie M., McNeil, Galen D., and Keltner, Dacher. 2015. "Positive affect and markers of inflammation: Discrete positive emotions predict lower levels of inflammatory cytokines." *Emotion* 15 (2): 129.

Taylor, Andrea Faber, and Kuo, Frances E. 2009. "Children with attention deficits concentrate better after walk in the park." *Journal of Attention Disorders* 12 (5): 402–409.

Tibbitts, Felisa. 2005. "Transformative learning and human rights education: Taking a closer look." *Intercultural Education* 16 (2): 107–113.

United Nations. 2014. "International Decade for Action 'Water for Life,' 2005–2015" *United Nations Department of Economic and Social Affairs.* www.un.org/waterforlifedecade/human_right_to_water.shtml.

"Universal Declaration of Human Rights." *United Nations*, 1948, www.ohchr.org/EN/UDHR/Documents/UDHR_Translations/eng.pdf.

Warren, Karen, Roberts, Nina S., Breunig, Mary, and Alvarez, M. Antonio G. 2014. "Social justice in outdoor experiential education: A state of knowledge review." *Journal of Experiential Education* 37 (1): 89–103.

Zhang, J.W., Piff, P.K., Iyer, R., Koleva, S., and Keltner, D. 2014. "An occasion for unselfing: Beautiful nature leads to prosociality." *Journal of Environmental Psychology 37*: 61–72.

Zhang, J.W., and Keltner, D. 2016. "Awe and the Natural Environment." In H. S. Friedman (Ed.) *Encyclopedia of Mental Health* 2nd ed., Vol. 1: 131–134. Waltham, MA: Elsevier.

8 Conclusions and Resolutions

SUMMARY

This chapter is a summary of the Networking for Science Diplomacy Panel Discussion from the conference, chaired by Christopher Scott, with panelists Nélida Barajas Acosta, Cherry Murray, and Robert Varady. They share three different perspectives on the importance of networking in order to sustain science diplomacy efforts—from the perspective of a binational NGO (CEDO Intercultural), an international network of agencies (InterAcademy Partnership), and an academic research center (Udall Center for Studies in Public Policy)—followed by some Q&A with the audience. This is followed by the introduction to the final keynote speaker, Léna Salamé, a lawyer

DOI: 10.1201/9781003220503-8

and specialist in conflict management and mediation. Salamé provides insight on the importance of knowledge and capacity building in order to achieve the political will necessary to affect real change.

This is followed by the concluding remarks from the honorary chair of the conference, Dr. E. William Colglazier, as he shares conclusions drawn from the conference by the rapporteurs, and the resolutions to continue these efforts at The University of Arizona and around the world.

NETWORKING FOR SCIENCE DIPLOMACY PANEL DISCUSSION

Chair:
Christopher Scott
Director, Udall Center for Studies in Public Policy, The University of Arizona

Panelists:

Nélida Barajas Acosta
Executive Director, Intercultural Center for the Study of Deserts and Oceans (CEDO Intercultural)

Cherry Murray
Deputy Director of Research, Biosphere 2 Institute, The University of Arizona

Robert Varady
Research Professor of Environmental Policy, Udall Center for Studies in Public Policy, The University of Arizona

Introduction of Panel Members

Christopher Scott
Director, Udall Center for Studies in Public Policy, The University of Arizona

My name is Christopher Scott and I am on the organizing committee for this conference. This is a panel format, so it is not a session with formal presentations, but rather a discussion to understand the opportunities, limits, constraints, and ways forward with networks in science diplomacy. We often conceive of a set of arrangements of scientists with stakeholders and policymakers in a diplomatic sense, or a broader conception of public engagement around science to consist of scientists (or the scientific establishment) with colleagues and counterparts. But of course, there is a lot of lateral networking that supports a lot of these initiatives for lessons learned, for actually conducting the science that is the most policy relevant.

We have three distinguished panelists who are going to offer a few words each sharing their views and experiences, and also charging us with thinking in some new directions.

First we have Dr. Nélida Barajas Acosta, who is the new Executive Director of CEDO, the Center for the Study of Deserts and Oceans, based both in Rocky Point, Sonora and Tucson, Arizona. She has a biology degree from the National Autonomous

University of Mexico (UNAM), and both Master's and Ph.D. in Natural Resource Management from the Autonomous University of Chihuahua (UACH), so very much from our broader region. She is very well-networked across the U.S.-Mexico borders and beyond.

Our second speaker, who has presented on one of our earlier panels, is Dr. Cherry Murray. She is the newly appointed Deputy Director of Research for the Biosphere 2 Institute here at The University of Arizona, which has a strong mission associated with building networks of scientists for international engagements.

Our third speaker is Dr. Robert Varady, my colleague at the Udall Center, who has extensive experience in science and diplomacy as a topic. He helped organize the panel on Hydrodiplomacy as well as written special issues of journals that are looking at that very question. Dr. Varady has his Ph.D. in modern history from The University of Arizona. He has spent years at the Udall Center working on bilateral cooperation with Mexico, but also has very interesting set of links with other regions as well, including Europe, and is just back from some related travel in Budapest, Hungary.

Without further ado, I'm going to pass the microphone to Dr. Barajas Acosta, and then we'll go to Dr. Murray, and finally to Dr. Varady. I've asked each of them to provide brief remarks, and then we will come back from an open type of discussion, followed by some Q&A.

Dr. Nélida Barajas Acosta: CEDO Intercultural

Thank you everybody for being here. It is my great pleasure that this is my first opportunity to be a guest speaker as the new executive director of CEDO. CEDO is a nonprofit international organization based, as Dr. Scott mentioned, in Puerto Peñasco and here in Tucson. Our region is creating conservation communities, sustainable livelihoods, and healthy ecosystem. I guess that this vision clearly resembles what networking is for us. We network between two countries, though we say that there is not a division between us, because we have a shared cultural heritage. Most of our work is being done as a grassroots organization. We work with local fisherman, the local schools, and with local authorities in order to develop big agreements, and that is how my work resembles science diplomacy. There are many different ways to do science and diplomacy and this is one way that helps us a lot with building credibility among the communities and other stakeholders.

Some of you may know that our work has been recognized internationally, because we are the first and only organization to have been working on the Gulf of California for over 40 years. We are working on a long-term commitment with the people and with the environment and getting these interlinkages where fostering communities and regenerating ecosystems can provide real solutions to big problems. That, to me, is science and diplomacy. That is something that we need to learn by doing. By "learning by doing" I mean that, for example, I am not a diplomat, but I have to work as a diplomat because of the need for scientific knowledge to be provided to policymakers in order for them to make the decisions which their jobs entail.

What I think has been our biggest challenge in our history is political will. Some years have been amazing, and we got amazing results and learned many things to share. But others, like the one that we are living, it seems that all our work has gone unnoticed and unheeded. So how can we jump to the next level, where you have reliable science that promotes peace, well-being, and knowledge among vulnerable populations? We talk about doing science and diplomacy, but by implementing diplomacy into our work and our programs we achieve science and diplomacy.

We are not offering a brand new discourse about how to do things. We are simply recalling the United Nations' way: we need to work together because we don't have to leave anyone behind. And the people that we're not leaving behind are those fishermen and local people that are working with us. It reminds me of the focus for the UN's upcoming Convention on Biological Diversity, which is how biodiversity preservation contributes to "the nutrition, food security, and livelihoods of people, especially for the most vulnerable" (UN n.d.). It is important because not only should we work to preserve nature, but remember that people are also part of nature. This is why CEDO is a good example of an NGO working on science and diplomacy.

Dr. Cherry Murray: InterAcademy Partnership

I would like to talk a little about the InterAcademy Partnership (IAP) and what that does. I was elected in April in Songdo, South Korea to be one of the six co-chairs of the IAP. So what is that?

In 1993, the partnership was established under the name InterAcademy Panel, mostly because of Bruce Alberts. He was the president of the National Academy of Sciences at the time, and the U.S. had just received a Gates Foundation grant to help form 12 National Academies of Science in Africa. He wanted to form a partnership among science academies across the world, and called it capacity building (Alberts 2016, 4–5). Many Academies exist, but they are more or less honorific—that is to say, they get elected to the Academy and that's all that happens. The National Academy of Sciences of the U.S., or as they now call themselves, the National Academies of Science, Engineering and Medicine, does quite a bit of policy advice, and the capacity building is to build the capacity of Academies of science, engineering, and medicine worldwide, in order to provide science-based advice to their governments. There are now 26 African Academies, so real evidence of capacity building in Africa.

Two other networks were established back in 2000. What was called the InterAcademy Panel morphed into the InterAcademy Council. The InterAcademy Partnership for Health, which were the medical academies, have all merged into a network of more than 140 Academies. This includes the World Academy of Sciences, the Global Young Academy, National Academies of Science, Engineering and Medicine, and four very active regional networks which actually do regional studies in the Americas (IANAS), Asia, Europe, and Africa. The European Academy network, called EASAC, is the body that provides advice to the European Commission. The African group of Academies, NASAC, provides advice to the African Union. So these are very serious regional networks.

As I said, we are now consolidated into a single network because global problems are multidisciplinary. You can't just have science, engineering, or medicine, because every problem is much more global than that. Science includes social science. There are two secretariats: one funded by UNESCO at the World Academy of Sciences in

Trieste, Italy, and one hosted by the U.S. and funded by the National Academy of Sciences in Washington, D.C. That allows the network to get grants to conduct policy studies, in either Europe or the U.S. For any of you who have had Foundation Grants, you are aware that the U.S. foundations like to go through a U.S. entity; likewise, European foundations want to go through a European entity, so this works out very well. The Swedish development authority, Sida, developed one of these entities, and certainly they exist among the African Academies as well.

The goals of the network are, as I said, capacity building to provide science-based policy advice, to enhance STEM education, active learning, and diversity in STEM, specifically by including women. As I pointed out, global challenges must be solved not only in this multidisciplinary way, but locally, and the advantage of the IAP is that the Academies can work together and then they can each take the relevant recommendations for their nation and provide advice locally. This has turned out to be extremely important because no two countries are the same.

One of the first things that the United Nations Secretary General requested of the IAP was an external science body to peer review the Intergovernmental Panel on Climate Change (IPCC) procedures, management structure, and reports. There have been very large national, regional, and global studies on food and nutrition security, which is really quite impressive. Just recently, it funded a new climate change and health study that is also at the national, regional, and local levels (InterAcademy Partnership 2019a; EASAC 2019).

The IAP also provides statements on things, like the freedom of scientists in tropical forests. As you know tropical forests are burning at an alarming rate at the moment, so the IAP has released a statement on that (InterAcademy Partnership 2019b). And lastly, the Partnership collaborates with other international science bodies, such as the International Science Council and other UN entities, on global issues such as the sharing of data, ethics of research, refugee scientists and how to manage them, and how the Academies can provide science advice to governments for the Sustainable Development Goals (InterAcademy Partnership 2020).

Dr. Robert Varady: Udall Center for Studies in Public Policy

As Dr. Scott said, I'll be talking about the networking that we do at the Udall Center at The University of Arizona, but I have to tell you that I'm also part of the organizing committee of this conference, and am on the board of CEDO. You heard just now from these two presentations some good examples of networking and why they are so important and effective. I'm going to tell you about some of the work that my colleagues and I have done, and why networking is important for a place like the Udall Center, which is an interdisciplinary public policy center. We are not interdisciplinary for everything, as we focus just on public policy, and only in some parts of the policy—in this case I'll be talking about environmental policy.

For a center like ours, networking is really our lifeblood. We couldn't do what we do if we didn't have strong networks in which we are engaged, and on which we rely. For us, as an interdisciplinary center, we can't fulfill our mandate without cooperating with others and teaming with them in the work that we do.

This was especially important when the center began, a little more than 30 years ago, and I have been there for 30 years now, so in a way this is not just about the center, but also about my career in a way. We are only about 100 kilometers from Mexico, so when we began at the programs at the Udall Center, it was very important for us to talk about some of the issues that were bilateral and cross-border. We began right away in 1989 receiving a stream of long-term funds from the Ford Foundation to look at transboundary water issues, especially in the region just south of us.

Much of the early work was on transboundary water management and policy and governance. That has allowed us to continue building on that very early work, and we have expanded from that. It was always pinned on the role of people and institutions, or communities and institutions. Eventually, by the late 1990s, we realized that climate is an important part of the water equation, and we began getting some funding to work on water and climate, initially in this border region. So we received some grants from NOAA way back in 2000, and we continued getting a string of grants from NOAA that allowed us to build our networks all the way until about 2015.

The NOAA grants led to a whole series of other grants. Starting right about the time that Chris Scott joined the Udall Center in 2006, we began getting funding from an organization called the Inter-American Institute for Global Change Research, or IAI, which is a consortium of NSFs across the Americas; 19 countries are members, and each of their NSFs sits on the board of that organization. IAI decided that the way to promote Global Change Research was to foment or instigate multi-party research that involved in these three countries. So they created a system of grants mostly funded by NSF, with 80–90% of the money coming from the U.S. side. These were called Collaborative Research Network (CRN) grants (IAI n.d.). The word "network" was an integral part of what they were doing, because they were attempting to get scientists from different countries to work together toward common problems that involve their regions (Pittman, Tiessen, and Montaña 2015).

When we first applied for a CRN grant in 2007, we did not receive one. But we did get a smaller grant, because they realized that social sciences were also an important part of what they did, and they were funding almost exclusively physical sciences on climate and ecology. So they created a special set-aside grant for people to apply for social sciences and human dimensions grants related to that. That allowed us to lead into applying for a full-fledged "next phase" of their Collaborative Research Network grants, which was the third phase of it, or CRN3. As we were doing that, the IAI decided that we would be their first Center of Excellence in the theme of water security for the arid Americas, called AQUASEC, and that led us to get the networking grant. We did that, in part, by establishing partnerships with countries—Mexico, first of all, because we had such a long-standing set of relationships with Mexican colleagues, but then we expanded and brought in partners from Peru, Argentina, and Brazil—all of which have arid regions.

As we embarked on that project, almost simultaneously, again through networking activities, a colleague who is in England at the University of the West of England (UWE) approached us to say that he was asked by a brand-new British foundation, called Lloyd's Register Foundation, if we would be interested in submitting a proposal for their first round of brand-new grants. So my English colleague and I got together with a couple other people, we gave them a proposal, and in very short order they approved a grant that was worth about $4 million to UWE. We became the leads for the Americas portion of what we call the International Water Security Network (IWSN), and we focus on Transboundary Water Security (International Water Security Network n.d.). We were clever about this because since it is a network, and we needed partners, we used the same partners that we used in the AQUASEC configuration and in the Collaborative Research Network program that was funded by IAI. This is how the Udall Center operates, by continually growing and building upon networks.

We continue to do this, but in the course of establishing partnerships like this, some are more effective than others. I am not going to say that every single one was as successful or as durable as the others. So the question is, what are the key factors that make one set of partnerships with networks work better than others? For me, the first key is chemistry. Not molecular chemistry, but rather, human chemistry. Without that it's a non-starter; you cannot get a successful network. You can't work with partners unless you can establish that level of trust, and that's a *sine qua non*. Trust must be the foundation. Beyond that, there are things like the alignment of existing research objectives; having similar environments, meaning physical environments—we deliberately work with places that have issues and environments similar to the ones that we face here; solid scientific capability, so we want partners that we know conduct good research and good science; a desire for long-term collaboration, on their part as well as ours; and finally, official interests, meaning government interests in condoning at least, if not supporting, the kind of work that we are doing.

So in this sense, obviously, our relationships with Mexico have been incredibly successful and long-lasting. I just came from the panel called "Hydrodiplomacy in Ambos Nogales," and it turns out that the center point of several of those presentations was the precursor for the Ford Foundation work that we did 30 years ago. I can't think of a better way to illustrate how networking over a very long period of time with some of the same people is actually instrumental in how we get things done at the Udall Center. Also, it turns out that the partnership that we had in Mendoza, Argentina, which has a very similar climate, layout, and infrastructure, is a good example of a partnership that has worked and that we expect to continue fomenting and expanding.

This is our approach to hydrodiplomacy; all of this has to do with water and with water-short areas. At the Udall Center, we are firm believers that this is the way to do diplomacy while conducting research.

Questions and Answers

Question

Thibodeaux-Yost: My name is Singleton Thibodeaux-Yost. I am a staff member here at The University of Arizona working in spacecraft operations, but also an early scientist trying to come from that perspective of just getting started in this. It's clear you all three showed examples of efforts in many different capacities that maybe didn't bear fruit, or didn't fully come to have a large effect until much later on. Something that I am encountering in my early work is not really being certain where to put efforts, because a lot of this kind of advocacy and involvement that we do, really on top of the jobs that we are paid for, and we're spending a lot of our "extra" time on these efforts, but burnout is a real thing. Can each of you speak to how you knew, throughout your careers, or where you began to feel that you started to understand what efforts were more fruitful than others, or how you learned to deal with not always getting immediate results, but seeing those results come later?

Answer

Acosta: In my case, it was understanding the interlinkages of the science and policy interface. So I have been working and have been trained as an academic, and have a solid scientific basis. But I realized when I started working with policymakers that they weren't understanding anything I was saying. So in the beginning, I needed a translator. Sometimes it is a difficult matter to get a clear understanding, but you must be certain that people understand what you're trying to say. So when you get to that point, you may open the gate for a very fruitful relationship.

Murray: I completely agree with that, which is, know who you are talking to and speak to something that they are interested in, and at their level, wherever it is. I am going to talk about Academies working with other Academies, because I think this is a relevant point. It is much easier, just like in bilateral relations with countries, to just work with another Academy you are used to working with. So, the National Academy of Sciences and the Royal Academy of Sciences—No problem. Start working with 140 National Academies and it takes a long time to come to some agreement, to even have a statement, let alone action. It's like a U.N. process almost, except that they are not politicians, they are scientists and engineers and medical doctors, and they more or less speak the same language, so that makes it a little bit easier than a U.N. deliberation. But I am pushing the National Academy of Sciences to broaden out and be more inclusive in who they partner with, and not just work with those most comfortable ones. We need to be inclusive, and that means it takes more work, that it will take longer, but that we will get a better outcome.

Varady: I am going to come at this from a different point, because I understood your question to say, well, here you are, and there are all these things you could be doing, but how do you decide whether to do this or that? How do you know in advance, or how much confidence can you have that this one is not a dud? Well, we are a center that relies primarily on external funds. So for us, that is a choice that is very similar to deciding where are you going to put your effort to apply to get money. I have always found that the best way to get funded is through people—through connections, networks, and people that you know. After a few years of this, you begin to develop a really good olfactory sense for it. So some of this is intuitive, and we have benefited from our networks. We have a really good record of percentage of getting funded for applications. We tend not to go for things that we don't think we could accomplish, and we already know who we are going to work with. So a lot of that is based on experience and knowing the field in which you're operating, and the other characters with whom you're going to join forces.

Question

Cuello: Thank you for the insights you shared. I'm Dr. Joel Cuello, professor of biosystems engineering. This is more of a remark rather than a question, but it echoes and crystallizes a lot of the insights you've shared. To me, at least in a national level, I find that the most effective networking for science diplomacy is one that results in a particular outcome, and that outcome is the government of a public-private academic/non-academic partnership. Dr. Barajas Acosta spoke about political will, and I find that political will is engendered and sustained whenever you have a cohesive public-private academic/non-academic partnership. And this could be tied to other efforts, especially when there is a national challenge—be it food security or climate change adaptation or water issues or job creation. Particularly that is when every party becomes really eager to work together to come up with solutions. And that's where science diplomacy really is most effective.

Answer

Murray: I totally agree. What I think is really important is having quality outputs of the networks, because you need to get funded; all of these things need to get funded. For example, by quality output in policy advice (which is what the Academies do mostly), the proposed solution or the proposed recommendations must be actionable. That is incredibly important. There are processes to review reports—very much like the National Academy of Science does, but this is on the international level—to make sure that whatever advice is being given is science-based and is also actionable by whoever it's targeted to. This is how real change is made.

INTRODUCTION TO LÉNA SALAMÉ

Elizabeth Cantwell

Senior Vice President for Research and Innovation, The University of Arizona

Before I introduce our final speaker for our Science, Health, and Engineering Policy and Diplomacy conference, I'd like to thank everyone for attending and participating in discussions to facilitate greater diplomacy among scientists and professionals. This conference was focused on "Sustainable Development for the Americas," but in reality, it extends beyond that, and is truly about sustainable development for the world.

As the Senior Vice President for Research and Innovation, I get the privilege of learning about all the incredible and innovative research being conducted here at The University of Arizona, and I can honestly say that I am amazed at the variety and breadth of applied research conducted by our faculty, staff, and students. But it would not be half as rich if we did not collaborate with our partners and colleagues outside of the university, both nationally and internationally.

The real strength to our research and development is its diversity—the diversity of our scientists and researchers, the diversity of perspectives from different levels, and the diversity of our collaborators. As our diversity grows and we are more inclusive and cooperative, so does our research portfolio and our impact, as you have heard from various presentations throughout this conference.

Our final speaker, Léna Salamé, is a lawyer specialized in conflict management. She has served in the United Nations' system for 17 years as the strategic and operational coordinator of its program on water conflict and cooperation (PCCP). She has played a central role in the development and promotion of the modern concept of hydrodiplomacy, or water diplomacy. She is now leading the development of the Universities' Partnership for Water Cooperation and Diplomacy (UPWCD), which is a platform for partners in the field of water cooperation and diplomacy to share and access information, co-develop research and education tools, and jointly support professional training in this field.

Without further ado, please let me turn the time now over to Léna Salamé.

DETERMINE "X" IN THE FOLLOWING FORMULA: (TRUST + X^2) × PERSEVERANCE = POLITICAL WILL

Léna Salamé

Lawyer, Specialist in conflict management and mediation, Water Ethics Network

My name is Léna Salamé. Today I am speaking to you, not because I'm an academician, because I am not; I am a "pracademician." This is a person who tries to take research and knowledge into practice, for instance, some science into diplomacy. I'm also not here because I'm a scientist. I am not a scientist; I am a lawyer. So, do not expect that the explanation of my formula would be very rational and Cartesian, because there is nothing rational about political will. Actually, I chose to explain political will with this crazy formula because at one point, about a year ago, I was sitting with the former president of Slovenia, Danilo Turk, who has been very much involved in water diplomacy these days (Ooska 2019). He was wondering if there was a secret formula to achieve political will at the surface of water diplomacy. I started scribbling things on my notebook, because I felt that with all my years of experience, I must come up with something to respond to him. And I came up with this:

$$\textbf{Political will} = (\textbf{Trust} + X^2) \times \textbf{Perseverance}$$

WHAT IS WATER DIPLOMACY?

In order to explain this formula, I will first try to explain what I understand by water diplomacy. There are tons of definitions of water diplomacy. Some people say "this is too simplistic," "this is too complicated," "this is too lawyer-ish," "this is not complicated enough," and whatever. So for me and a couple of my colleagues, with whom I published a paper, we have defined water diplomacy as referring to all positive interactions among actors and stakeholders that are concerned with good water management, with sound water management, and peaceful water management (Marshall, Salamé, and Wolf 2017). As simple as that. In its large understanding water diplomacy, the actors and stakeholders include all kinds of actors from international, regional, national, or governmental players to non-governmental players, faith representatives, activists, and social and civil society actors on different sides of a water body or sharing the same water resources.

The water diplomacy process can be a formal process, what we call a "Track I" process, which is the one involving governmental actors, or it can be an informal process involving non-governmental actors, or what we call "Track II." The most effective process though is one that involves the Track II and Track I, a process which puts both together, because they can complement each other (Patrick et al. 2014). Actually, they can complete each other, and this is the best to make a real difference.

When we start building trust and building relationships at a lower level with people who have the same stakes at the field level at the grassroots level, then we can hope to build trust at a higher level, where other problems occur, and we will get back to that. So that was, in a couple of words, the simple definition of water diplomacy.

What Is Political Will?

Now that we have defined water diplomacy, let's look at political will. Throughout my 20 years of experience in water diplomacy, the most common excuse not to achieve water cooperation or efficient transboundary water cooperation is the lack of political will. People are fond of saying, "we have all the technology, we have the solution, but we lack political will." I've heard it so much that it became almost a joke in our circle; we have everything ready and we are only missing this tiny, little, insignificant element which is political will, and then we're all set!

But then, what exactly is political will and how do we make it, or how do we achieve it? A decade ago, Hammergren characterized political will as "the most slippery element of the policy lexicon," and he says that it is "the *sine qua non* of policy success, which is never defined except by its absence" (Hammergren 1998, 12). So we always are wondering, what is political will?

Now, the literature has surfaced four elements to define political will, or to say that we have put together the four pieces of the puzzle of political will, and they are (1) a set of sufficient set of decision-makers who (2) have a common understanding of a particular problem on the formal agenda, who are (3) committed to (4) supporting the commonly perceived, potentially effective policy solution (Post, Raile and Raile 2010).

If these are the different pieces of the puzzle, then to achieve this, you obviously need to build trust among those decision-makers, who are on different sides of the border, or on different shores of a water body, or from different sectors of activities. And trust is one very, very important ingredient that is also very difficult to build. And, of course, it's like an eraser; for every mistake, it becomes smaller, and when you break it, it is almost impossible to build again. So, you need trust, and you need to build it among *all* the stakeholders, all the actors of the process of water diplomacy, whoever they are, and whatever is at stake.

You also need a lot of perseverance. This is why I wrote it in my formula:

$$\textbf{Political will} = (\textbf{Trust} + \mathbf{X^2}) \times \textbf{Perseverance}$$

You need you need a lot of perseverance, and you need a long-term commitment. This is very difficult to achieve when you're talking about governmental institutions where high-level actors are changing over time and changing with governments and you have to start all over again every few years. So, this really poses a problem in building political will. If you don't have perseverance and long-term commitment, then you don't really have everything for this equation.

So, you need trust, which is difficult to build, you need perseverance, which is complicated, and long-term commitment, which is complicated to maintain throughout changes in the institutional and governmental system.

This sounds already like the hardest thing to achieve, right? But actually, we're not even there yet, because you need to also add something that is essential for achieving real political will, and this is our "X." This variable is knowledge and capacity building, and it has to be squared. It has to be squared because without knowledge and without capacity, no matter what else you have, it will not work. And you need it at every level of the society.

A few years ago, we published to the World Water Development Report the studies that we had done, which showed that in order to meet the eight Millennium Development Goals (this was prior to the Sustainable Development Goals), Africa needed an estimate of 300% increase in the number of trained water professionals; in Asia we needed a 200% increase in the number of trained water professionals, and Latin America and the Caribbean needed a 50% increase of the trained water professionals (WWAP 2016). And I'm telling you these figures, first of all, because the trend hasn't been reversed; and second, because there are no new figures.

We recognize that the lack of trained professionals is a real obstacle to achieving the Sustainable Development Goals. If you really want to make a change, you don't come with your money and then just go away. You really have to build the capacity of the people who will have to do things on their own. You don't only want to intervene when they are already politicians or decision-makers. You need to intervene at every level of the education ladder. You need to start from the kids to the young professionals to the mid-level professionals to the high-level professionals, and even then to the Advisor of the Minister and then to the Minister. If you start from the beginning, from the lowest level of the ladder, then you might hope to have the future generation of political leaders who can make the difference. As Confucius said, "If your plan is for one year, plant rice. If your plan is for ten years, plant trees. If your plan is for one hundred years, educate children."

However, our system today is a system that is not very much adapted to the complex issues of water. Water is an interdisciplinary topic, though cross-disciplinary studies are not widely or readily supported by many institutions. Beyond this, the process of water diplomacy is also an interdisciplinary process. So we have to adapt our educational system, our capacity, and knowledge development system to this new aspect, and then we can hope to make a contribution to the formula.

Universities Partnership for Water Cooperation and Diplomacy (UPWCD)

With something called the Universities Partnership for Water Cooperation and Diplomacy (UPWCD), we tried to stretch beyond the usual frames and to bring together like-minded institutions that are committed to knowledge and capacity development. We started only with universities, but now we have included institutions that are not necessarily universities, but that are committed to make a difference in water, cooperation, and diplomacy through capacity development and knowledge development. So it is a network of institutions across the globe. So far, we have 20 partners. The coordination unit is at the University of Geneva, with which I'm affiliated, and the objective of this partnership is to link together all these universities and institutions and help them by exchanging and by being in touch with each other to make a real difference and impact the global discourse related to water diplomacy; to exchange knowledge and experience and expertise; and most importantly, to bring into the global dialogue, and the diplomatic dialogue, a little bit of science and a little bit of knowledge and a little bit of data and a solid basis for the negotiations. This is because very often politicians are negotiating without knowing what data they are negotiating on—what the water transponder basin is or what a conflict is, for that matter. So these are the aims of the partnership, and all are welcome to recommend an institution or to have your institution join it.

My last message is that the international community as a whole has a responsibility to play in building the capacity of the next generation of leaders who will have to achieve water cooperation and face the challenges that are lying ahead. If we want to do that, we don't have to only think outside of the water box, but we just have to think as if there was no box at all. Thank you.

CONCLUDING REMARKS AND RESOLUTIONS

E. William Colglazier
Honorary Chairman of the Conference

First of all, let me thank Léna Salamé. That was another wonderful talk. I will remember well the X-squared factor. She got it exactly right: knowledge plus capacity building. Those are two essential ingredients if we are going to be able to provide wise advice and help achieve the political will for a better future. And achieving all of that will take a lot of perseverance. I also like her definition of preventative diplomacy—educating and training the next generation—which is also absolutely crucial.

I am not going to try to summarize the conference, but what I want to do is to review the themes that emerged for our rapporteurs. We had some very able rapporteurs who wrote down and summarized what they learned from the talks they did a superb job covering many of the things that we care about, so I would like to share what they wrote.

First was the underlying theme of the importance of collaboration. It is hard to do on your own. Building a network of people, organizations, and countries engaged in science diplomacy is crucial. Doing so is essential for mutual growth and to helping thaw icy political relations. In times of political and international uncertainty, science diplomacy can serve as a strong diplomatic tool.

Just as important is building the infrastructure and communication networks to share information and start partnerships. Sustainability efforts require a lot of engagement, from local communities to state departments. Science diplomats, from experts to spokespersons, can help grease those wheels at all levels and bring in valuable resources from technology sharing, capacity building, and knowledge exchanges. Science diplomatic efforts can influence myriad issues and negotiations, from reducing overfishing to protecting human rights.

Short-term research collaboration can morph into long time international partnerships, especially between national health agencies. These long-term relationships can survive through shifting political landscapes and need to be prioritized.

It is crucial that we bring younger generations into science diplomacy earlier, with robust interdisciplinary training programs. I truly believe that it is important for training the next generation. One of the contributions of conferences has been to bring together very able young people from many different disciplines. They want to use their expertise to make the world a better place. But I want to emphasize that it is also important to train and educate the older generation—people like me. I learned a lot from attending the sessions. Our keynote speakers, also, were not only very informative, but very insightful and very inspiring. One of the many things I learned well from my colleague, John Hildebrand, is that in difficult times such as this, with our current challenges from our government, all Americans—including scientists—have to become diplomats for our country when they engage around the world. Other countries really admire and respect American science, engineering and health expertise and capabilities. John and I have traveled around the world, and we have always been treated warmly. We are just two of many Americans who are involved, not only trying to carry on research in our disciplines, but also trying to build relationships between scientific colleagues and citizens in other countries.

The title of this conference is Sustainable Development for the Americas: Science, Health and Engineering Policy and Diplomacy. For me, the title captures the two things that I have cared about most of my own life, and certainly for the latter part of my career. First, the issue of sustainable development is critical not only for the Americas, but for the whole world. Second, the science, health, and engineering policy and diplomacy, as emphasized in Léna Salamé's equation, are vitally important. Right now the missing ingredients for making progress on sustainable development are, in part, capacity building and expertise in knowledge. But in addition, I another essential aspect, emphasized in several of our talks and where I spend much of my time these days, is the effort at the UN trying to harness the capacities in science, engineering, health, and innovation to accelerate progress on the 17 broad Sustainable Development Goals (United Nations 2015).

As I said previously, the United Nations is great with aspirational rhetoric, but the real challenge now how we stimulate action. I am involved in help helping emerging countries to develop their "STI for SDGs" roadmaps for how they can marshall their capabilities in science, technology, engineering, and innovation to help achieve their highest priority social, economic, and environmental goals. There are several key ingredients to generate real action, but I will just mentioned two. First, these these plans have to be built up by engaging with all stakeholders in society. That goes from local communities and individuals through institutions through decision-makers in national governments. Second, as we heard in the talk by Lauren Alexander Augustine of her efforts toward resilience, a lot can be accomplished if the scientists who are willing to engage with local communities and listen to their needs, trying to help and work with citizens to create solutions to the problems important to them, as communities try to improve their own situation, and help solve problems for their society as a whole.

I would like to applaud The University of Arizona. In these difficult times for our country, with our government absent from a number of these discussions right now, at least for science diplomacy, it is essential for our non-governmental institutions to take the lion share of being science diplomats, and indeed serving as diplomats in general around the world. I had the pleasure in September of visiting the University of Idaho, just like going today at the University of Arizona. On both campuses, there is a strong commitment at the university level to engage with expertise across all faculty members and to provide opportunities for their students to learn about how they can address some of these really important issues that face our communities, our nations, and globally. This commitment was displayed at The University of Arizona two years ago with the first conference of this series, and reinforced with the conference this year. I certainly hope the University can sustain the commitment and continue this effort.

In building those bridges, not only between disciplines and providing new opportunities for students, but also providing linkages with the other universities trying to do the same thing. I would like to thank very much the people that made this conference possible. We have been fortunate so many of the leaders of the University were here, which is great. We had the provost, the dean of engineering, the senior vice president for research, and directors from many institutions across campus, which shows that the university is committed at the highest levels. I then look at the people that really made this happen: Kevin Lansey, John Hildebrand, Jeff Goldberg, and my good friend, Hassan Vafai. This conference would not have happened if it had not been for all of them. Additionally, I want to thank the students and rapporteurs, the great keynote speakers, and the people who contributed papers to these sessions. I hope this type of initiative can not only continue at The University of Arizona, but can be a spark and inspiration for other universities who aspire to do the same.

The last part of my role here is to present three award certificates of recognition to the young people who are here and who did a magnificent job in their talks. The first goes to Saffeyya-Grace Ibrahim for "Best Student Presentation." The second is to Samragyee Gautam for "Best Young Investigator Presentation." And lastly, "Best Overall Presentation" is awarded to Mariana Rivera-Torres for her excellent research and presentation. Congratulations to these wonderful young scholars, and we hope to see their work continue to grow.

Again, I want to thank everyone for participating. I also want to thank the session organizers, who were a big help to make sure people came and that we had something really strong to show. Also, thanks for all those who presented a paper. The rapporteurs have been fabulous; we cannot thank them enough for their service. They showed a huge amount of devotion to the conference and in being here.

Lastly, I do want to go back and also thank the sponsors, who are also listed on the back of the program: the Richard Lounsbery Foundation provided valuable seed funding, CNRS and the World Science Foundation have been very supportive, and various departments across The University of Arizona have been of great assistance. We want to thank each one of them for a huge effort getting this conference underway and seeing that the University has this opportunity to make a mark in science diplomacy. Thank you.

REFERENCES

Alberts, Bruce. 2016. "An Addiction to Education at All Levels." *Cell* 167, September 22, 2016. Retrieved from https://brucealberts.ucsf.edu/previous-work/interacademy-council/.

EASAC (European Academies' Science Advisory Council). 2019. "Urgent Action Is Needed to Protect Human Health from the Increasing Effects of Climate Change." *The Lancet Planetary Health*, August 2019. Retrieved from www.interacademies.org/sites/default/files/2020-06/Lancet%20Planetary%20Health_CCH.pdf.

Hammergren, Linn. 1998. "Political Will, Constituency Building, and Public Support in Rule of Law Programs." Center for Democracy and Governance; Bureau for Global Programs, Field Support, and Research; U.S. Agency for International Development (August). Available at http://pdf.usaid.gov/pdf_docs/PNACD023.pdf.

IAI. n.d. "IAI's Science Program." Inter-American Institute for Global Change Research. Retrieved from www.iai.int/en/research/program.

InterAcademy Partnership. 2019 (a). "Climate Change and Health." Project Overview. Retrieved from www.interacademies.org/project/climate-change-and-health.

InterAcademy Partnership. 2019 (b). "Communique on Tropical Forests from the InterAcademy Partnership (IAP)." Retrieved from www.interacademies.org/sites/default/files/publication/iap_tropical_forests_communique.pdf.

InterAcademy Partnership. 2020. *IAP Annual Report 2019*. Gorizia, Italy: The InterAcademy Partnership. Retrieved from www.interacademies.org/sites/default/files/2020-10/IAP_AR19_web.pdf.

International Water Security Network. n.d. "Transboundary Water Security." Retrieved from www.watersecuritynetwork.org/.

Marshall, Dena, Salamé, Léna, and Wolf, Aaron T. 2017. "A Call for Capacity Development for Improved Water Diplomacy." In: Shafiqul Islam and Kaveh Madani, eds. *Water Diplomacy in Action: Contingent Approaches to Managing Complex Water Problems*, 141–154. London/New York: Anthem.

Ooska News. 2019. "Water Diplomacy Talks" – Danilo Turk, Chair of the Global High-Level Panel for Water and Peace; "We Are All Water Diplomats"! *The Water Diplomat*. June 19, 2019. www.ooskanews.com/story/2019/06/water-diplomacy-talks-danilo-turk-chair-global-high-level-panel-water-and-peace-we-are.

Patrick, M.J., Komakech, H., Mirumachi, N., Moosa, H., Prakash, A., Salame, L., Shubber, Z., van der Zaag, P., and Wolf, A.T. 2014. "Building Bridges between the Sciences and the Arts of Water Co-Operation." *Aquatic Procedia* 2: 48–54.

Pittman, Jeremy, Tiessen, Holm, and Montaña, Elma. 2015. "The Evolution of Interdisciplinarity over 20 years of Global Change Research by the IAI." *Current Opinion in Environmental Sustainability* 19 (April): 87–93. https://doi.org/10.1016/j.cosust.2015.12.004.

Post, L.A., Raile, A.N.W. and Raile, E.D. 2010. "Defining Political Will." *Politics & Policy* 38: 653–676. https://doi.org/10.1111/j.1747-1346.2010.00253.x.

United Nations. 2015. "Transforming Our World: The 2030 Agenda for Sustainable Development." https://sdgs.un.org/2030agenda

UN. n.d. "UN Biodiversity Conference." UN.org. Retrieved from www.un.org/en/food-systems-summit-2021-en/un-biodiversity-conference.

WWAP (United Nations World Water Assessment Programme). 2016. *The United Nations World Water Development Report 2016: Water and Jobs*. Paris, France: United Nations Educational, Scientific and Cultural Organization (UNESCO).

Index

For Product Safety Concerns and Information please contact our EU representative GPSR@taylorandfrancis.com
Taylor & Francis Verlag GmbH, Kaufingerstraße 24, 80331 München, Germany

www.ingramcontent.com/pod-product-compliance
Ingram Content Group UK Ltd.
Pitfield, Milton Keynes, MK11 3LW, UK
UKHW022209190926
475702UK00003B/272
9781032115719